ALTERNATIVE USES
FOR SEWAGE SLUDGE

Related Pergamon Titles of Interest

BOOKS

CHIN
Industrial Water Technology, Treatment, Refuse and Recycling

LIJKLEMA
Water Pollution Research and Control (4 Vols)

MATSUMATO
Treatment, Disposal and Management of Human Wastes

SMITH
Design of Water and Wastewater Services for Cold Climates

TOFT
Drinking Water Treatment: Small System Alternatives

VAN HAUTE
Advanced Treatment Technologies for Removal and Disposal of Micropollutants

JOURNALS

Environment International

Environmental Toxicology and Chemistry

Water Quality International

Water Research

Water Science and Technology

Free details of all Pergamon publications/free specimen copy of any Pergamon journal available on request from your nearest Pergamon office.

ALTERNATIVE USES FOR SEWAGE SLUDGE

Proceedings of a conference organised by WRc Medmenham and held at the University of York, UK on 5–7 September 1989

Edited by

J. E. HALL
WRc Medmenham, U.K.

PERGAMON PRESS
Member of Maxwell Macmillan Pergamon Publishing Corporation
OXFORD · NEW YORK · BEIJING · FRANKFURT
SÃO PAULO · SYDNEY · TOKYO · TORONTO

U.K.	Pergamon Press plc, Headington Hill Hall, Oxford OX3 0BW, England
U.S.A.	Pergamon Press Inc., Maxwell House, Fairview Park, Elmsford, New York 10523, U.S.A.
PEOPLE'S REPUBLIC OF CHINA	Pergamon Press, Room 4037, Qianmen Hotel, Beijing, People's Republic of China
FEDERAL REPUBLIC OF GERMANY	Pergamon Press GmbH, Hammerweg 6, D-6242 Kronberg, Federal Republic of Germany
BRAZIL	Pergamon Editora Ltda, Rua Eça de Queiros, 346, CEP 04011, Paraiso, São Paulo, Brazil
AUSTRALIA	Pergamon Press (Australia) Pty Ltd, P.O. Box 544, Potts Point, N.S.W. 2011, Australia
JAPAN	Pergamon Press, 5th Floor, Matsuoka Central Building, 1-7-1 Nishishinjuku, Shinjuku-ku, Tokyo 160, Japan
CANADA	Pergamon Press Canada Ltd, Suite No. 271, 253 College Street, Toronto, Ontario, Canada M5T 1R5

Copyright © 1991 Pergamon Press plc

All Rights Reserved. No part of this publication may be reproduced, stored in a retrieval system or transmitted in any form or by any means: electronic, electrostatic, magnetic tape, mechanical, photocopying, recording or otherwise, without permission in writing from the publisher.

First edition 1991

Library of Congress Cataloging in Publication Data
(applied for)

British Library Cataloguing in Publication Data
Alternative uses for sewage sludge.
1. Sewage sludge
I. Hall, J. E.
628.3

ISBN 0-08-040271-2

Printed in Great Britain by BPCC Wheatons Ltd, Exeter

Contents

Editor's Preface	viii

Session 1: Land Reclamation

The potential value of sewage sludge in land reclamation
   K. L. BYROM (*West Yorkshire Waste Management, Wakefield, UK*)
   A. D. BRADSHAW (*University of Liverpool, UK*)   1

Utilisation of sewage sludge in the United States for mine land reclamation
   W. E. SOPPER (*The Pennsylvania State University, USA*)   21

Sewage sludge as an amendment for reclaimed colliery spoil
   I. D. PULFORD (*University of Glasgow, UK*)   41

Reclamation of acidic soils treated with industrial sludge
   D. DAUDIN, J. F. DEVAUX, D. FULCHIRON,
   M. C. LARRE-LARROUY and P. LORTHIOS
   (*Chambre d'Agriculture de Vaucluse, Avignon, France*)   55

Experiences of the usage of heavy amounts of sewage sludge for reclaiming opencast mining areas and amelioration of very steep and stony vineyards
   W. WERNER, H. W. SCHERER and F. REINARTZ
   (*Rhein. Friedr.-Wilhelms-University, Bonn, FRG*)   71

Consolidated sewage sludge as soil substitute in colliery spoil reclamation
   B. METCALFE (*Yorkshire Water, Dewsbury, UK*)
   J. C. LAVIN (*West Yorkshire Ecological Advisory Service, Keighley, UK*)   83

Discussion   97

Session 2: Forestry

The potential for utilising sewage sludge in forestry in Great Britain
   C. M. A. TAYLOR (*Forestry Commission, Roslin, UK*)
   A. J. MOFFAT (*Forestry Commission, Farnham, UK*)   103

Sewage sludge utilisation in forestry: the UK research programme
  C. D. BAYES (*Water Research Centre, Stirling, UK*)
  C. M. A. TAYLOR (*Forestry Commission, Roslin, UK*)
  A. J. MOFFAT (*Forestry Commission, Farnham, UK*) 115

Operation experiences of sludge application to forest sites in Southern Scotland
  J. M. ARNOT (*Strathclyde Regional Council, Hamilton, UK*)
  J. D. MCNEILL (*Forestry Commission, Roslin, UK*)
  B. F. J. WALLIS (*Borders Regional Council, Melrose, UK*) 139

US forestry uses of municipal sewage sludge
  C. G. NICHOLS (*County Sanitation Districts of Orange County, California, USA*) 155

Utilisation of dehydrated sludge from Marseille's purification station in forestry
  G. LAVERGNE (*Société du Métro de Marseille, France*) 167

Long-term effects of sewage sludge application in a conifer plantation on a sandy soil
  S. E. OLESEN and H. S. MARK (*Hedeselskabet, Danish Land Development Services, Viborg, Denmark*) 177

Discussion 199

Session 3: Landfill and incineration

Co-disposal of sewage sludge and domestic waste in landfills: laboratory and field trials
  N. C. BLAKEY (*Water Research Centre, Medmenham, UK*) 203

Co-disposal in MSW landfills of coal fly-ash and domestic sludge
  R. COSSU, R. SERRA, P. CANNAS and G. CASU
  (*University of Cagliari, Cagliari, Italy*) 215

The co-dosposal of controlled waste and sewage sludge - some practical aspects
  C. P. HILL (*West Yorkshire Waste Management, Wakefield, UK*) 233

Environmental aspects of landfilling sewage sludge
  D. BEKER (*RIVM, Bilthoven, The Netherlands*)
  J. J. VAN DEN BERG (*Grontmij, De Bilt, The Netherlands*) 243

Physical aspects of landfilling of sewage sludge
  J. J. VAN DEN BERG (*Grontmij, De Bilt, The Netherlands*)
  P. GEUZENS (*Nuclear Research Institute, Mol, Belgium*)
  R. OTTE-WITTE (*University of Bochum, FRG*)   263

The use of sewage sludge as a fuel for its own disposal
  P. LOWE (*Yorkshire Water, Wakefield, UK*)
  J. BOUTWOOD (*Yorkshire Water, Leeds, UK*)   277

Discussion   285

Session 4: Other uses

Economics and marketing of urban sludge composts in the EEC
  J-L. MARTEL (*Agro Developpement SA, St. Quentin, France*)   291

Compost - a sewage sludge resource for the future
  P. J. MATTHEWS (*Anglian Water, Cambridge, UK*)
  D. J. BORDER (*Hensby Biotech Ltd., Huntingdon, UK*)   303

The manufacture of a quality assured growing medium by amending soil with sewage sludge
  K. M. PANTER (*Thamesgro Land Management Ltd., Slough, UK*)
  J. E. HAWKINS (*Thames Water, Ware, UK*)   311

Energy from sludge
  R. C. FROST and A. M. BRUCE
  (*Water Research Centre, Swindon, UK*)   323

Resource recovery through unconventional uses of sludge
  M. D. WEBBER (*Environment Canada, Burlington, Canada*)   343

The development of a sludge treatment and disposal strategy
  C. POWLESLAND (*Water Research Centre, Medmenham, UK*)   359

Discussion   375

List of Participants   377

Editor's Preface

The treatment and disposal of sewage sludge is an expensive and environmentally sensitive problem. It is also a growing problem worldwide since sludge production will continue to increase as new sewage treatment works are built and environmental quality standards become more stringent. With some traditional disposal routes coming under pressure and others such as sea disposal in the UK, being phased out altogether, the challenge facing the sludge disposal authorities is to find cost-effective and innovative solutions whilst responding to environmental and public pressures.

This book records the proceedings of an international conference organised by WRc and attended by 159 delegates from 17 countries, the purpose of which was to review the range of alternative disposal/recycling options for sewage sludge. The alternative uses considered are in land reclamation, forestry, compost and soil production, landfill, incineration, and energy and resource recovery. The papers presented include recent research findings and technological developments, as well as operational implementation of schemes. They show that beneficial re-use of sludge can be entirely compatible with a cost-effective and environmentally sensitive approach to sludge disposal, and the conference struck a new note of optimism in how the sludge problem may be tackled in the future.

The conference was divided into four main sessions and a poster session; papers from the latter are published separately by WRc [1]. Several technical visits were also made, and the help and support of Yorkshire Water Services Ltd, West Yorkshire Waste Management and Wakefield Metropolitan District Council are gratefully acknowledged. Special thanks are also due to Mr J Taylor and Prof. R Edwards for respectively opening and closing the conference, and to Dr P J Matthews, Mrs L. Duvoort-van-Engers, Mr R C Ramsay and Prof. R Edwards for chairing the sessions.

J. E. Hall
WRc Medmenham

---

[1] Report No. CP 596 available from WRc, Henley Road, Medmenham, PO Box 16, Marlow, Buckinghamshire, SL7 2HD, UK.

SESSION 1

Land Reclamation

The potential value of sewage sludge in land reclamation

K. L. BYROM[1] and A. D. BRADSHAW[2]

[1]West Yorkshire Waste Management, Wakefield, West Yorkshire, UK.
[2]Department of Environmental and Evolutionary Biology,
University of Liverpool, UK.

SUMMARY

Derelict and degraded land is a persistent feature of present-day society. Much of it must be reclaimed to a soft, green, end use, but the materials on which vegetation is to be established are nearly always extremely deficient in organic matter and nutrients, especially nitrogen. To maintain an adequate supply of nitrogen, a capital of over 1,000 kg N/ha is required in soil. This can be readily provided by a single large dose (200 t/ha) of sewage sludge. Experimental studies show that this will release nitrogen to plants over a period of several years, and yet will retain it against leaching. Large amounts of phosphorus and organic matter are also contributed, sufficient to restore these to normal soil levels. Field experiments confirm the superiority of such sludge treatments over more traditional approaches using fertilisers. Problems due to potentially toxic elements do not arise because the amounts contained in a single, albeit large, dose do not even approach maximum permissible amounts. However, practical problems of transport, application and user antipathy have to be overcome.

1. INTRODUCTION

Derelict and degraded land is a persistent feature of almost every country of the world. It arises from the many activities which provide the comforts of civilization - particularly the production of fuel, building materials and other products which come from the ground. It is also the outcome of the damage caused by the disposal of solid wastes and by industrial dereliction. A great deal of effort has recently been put into restoration. In England, where up-to-date surveys are available, the total area has diminished very little. As fast as derelict land has been reclaimed, more has been produced (Table 1).

Recent policy in many countries has been to encourage hard end uses, by the building of new factories and houses. However, the size of the total area remaining, as well as its situation, means that much of it must have a soft, i.e. green, end use. Such restoration has the particular advantage that it is relatively cheap, which is important when there is far more land to be reclaimed than can possibly be paid for (1).

Derelict and degraded land can have many origins (Table 2). The conspicuous feature is that it no longer possesses the normal layers of surface soil which are crucial for sustaining plant growth. As a result there are both physical and chemical deficiencies. Some materials may also have specific toxicity problems, connected with their origin and subsequent history.

Table 1. Changes in the areas of derelict land in England 1974-88.

1974	43,273 ha	16,952 ha reclaimed
1982	45,683 ha	13,158 ha reclaimed
1988	41,456 ha	

The most common problems are extremes of soil texture and structure, and a major shortage of plant nutrients. These will be related, particularly, to a complete absence of organic matter, which would have accumulated in normal soils as a result of plant growth and decay.

The physical value of organic matter to soils is well understood. It makes a direct contribution to water holding, by having an available water capacity (awc) at least four times that of mineral materials. It also makes an indirect contribution as a substrate for the growth of the soil micro-organisms which affect soil structure, particularly ensuring the maintenance of crumb structure and pore spaces.

The main nutrient value of organic matter lies in its nitrogen content. The only major way nitrogen is held in soils is in organic form. Nitrogen supply to plants depends on the slow but continuous breakdown of this organic matter, releasing nitrogen by mineralisation. This is well understood, but to appreciate its significance for derelict land reclamation some quantification is necessary (2).

The annual nitrogen requirement of a temperate ecosystem is about 100 kg N/ha. A typical organic matter decomposition rate (k) is about 1/16 per year. To provide the necessary supply of mineral nitrogen therefore requires a total capital of about 1,600 kg N/ha. Direct observations suggest that 750 kg N/ha is the minimum requirement. A derelict or degraded soil is likely to contain 0-500 kg N/ha. To raise it to give the level of function of a normal soil, an addition of about 1,000 kg N/ha is therefore required. Without a full soil nitrogen capital, reclamation schemes show rapid regression after initial treatment. This usually occurs within a single season. Experiments indicate that it is due to acute nitrogen deficiency (4). Since initial treatments usually provide only 100-200 kg N/ha, this is scarcely surprising.

Table 2. The characteristics of different derelict land materials (3).

Materials	Texture and structure	Stability	Physical		Chemical				
			Water supply	Surface temp.	Macro-nutrients	Micro-nutrients	pH	Toxic materials	Salinity
Colliery spoil	OOO	OOO/o	O/o	o/***	OOO	o	OOO/o	o	o/**
Strip mining	OOO/o	OOO/o	OO/o	o/***	OOO/o	o	OOO/o	o	o/**
Fly ash	OO/o	o	o	o	OOO	o	*/***	**	o
Oil shale	OO	OOO/o	OO	o/**	OOO	o	OO/o	o	o/*
Iron ore mining	OOO/o	OO/o	O/o	o	OO	o	o	o	o
Bauxite mining	o	o	o	o	OO	o	o	o	o
Heavy metal wastes	OOO	OOO/o	OO/o	o	OOO	o	OOO/*	*/***	o/***
Gold wastes	OOO	OOO	O	o	OOO	o	OOO	o	o
China clay wastes	OOO	OO	OO	o	OOO	o	O	o	o
Acid rocks	OOO	o	OO	o	OO	o	O	o	o
Calcareous rocks	OOO	o	OO	o	OOO	o	*	o	o
Sand and gravel	O/o	o	o	o	O/o	o	O/o	o	o
Coastal sands	OO/o	OOO/o	O/o	o	OOO	o	o	o	o/*
Land from sea	OO	o	o	o	OO	o	o/*	o	***
Urban wastes	OOO/o	o	o	o	OO	o	o	o/**	o
Roadsides	OOO/o	OOO	OO/o	O/o	OO	o	O/o	o	o/**

Notes:
Deficiency			Adequate	Excess		
severe	moderate	slight		slight	moderate	severe
OOO	OO	O	o	*	**	***

Source: from Bradshaw and Chadwick (1980) with kind permission of Blackwells

It would be possible to provide the capital by the addition of fertiliser nitrogen, but to avoid toxicity and wastage by leaching, a large number of small doses would have to be given over a number of years, which would be troublesome and expensive. A well-established alternative is to incorporate legumes in the vegetation. These can accumulate through their root-nodule bacteria as much as 100 kg N/ha per year. The difficulties of this approach are the problem that may be experienced finding legume species tolerant of difficult conditions, the attention that has to be paid to the other nutrient requirements of the legume, and the fact that the accumulation takes time (5).

2. THE POTENTIAL OF SEWAGE SLUDGE

2.1 Nutrient and organic matter capital

Sewage sludge provides a very obvious alternative. Its outstanding characteristic is that it contains high levels of nitrogen in an organic form, ranging from 1 - 7% total N (5) (Table 3). In digested sludge up to 70% may be in the liquid fraction, which can mean an almost excessive availability equivalent to fertiliser. While this can be valuable, what remains in the organic fraction is really more important for long term reclamation, since it is this which will contribute to the long term soil capital.

Table 3. Major nutrients in three common types of sewage sludge (6).

	Liquid undigested	Liquid digested	Digested cake
Dry solids (%)	2 - 7	2 - 8	20 - 50
Organic matter (% ds)	50 - 70	50 - 60	50 - 70
Total N (% ds)	2.1 - 7.6	0.9 - 6.8	1.5 - 2.5
Total P (% ds)	0.6 - 3.0	0.5 - 3.0	0.5 - 1.8
Total K (% ds)	0.1 - 0.7	0.1 - 0.5	0.1 - 0.3
Total Ca (% ds)	1.4 - 2.1	1.5 - 7.6	1.6 - 2.5
Total Mg (% ds)	0.6 - 0.8	0.3 - 1.6	0.1 - 0.5

Different types of sludge obviously have different water contents and different amounts of available and total nitrogen. In land reclamation, so long as a sludge is accessible, quite large amounts can be applied, even up to 300 t/ha of liquid or semi-solid material. The amounts of nitrogen which can be added are consequently very considerable, equal to or in excess of the soil capital requirements already discussed (Table 4). Thus in one step, in the initial operations, it should be possible to overcome permanently the over-riding problem of nitrogen deficiency.

The organic matter itself must not be forgotten; while derelict and degraded materials may have none, normal soils have 2.5% or more, concentrated mainly in the top 100 mm. If for simplicity the bulk density is assumed to be one, the total amount of organic matter in a normal soil will be about 25 t/ha, an amount approached or exceeded by the amounts contributed by the treatments suggested in Table 4. It is unlikely that this can ever become incorporated as completely as it would be in normal soils, but it nevertheless represents a potentially remarkable one-step restoration of the soil organic matter.

Table 4. The value of addition to land of two different types of sludge applied at 200 t/ha, assuming values of Hall et al. (7).

Sludge type	Dry solids		Total nitrogen		Total phosphorus	
	%	t/ha	N %ds	kg N/ha	P %ds	kg P/ha
Liquid digested	4	8	5	400	1.7	140
Digested cake	25	50	3	1,500	1.5	785

Sewage sludge contains many other plant nutrients which can be considered as a bonus in the restoration process. There is one nutrient, phosphorus, of special importance; firstly because its concentration in sewage sludge is much higher than normally occurs in organic materials, and secondly because it is often an important nutrient which is deficient in derelict land. Additions of 50 kg P/ha (115 kg P_2O_5) are often required in reclamation, and in some situations, such as in the reclamation of colliery spoil where there is severe binding of phosphorus, 250 kg P/ha may be necessary (8). These amounts can be readily provided or exceeded by the rates of application suggested in Table 4.

2.2 Nutrient release

Additions of nutrient capital to a soil are only effective to the extent that the nutrients are available. This is particularly important for nitrogen, where the availability (beyond that which is already in a soluble form) is entirely dependant on the rate of organic matter decomposition. C/N ratios in sludge normally fall well below the threshold of 25:1, above which nitrogen supply may limit decomposition (9). Incubation studies suggest rapid mineralisation, from 6% over 9 weeks (10) to >30% over 18 weeks (11).

What is perhaps most significant is the continued rate of release over a long period. This was demonstrated by measuring the production of grass (*Lolium perenne*) established on a derelict land material (colliery spoil) amended with different amounts of sewage sludge, in a replicated pot experiment (12) (Figure 1). As might be expected, there was a major effect on growth in the first year. In contrast to the potting compost control, however, there was a very satisfactory continuation of growth into the second year, particularly at the higher levels of sludge addition. Growth without sludge was negligible. Such a long term release is borne out by other experiments. A large scale experiment on the release of nitrogen from sludge applied to a clay sub-soil (13) showed that release was still occurring after four years, at appreciably higher levels from undigested rather than digested material (Figure 2).

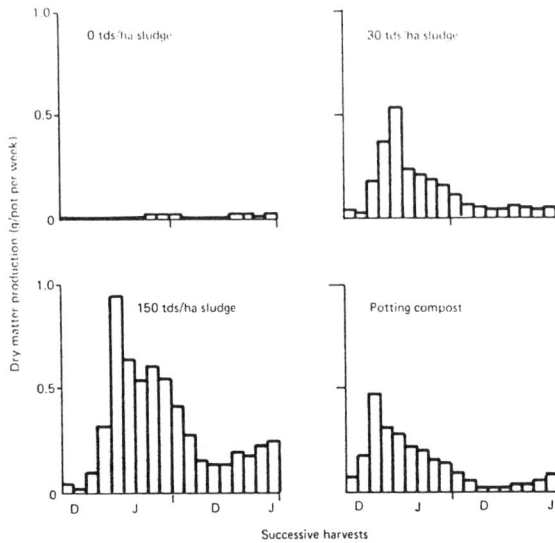

Fig. 1. Dry matter production of grass *(Lolium perenne)* on colliery spoil amended with different amounts of sludge, compared with production of potting compost, over 18 monthly periods.

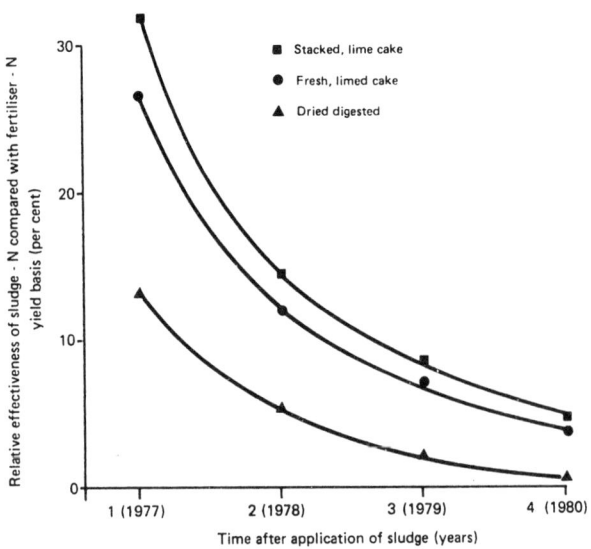

Fig. 2. The influence of time after application on the effectiveness of sludge application (13).

The general picture of decay rates is that release of nitrogen decreases by about half each year (14, 15). This of course suggests $k = 1/2$, which is considerably higher than that typical for normal organic matter. Herein is perhaps both an advantage and a disadvantage. The higher rate will ensure good release, but it will tend to allow too much to be released early on and rather too little to be retained for later years. However, the rate of release is still completely different to that of fertiliser. At the same time, in an actual reclamation scheme on which vegetation has been established, as the vegetation grows, there will be a progressive transformation of the nitrogen from sludge nitrogen with its high rate of breakdown to plant nitrogen with a much lower C/N ratio (<30) and much lower rate of mineralisation.

When high levels of sludge are applied to land there may be large losses of nitrogen to ground water by leaching. With liquid digested sludge, in which there are high concentrations of nitrogen in solution, this must be expected (16). With solid sludge the same effect should be less, but the relatively high rate of release of the organically bound nitrogen just discussed could lead to similar problems.

A lysimeter study was therefore undertaken of the effects on the nitrogen budget of adding, in comparison with fertiliser, different amounts of solid digested sludge to colliery spoil with either a grass cover (*Lolium perenne*) or none (12). The spoil, initially very acidic, was heavily limed to achieve a nearly neutral pH, before the incorporation of sludge to 15 cm. There was almost no leaching of ammonium ions, except for the first two months, from the lysimeters treated with fertiliser. The losses of nitrate were very interesting (Figure 3).

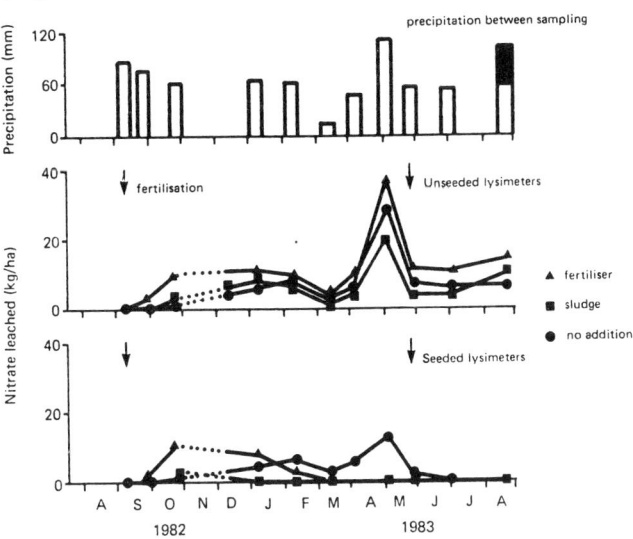

Fig. 3. Leaching of nitrate from colliery spoil in lysimeters treated with sludge cake (90 tds/ha), fertiliser (2 x 100 kg N/ha), or nothing, with or without a grass cover (*Lolium perenne*).

In the seeded lysimeters losses of nitrogen were negligible from those treated with sludge (about 3 kg N/ha), but rather more from those given nothing or fertiliser (Table 5). In the unseeded lysimeters there were appreciable losses, both initially and thereafter, which were highest from the fertiliser treated and less from the sludged lysimeter. In the highest sludge treatment (3,000 kg N/ha), despite the very high nitrogen loadings, the losses were lowest (72 kg N/ha). However, this nitrogen must have been available since this treatment gave the greatest growth and nitrogen uptake in the seeded lysimeters, equivalent to what could be expected from a pasture of normal productivity.

Table 5. Total amounts of nitrogen added, taken up or leached (kg N/ha) from colliery spoil in lysimeters over one year, treated with sludge cake, fertiliser or nothing, with or without a grass cover (see Fig. 3).

	Control		Fertiliser		Sludge	
	Unseeded	Seeded	Unseeded	Seeded	Unseeded	Seeded
Added in fertiliser or sludge	0	0	200	200	3,006	3,006
Uptake into grass	-	128	-	332	-	588
Leached	76	35	123	23	72	3

This suggests a very large amount of nitrogen can be added, without over-loading the soil part of the ecosystem, by a single heavy application of solid sludge. The nitrogen is relatively fixed, i.e. resistant to leaching. Yet it is labile and available to plants. The nitrogen must be released by mineralisation and able to be intercepted by plants, before being bound again by other micro-organisms using other materials in the sludge as a substrate. Simultaneous mineralisation and immobilisation is known to occur in normal soils (17,18).

These results also support the importance of a vigorous intact vegetation cover in preventing leaching losses, suggested by other work (19, 20). The experiment also showed that the spoil itself released some nitrogen, presumably from the coal organic fraction, as has been found previously (21).

2.3 Physical effects

The large store of nitrogen and other nutrients, especially phosphorus, will have important effects on the general vigour and root growth of newly established vegetation, and thus help the restoration of soil structure by root activity and subsequent incorporation of organic matter. However, this will take time. Sludge organic matter can, however, have immediate and direct effects on soil physical characteristics. To test this critically a simple pot experiment was carried out in which different amounts of solid sludge were mixed with the same colliery spoil used in the previous experiments (Figure 4).

The pots were left outside for three months, exposed to the sort of winter weather conditions which cause substantial collapse of soil structure in reclaimed colliery spoil (12). The effects on both bulk density and water content at field capacity were quite marked, although only to a really useful extent at the highest level of sludge addition. Similar direct effects on soil structure, but again only at high levels of application, have been recorded in several field experiments (7).

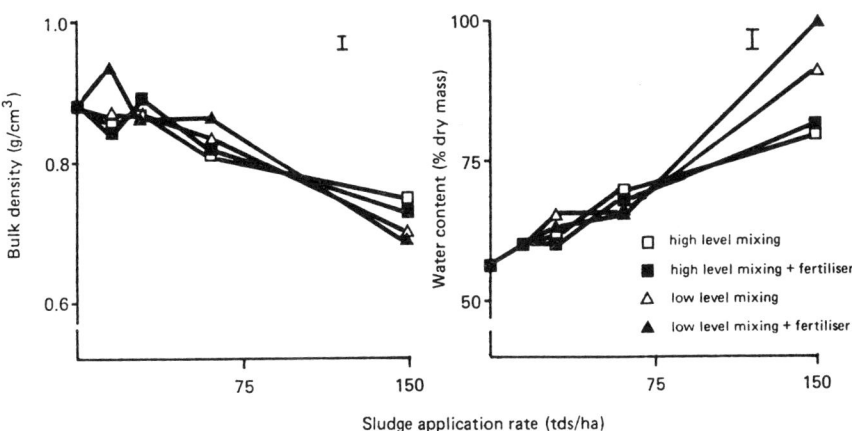

Fig. 4. Bulk density and water content at field capacity of colliery spoil amended with sludge cake, with two mixing regimes and two fertiliser levels, after three months.

The need for high application rates (>50 tds/ha) is perhaps to be expected, since from a physical standpoint sludge has a labile amorphous structure. Such high rates are in the range required, anyway, to give adequate nitrogen capital. At lower rates the useful influence of sludge will therefore be more due to its indirect effects on soil structure through its direct effects on plant and root growth.

3. POSSIBLE PROBLEMS

3.1 Heavy metal toxicity

Whenever the use of sewage sludge, particularly in large quantities, is proposed, the question of its heavy metal content is raised. This is indeed a problem which must always be addressed when sludge is applied to land. However, its critical significance is when sludge is being applied continuously over a long period, allowing a substantial accumulation to occur. When only a single application is given, the situation is rather different.

The levels of toxic metals found in sludge are markedly affected by the source of the waste water being treated, and very contaminated samples must be rejected. Some typical values are given in Table 6.

Table 6. The concentrations in sludge and amounts of potentially toxic elements contributed to land by two different sludge treatments of 200 t/ha, as in Table 4 (assuming values of Hall *et al.* (7)) - the amounts are also the resulting soil concentrations if sludge is mixed to a depth of 100mm in soil.

Element	Concentration in sludge (mg/kg ds)		Amount (kg/ha) or soil concentration (mg/kg)		Maximum permissible addition (kg/ha)
	Range	Common value	Liquid digested	Digested cake	
Cd	2 - 1,500	20	0.16	1.0	5
Cu	200 - 8,000	650	5.2	32.5	280
Ni	20 - 5,000	100	0.8	5.0	70
Zn	600 - 2,000	1500	12.0	75.0	560
Pb	50 - 3,600	400	3.2	20.0	1,000
Hg	0.2 - 18	5	0.04	0.25	1
Cr	40 - 14,000	400	3.2	20.0	1,000
Mo	1 - 40	6	0.05	0.3	4
As	3 - 30	20	0.16	1.0	10
Se	1 - 10	3	0.02	0.15	5
B	15 - 1,000	50	0.4	2.5	4
F	60 - 4,000	250	2.0	12.5	600

From these, the total amounts which would be contributed by single sludge applications of the magnitude given in Table 4 can be determined, and compared with recommended limits for additions given over a 30 year period. The resulting increases in concentration if these loads were incorporated into the surface 10 cm of a soil can also be calculated.

This shows very clearly that recommended limits are not reached even by heavy sludge applications. A similar viewpoint has been expressed in the extensive review by Hall *et al.* (7). The crucial point is that only a single, although large, and not repeated, application is being considered.

3.2 Other phyto-toxic effects

When sewage sludge is used in a seed bed, applied immediately before sowing, significant toxic effects can occur. These are easy to see in pot experiments. They are also visible when sludge is used as a seed-carrying mulch, a technique similar to hydraulic seeding, of value in the establishment of vegetation on quarry rock faces and other inaccessible sites (22).

The noticeable results are a delay, or permanent reduction, in germination, which is related to sludge concentration. The effect is, however, not permanent, but disappears with time. The most likely culprit is ammonium ions or ammonia, since these are known to have distinct inhibitory effects on seed germination. As the level of ammonium ions falls in fresh sludge, so its phyto-toxic effects disappear (12) (Figure 5).

3.3 Other problems

There are a number of practical and economic problems which have to be faced when sludge is being used in land reclamation. These are also found in other situations.

One of the crucial points in its favour is that sludge is normally available free from most sewage treatment plants. However, sites requiring reclamation may often be a long way from the nearest treatment plant. There are no ways in which this problem of distance can be overcome. Whether it is economic to use sludge is therefore a matter of the economics of transport. This has been been discussed by Ellis and Brade (23). Their model shows that the most important factors affecting cost are journey distance and tanker size. To assist disposal, most treatment plants are prepared to deliver sludge some distance without charge. The maximum distance sludge will be transported free is then important. This obviously varies with proximity of other competing users.

A survey of Regional Water Authorities in England (12) showed a lack of common policy, and that the maximum distance could vary from 1-2 km to 55-60 km, or "reasonable", or "short as possible"; the average was about 12.5 km. With costs beyond this distance being variable, but about 6p/km per tonne, it is difficult to generalise, but if the minimal assumption is made that 200 m^3 of liquid digested sludge will give growth equivalent to 100 kg N and 100 kg P, it would be economic at present day fertiliser costs

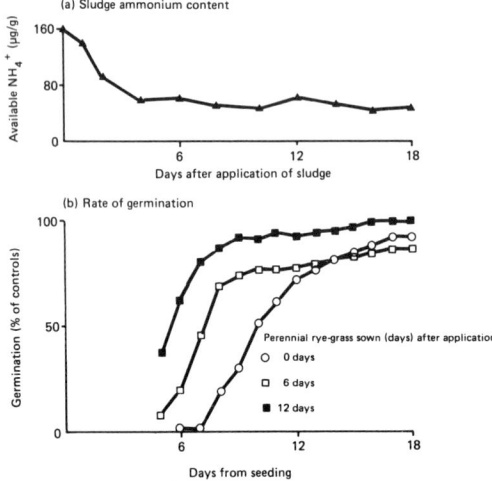

Fig. 5. Changes in the inhibitory effect of sludge on germination of *Lolium perenne* in relation to weathering and ammonium ion content (*Lolium* sown 0, 6, and 12 days after sludge application).

for the sludge to be transported a further 10 km (12). Providing the economic value is realised and users are prepared to contribute to transport costs, the aggregate distance of 22.5 km does then put most reclamation sites within economic distance, since they commonly occur in industrial areas where treatment plants are also sited.

There is then a problem of availability. A reclamation scheme may require a large volume of material over a short period. This may be difficult to deliver against competing users. The best solution is to design schemes so that sludge can be taken over a longer period, which should not be difficult.

Handling may be a source of problems. Compared with fertiliser, sludge is very heavy and large volumes have to be applied. Effective access and spreading routes have therefore to be devised which are relatively resistant to problems caused by weather. This is not always easy, although some problems can be overcome by the use of irrigation pipes. This however adds to costs and is not applicable to semi-solid material. Spreading the latter is particularly susceptible to problems due to weather. After spreading the sludge has to be incorporated. This can be done by ripping and cultivation; it can also be helped considerably by the application of a suitable ripping treatment beforehand. If sludge is to be used effectively, it is essential that its use is carefully planned in advance of actual field operations. In view of its potential benefits this is not unreasonable to suggest. Injection, although valuable later (24), is not really necessary in the primary stages of reclamation.

Finally, there can be a problem of user resistance, in both management and workforce. This can arise from views about smell, for example, but it is more often related to an unease in using a material which is seen as being sewage, even if it has been properly treated. There are worries about disease and potential toxicity (from heavy metals). The only solution to this is education and a greater familiarity with the benefits of sludge use.

4. FIELD EXPERIMENTS

4.1 China clay sand waste

The china clay industry in Cornwall produces large volumes of a coarse sand waste which, because the clay deposits are bottomless, cannot be returned to the pits. It is therefore deposited in large steep-sided heaps. The material contains no toxic elements but is almost completely devoid of plant nutrients. Successful vegetation establishment depends on the development of an adequate nitrogen capital. This is normally achieved by a combination of fertiliser and legumes (25).

It is an obvious situation for the use of sludge. Digested drying bed sludge (24% dry solids; 5.1% N, 1.2% P in the dry solids) was incorporated in different amounts to a depth of 10 cm, and compared to various combinations of peat and slow release fertiliser (11) (Figure 6).

Growth was poor in the first year because of drought. The results for the second year showed that the yield of the highest sludge treatment (given in the previous year only) was equal to that of the fertiliser treatments (repeated in the second year). In the third year, yields of the fertilised plots (not retreated) collapsed even where slow release material had been used, while the highest sludge treatment continued to show satisfactory, although somewhat reduced, growth. The plots without fertiliser made no growth at all, and the peat had no effect. The slow release fertiliser (sulpher-coated urea) was little better than the normal material.

The total amount of nitrogen applied by the highest sludge treatment was 3,670 kg N/ha. At the end of the second season this treatment still had 1,250 kg N/ha, providing a good indication as to why growth continued in the third year. The levels in the fertilised plots were about 200 kg N/ha, clearly insufficient to sustain further growth.

4.2 Colliery spoil

Colliery spoil may have severe acidity problems which have to be treated by the application of lime. In other respects it is less extreme than china clay waste, although it also has serious nutrient deficiencies, especially nitrogen and phosphorus. Successful reclamation depends on the use of fertilisers and the inclusion of clover (*Trifolium repens*) as a nitrogen accumulator.

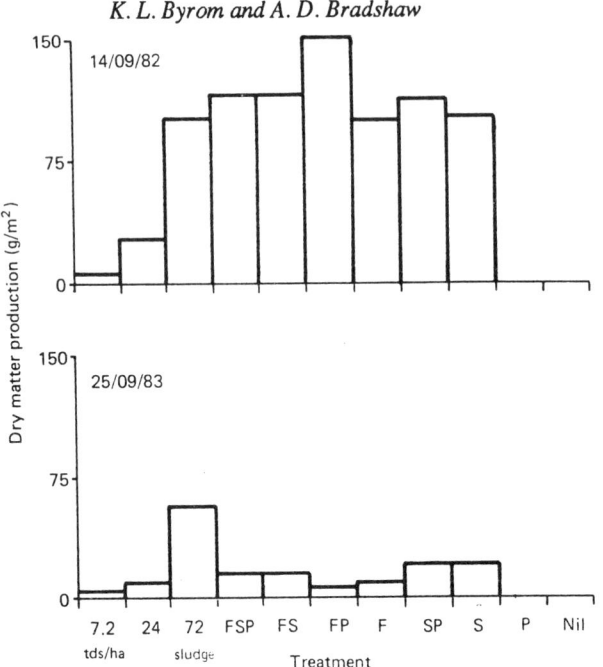

Fig. 6. The yields at the end of the second and third years of grass established on china clay sand waste amended with different amounts of sludge compared with fast (F) and slow (S) release fertiliser with and without peat (P) (see text).

Two types of sludge, polyelectrolyte stabilised (ps)(3.3% N, 0.9% P) and lime pressed cake (lp) (1% N, 0.8% P), were incorporated in different amounts and compared again with various fertiliser and peat combinations. The most significant results are the two harvests of the second year (Figure 7).

In the first harvest in July, the high ps-sludge treatments (given one year previously) gave excellent yields, surpassing those of the fertiliser treatments (re-treated 3 months previously). In the second harvest in October any effects of the fertiliser treatments had disappeared, but the high ps-sludge treatment (and to a lesser extent the high lp-sludge) still gave marked yield improvements. Since nitrogen and phosphorus were applied separately it was possible to see the over-riding importance of nitrogen but also some effect of phosphorus. Peat had no effect.

Because of the presence of fossil nitrogen in the "coal" fraction of the soil, it was not possible to determine the contribution of the sludge to the soil nitrogen directly. But plant analysis gave evidence of a very satisfactory supply of plant-available nitrogen in the sludge treatments through to the last harvest, in complete contrast to the fertiliser treatments. The same was true for phosphorus.

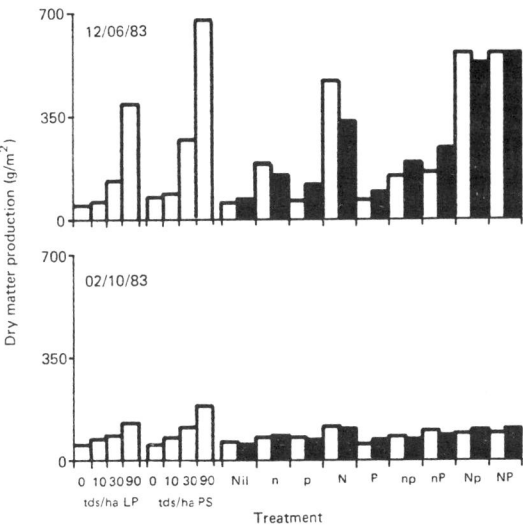

Fig. 7. The yields at the middle and end of the second year of grass established on colliery spoil amended with different amounts of sludge compared with combinations of low and high levels of nitrogen (n,N) and phosphorus (p,P), with and without peat (black, white).

4.3 Other sites

Sludge can be of great value in the reclamation of many difficult sites where the starting materials are low in nutrients and have poor physical characteristics. It has been used with success on metal contaminated materials, where the organic material temporarily relieves the metal toxicity (26), on slate waste and quarry faces (22), and on clay subsoils (13). It is outstandingly successful in the restoration of eroded sand-dunes, where its abilities to stabilise sand surfaces and give long-term nutrient release are invaluable (3).

5. CONCLUSIONS

All derelict land soil materials are very deficient in nitrogen. A major problem is, therefore, to find a way in which this capital can be restored. If this is not done, failure is inevitable. Fertiliser nitrogen is not an effective solution since it is completely soluble and too readily available. Sludge provides a unique alternative.

Sewage sludge contains, apart from soluble nitrogen in the water fraction, large quantities of insoluble nitrogen and phosphorus which, although organically bound, are readily available to plants by mineralisation. This organic nitrogen fraction is the important component since it appears to be equivalent to organically bound soil nitrogen. It is possible to apply in one application the amount of sludge necessary to build up the total capital required, about 1,000 kg N/ha. The soluble nitrogen is also valuable, but for this purpose can be discounted, since it can be considered equivalent to normal fertiliser nitrogen.

However, available evidence suggests that the organically bound fraction in sludge is more readily mineralised than soil organic nitrogen. If the main virtue of sludge is the way in which it can restore a normal nitrogen capital, then this rather rapid breakdown could be seen as a disadvantage. However, two processes are important. Firstly, if a vegetation cover is present it will take up whatever nitrogen is released from the sludge, transforming it into plant organic nitrogen, from which it will be subsequently released at a slower rate. Secondly, from the lysimeter experiment reported it appears that mineralisation within the sludge is accompanied by immobilisation, and that most of the nitrogen which could be available to plants, if not taken up by them, is taken out of circulation again. So little nitrogen is lost, except perhaps by volatilisation, and the soil capital is genuinely restored.

However, to carry out a full restoration in one step does mean that sufficient sludge must be added. An example of what would be required to achieve 1,000 kg N/ha could be as follows:

Digested cake available	25% dry solids - containing 3% N (15% available in first year)
total N	7.5 kg N/t
available N	1.1 kg N/t
hence organic N	6.4 kg N/t

Therefore to provide 1,000 kg organic N/ha, 156 t digested cake/ha is required.

Such an amount is large, about six times a typical dressing of farmyard manure, but it only has to be applied once. Similar calculations can be made for liquid sludge, which will be easier to spread and to incorporate into a ripped surface but, because of dilution, the volumes will be greater. The relative values of different treatments are discussed further in the recently published draft code of practice (27). If large amounts prove difficult to apply it is, of course, possible to compromise and to use reduced amounts, relying on legumes to build up the rest of the nitrogen.

For legumes in particular the other benefits of sludge must not be forgotten, especially the phosphorus content. The above material might contain 2 kg available P/t, so that the treatment would provide over 300 kg available P/ha, a very substantial amount. This would have particularly beneficial effects on legumes, as observed in field trials. The potential benefits of the organic matter have already been noted.

Providing the problems of availability, transport and spreading can be overcome, sewage sludge is an invaluable and important material for derelict land reclamation.

ACKNOWLEDGEMENTS

Much of this work was carried out with the support of an SERC CASE research studentship to K. Byrom in collaboration with the Water Research Centre, with the supervision and advice of R. D. Davis and J. E. Hall. She is also grateful for the support of West Yorkshire Waste Management over the preparation of the paper.

REFERENCES

(1) Michael, N. and A. D. Bradshaw (1989). A hard future for derelict land. Landscape Design, 177, 37-40.
(2) Bradshaw, A. D. (1983). The reconstruction of ecosystems. J. Applied Ecology, 20, 1-17.
(3) Bradshaw, A. D. and M. J. Chadwick (1980). The restoration of land. Blackwell, Oxford.
(4) Bloomfield, H. E., J. F. Handley and A. D. Bradshaw (1982). Nutrient deficiencies and the aftercare of derelict land. J. Applied Ecology, 19, 151-158.
(5) Jefferies, R. A., A. D. Bradshaw and P. D. Putwain (1981). Growth, nitrogen accumulation and nitrogen transfer by legume species established on mine spoils. J. Applied Ecology, 18, 945-956.
(6) Institute of Water Pollution Control (1978). Sewage sludge III: utilisation and disposal. I.W.P.C.
(7) Hall, J. E., A. P. Daw and C. D. Bayes (1986). The use of sewage sludge in land reclamation. Report 1346-M. Water Research Centre, Medmenham, UK.
(8) Fitter, A. H. (1974). A relationship between phosphorus requirement, the immobilisation of added phosphate and the phosphate buffering capacity of colliery shales. J. Soil Science, 25, 41-50.
(9) Reddy, K. R., R. Khaleel, M. R. Overcash and P. W. Westerman (1977). Conceptual modelling of non-point source pollution from land areas receiving animal wastes. I. Nitrogen transformations. American Society of Agricultural Engineering, 77, 4046.
(10) Tester, C. F., L. J. Sikora, J. M. Taylor and J. F. Parr (1977). Decomposition of sewage sludge in soil: I. Carbon and nitrogen

transformations. J. Environmental Quality, 6, 459-463.
(11) King, L. D. (1973). Mineralisation and gaseous loss of nitrogen in soil-applied liquid sewage sludge. J. Environmental Quality, 2, 356-358.
(12) Byrom, K. (1984). The use of sewage sludge in land reclamation. Ph.D. thesis, University of Liverpool.
(13) Coker, E. G., R. D. Davis, J. E. Hall and C. H. Carlton-Smith (1982). Field experiments on the use of consolidated sewage for land reclamation: effects on crop yield and composition and soil conditions, 1976-1981. Tech. Rep. 183. Water Research Centre, Medmenham, UK.
(14) Rundle, H. L. (1984). The use of sewage sludge in land reclamation. Part I: Establishment of grass on colliery shale. Severn-Trent Water Research and Development Project Report No. RP84-065.
(15) Rimmer, D. L. and A. Gildon (1986). Reclamation of colliery spoil: the effect of amendments and grass species on grass yield and soil development. J. Soil Science, 37, 319-327.
(16) Hinesly, T. D., O. C. Braids and J. E. Molina (1971). Agricultural benefits and environmental changes resulting from the use of digested sewage on field crops. United States EPA, Report No. SW30D.
(17) Woodmansee, R. G., I. Vallis and J. J. Mott (1981). Grassland nitrogen. In: Terrestrial nitrogen cycles; processes, ecosystem strategies and management impacts (F. E. Clark and T. Rosswall, Eds.), pp. 443-462. Ecological Bulletins, Stockholm.
(18) Haynes, R. J. (1986). Mineral nitrogen in the plant-soil system. Academic Press, Orlando.
(19) Likens, G. E., F. H. Bormann, N. M. Johnson, D. W. Fisher and R. S. Pierce (1970). Effects of forest cutting and herbicide treatment on nutrient budgets in the Hubbard Brook watershed ecosystem. Ecological Monographs, 40, 23-47.
(20) Marrs, R. H. and A. D. Bradshaw (1980). Ecosystem development on reclaimed china clay wastes. III Leaching of nutrients. J. Applied Ecology, 17, 727-736.
(21) Cornwell, S. M. and E. L. Stone (1968). Availability of nitrogen to plants in acid coal mine spoils. Nature, 217, 768-769.
(22) Coppin, N. J. and A. D. Bradshaw (1982). Quarry reclamation. Mining Journal Books, London.
(23) Ellis, J. D. and C. E. Brade (1982). Efficient use of sludge tankers. Paper presented to joint meeting of East and West Branches of the Institute of Water Pollution Control.
(24) Hall, J. E., R. J. Godwin, N. L. Warner and J. M. Davis (1986). Soil injection of sewage sludge. Report ER 1202-M. Water Research Centre, Medmenham, UK.

(25) Roberts, R. D., R. H. Marrs, R. A. Skeffington, A. D. Bradshaw and L. D. C. Owen (1982). Importance of plant nutrients in the restoration of china clay and other mine wastes, Transactions of the Institute of Mining and Metallurgy, *91A*, 42-50.
(26) Goodman, G. T., C. E. R. Pitcairn and R. P. Gemmell (1973). Ecological factors affecting growth on sites contaminated with heavy metals. In: Ecology and reclamation of devastated land (R. J. Hutnik and G. Davis, Eds.), pp. 149-174. Gordon and Breach, New York.
(27) Hall, J. E. (1989). The use of sewage sludge in land restoration. Draft Code of Practice, Report PRS 1783-M. Water Research Centre, Medmenham, UK.

Utilisation of sewage sludge in the United States for mine land reclamation

W. E. SOPPER

School of Forest Resources, The Pennsylvania State University, University Park, PA 16802, USA.

SUMMARY

Millions of hectares of marginal and barren land exist throughout the United States as a result of some types of mining activity. These lands are a source of acid mine drainage, surface runoff, erosion, and sedimentation, all of which have created serious water pollution and land degradation problems. During the past decade, a considerable amount of research has been conducted that has shown that stabilised municipal sludge from secondary wastewater treatment plants is an excellent soil amendment and chemical fertiliser substitute. In general, the results of that research has shown that municipal sludges can be used to revegetate mined lands in an environmentally safe manner with no adverse effects on vegetation, soil, or groundwater quality and with little risk to animal or human health.

1. INTRODUCTION

Surface mining alone has disturbed 1.76×10^6 ha, much of which is located in the populated eastern half of the United States. Pennsylvania is rich in coal. Recoverable coal reserves of over 34.5 billion tonnes exist under 41 of the 67 counties in the Commonwealth. In addition, over 97,000 hectares have been disturbed by strip mining and have been abandoned or inadequately reclaimed. This represents 27% of all the abandoned mine land in the nation.

Drastically disturbed lands generally provide a harsh environment for establishing vegetation. Major deterrents to vegetation establishment are usually a lack of nutrients and organic matter, low pH, low water-holding capacity, toxic levels of trace metals, compaction, and poor physical conditions of the spoil material.

During the past decade, a considerable amount of research has been conducted that has shown stabilised municipal sludge from secondary wastewater treatment plants is an excellent soil amendment and chemical fertiliser substitute. There has been considerable interest in the use of municipal sludge for the production of agricultural crops. However, sludge may contain every conceivable element or compound found in wastes from human, domestic, commercial, and industrial sources. It may contain substantial quantities of organic matter, plant nutrients, trace metals, and some potentially hazardous compounds. Thus, some concern has been raised concerning the potential health hazard of using sludges on agricultural land and the potential introduction of these elements into the human food chain. A possible alternative, which may alleviate these concerns, is to utilise the sludge to reclaim and revegetate marginal, unproductive land or barren land disturbed by coal mining activities. While the benefits of using sludge to reclaim land seems obvious, there is still some reluctance on the part of land owners and local government officials to use it. Perhaps the greatest obstacle is the lack of knowledge on the part of the general public concerning the possible impacts of such use on soils and vegetation, on groundwater and surface water, on energy conservation, on animal and human health, and on the overall balance between benefits and risks.

2. FEDERAL AND STATE GUIDELINES AND REGULATIONS

If sludge is used in the reclamation process there are state and federal guidelines and regulations related to land application of sludge that must be followed. Most of these guidelines set limits on sludge application rates based on nitrogen and other plant nutrient requirements of the vegetation as well as trace metal loadings. For example, the Environmental Protection Agency (EPA) has issued recommendations for the maximum amounts of trace metals that can be applied to agricultural soils via sewage sludge (1). These maximum amounts are related to the soil cation exchange capacity. The criteria are given in Table 1.

Table 1. EPA recommended total trace metal loadings for agricultural land (kg/ha).

Metal	Soil cation exchange capacity (meq/100g)		
	<5	5-15	>15
Pb	560	1120	2240
Zn	280	560	1120
Cu	140	280	560
Ni	140	280	560
Cd	6	11	22

In addition, some states have even more stringent guidelines concerning sludge application on the land. For instance, in 1988 Pennsylvania issued "Interim Guidelines for Sewage Sludge Use for Land Reclamation" (2). These guidelines state that due to the high permeability of mine spoils and low retention of organic matter, sufficient nitrogen in excess of the crop requirement must be provided in order to establish growth. To provide sufficient nitrogen a maximum application rate of 134 t/ha may be utilised for land reclamation. In addition, the application is further limited according to the trace metal content of the sludge and application rates may not exceed the limits given in Table 2.

Table 2. Pennsylvania Department of Environmental Resources (PDER) recommended maximum trace element loading rates for land reclamation.

Constituent	Maximum loading rate for land reclamation kg/ha	Maximum loading rate land reclamation for farming kg/ha
Cd	5.6	3.4
Cu	140	84
Cr	560	336
Pb	560	336
Hg	1.7	1.1
Ni	56	33.6
Zn	280	168

The state guidelines further require that the soil pH must be adjusted to 6.0 during the first year of sludge application and must be maintained at 6.5 for two years following final sludge application. Liming is required to immobilise the trace metals in order to reduce their availability for plant uptake and to prevent their leaching into groundwater.

3. DEMONSTRATION PROJECTS

To introduce the concept of the use of municipal sludge for revegetation of mined land and to bridge the gap between available technical information and public understanding, a demonstration programme was initiated in 1977. It consisted of establishing a series of 4 ha demonstration plots throughout the anthracite (deep mines) and bituminous (strip mines) coal mining regions of Pennsylvania.

One of these demonstration projects will be discussed to illustrate the various ways sludge can be used to revegetate marginal land and the environmental impacts on soil, vegetation, and groundwater.

The site was representative of bituminous strip mine banks, which have been backfilled and recontoured after mining without top soil replacement. Several attempts had been made, about ten years earlier, to revegetate the area using lime, commercial fertiliser and seed, but without success. The surface spoil was compacted, stony, extremely acid (pH 3.8) and devoid of vegetation. The plot was scarified with a chisel plough to loosen the surface spoil material and then treated with agricultural lime (8.4 t/ha) to raise the spoil pH to 7.0.

Stabilised sludges for the project were obtained from three local wastewater treatment plants. Liquid digested sludge was transported to the site in tank trucks. Dewatered digested sludge was brought to the site in coal trucks. The 4-ha plot was subdivided into four 1-ha subplots for application of liquid digested sludge at two rates and dewatered sludge at two rates; only the two highest rates are discussed in this paper. After the installation of groundwater wells and the collection of background water and spoil samples, the liquid digested sludge was applied with a vacuum tank liquid manure spreader at 155 m^3/ha (equivalent to 11 t/ha). Dewatered sludge was applied at 90 and 184 t/ha. Sludge was applied in late May 1977. Samples of the sludge were collected as the sludge was applied to the plots. These samples were analysed to determine the amounts of plant nutrients and trace metals applied.

The amounts of trace metals applied at the highest liquid and dewatered sludge application rates are given in Table 3, together with the US EPA and Pennsylvania Department of Environmental Resources (PDER) interim guideline recommendations. It is obvious that the amounts of trace metals applied, even at the highest sludge application rate, were well below the recommended lifetime limits, except for copper, which slightly exceeded the Pennsylvania guidelines. The amounts of nutrients applied by each of the sludge application rates and the commercial fertiliser equivalents are given in Table 4. Potassium is the only plant nutrient deficient in all sludge application rates. The highest sludge application rate (184 t/ha) was equivalent to applying an 11.9.0 commercial chemical fertiliser at 22,400 kg/ha.

Immediately after sludge application and incorporation, the site was broadcast seeded with a mixture of two grasses and two legumes, including Kentucky-31 tall fescue (*Festuca arundinacea* Schreb.) (22 kg/ha), Pennlate orchardgrass (*Dactylis glomerata* L.) (22 kg/ha), Penngift crownvetch

(*Coronilla varia* L.) (11 kg/ha) and Empire birdsfoot trefoil (*Lotus corniculatus* L.) (11 kg/ha), then mulched with straw and hay at the rate of 3.8 t/ha.

Table 3. Trace metal loadings of the highest liquid and dewatered sludge applications compared with EPA and PDER recommendations.

Constituent	Loading at sludge application rates (kg/ha)		Recommendations (kg/ha)	
	11 t/ha	184 t/ha	EPA (CEC 5-15)	PDER
Cu	21	129	280	84
Zn	21	147	560	168
Cd	0.1	0.6	11	3.4
Pb	10	55	1,120	336
Ni	1	12	280	33.6
Cr	16	74	None	336
Hg	0.01	0.09	None	1.1

Table 4. Commercial fertiliser equivalents of the sludge application.

Sludge application rate (t/ha)	Fertiliser equivalent (fertiliser formula)			
	Amount (kg/ha)	N (kg/ha)(%)	P_2O_5 (kg/ha)(%)	K_2O (kg/ha)(%)
184	22,400	2,388(11)	2,103(9)	21(0)
11	2,240	284(13)	143(6)	6(0)

A complete monitoring system was installed on the demonstration plot to evaluate the effects of the sludge applications on water quality, vegetation and soil. One groundwater well was drilled up-slope to sample background water quality and two groundwater wells were drilled down-gradient of the sludge-treated plots to sample the effects of the sludge application on groundwater quality. Samples were collected from the wells with a submersible pump and a Kemmerer water sampler. In order to obtain control data, groundwater samples were collected from all wells prior to sludge application. After sludge application, groundwater samples were collected bi-weekly for the first two months and monthly thereafter. Samples were analysed for pH, nitrate-N by ion-selective electrode (3), dissolved Cu, Zn, Cr, Pb, Co, Cd and Ni by atomic absorption spectrophotometry (4), and total and faecal coliform bacteria by the methods described by the American Public Health Association (5).

Spoil samples were collected at the 0-15 cm depth and analysed for pH and dilute hydrochloric acid extractable Cu, Zn, Cr, Pb, Co, Cd and Ni by atomic absorption (6). Spoil samples were taken at 1-year intervals to determine the effect of the sludge application on the soil chemical status and to evaluate the fate of the heavy metals.

At the end of each growing season, vegetation growth responses were determined by measurements of percentage area cover, height growth and dry matter production. Twenty $0.2~m^2$ samples of the vegetative cover (including all above-ground organic matter) were collected and oven-dried to determine dry matter production in kg/ha. No crops were harvested over the 5-year period. Three samples of each species, tall fescue, orchardgrass, crownvetch, and birdsfoot trefoil, from each plot were collected for foliar analyses. Plant samples were analysed for Cu, Zn, Cr, Pb, Co, Cd and Ni by atomic absorption spectrophotometry (6).

3.1 Vegetation growth response and foliar analyses

During the first two years, the two grass species dominated the site, but by the third growing season, the two legume species predominated. The site had a complete vegetative cover by August 1977, three months after sludge application. First-year average height ranged from 32 to 35 cm, with the greatest growth being produced with the 184 t/ha sludge application. The greatest vegetation height increase occurred the second year after sludge application, and then levelled off. Samples of the individual grass and legume species were collected at the end of each growing season for foliar analyses. Results for tall fescue and birdsfoot trefoil for the highest sludge application rate are given in Table 5 for 1977 to 1981. Foliar trace metal concentrations generally decreased over the 5-year period. Overall, the trace metal concentrations were well below the suggested tolerance levels. These levels represent the level at which a yield reduction might occur and do not represent levels at which toxicity occurs. There were no phytotoxicity symptoms observed for any vegetation on the sludge-treated areas.

In general, the vegetation cover improved over the five growing seasons (1977 to 1981) following sludge application. No deterioration in vegetation quality or yield has been measured or observed even after ten years. In comparison, the remainder of the site, not treated with sludge, remains barren.

Table 5. Average concentration (mg/kg) of trace metals in the foliar samples collected from the 184 t/ha plot.

Species	Year	Cu	Zn	Cr	Pb	Co	Cd	Ni
Tall fescue	1977	9.4	44.4	0.8	4.5	1.5	0.20	9.8
	1978	8.6	44.4	0.8	4.5	1.6	0.41	3.7
	1979	9.2	72.5	0.5	1.8	0.6	0.08	2.5
	1980	3.5	41.9	1.1	3.8	1.8	0.73	7.3
	1981	12.7	34.8	0.1	1.9	1.0	0.50	0.7
Birdsfoot trefoil	1977	13.9	95.9	1.0	7.4	2.1	0.43	6.3
	1978	7.7	30.4	0.3	8.5	3.0	0.07	4.8
	1979	9.2	41.5	1.7	1.8	0.3	0.04	6.3
	1980	8.2	45.3	1.9	4.5	1.4	0.08	6.5
	1981	11.6	40.9	0.1	1.5	0.9	0.37	3.3
Suggested tolerance level (7)		150	300	2	10	5	3	50

3.2 Spoil chemical characteristics

To evaluate the effects of the sludge treatment on the chemical properties of the spoil, samples were collected at various locations and depths at the end of each year. Results indicated that the lime and sludge applications did raise the spoil pH significantly, from 3.8 to 6.9 in about four months (1977). Average spoil pH was 6.3, 7.4, and 6.9 in 1978, 1979 and 1981, respectively and the higher pH was maintained. Under Pennsylvania guidelines, surface spoil samples must be collected at the end of the first and second year following sludge application to document that the pH has not dropped below pH 6.5. Should the pH drop below this level, lime must be applied to raise it to at least 6.5.

Spoil samples were also analysed for trace metals. A comparison of trace metal concentrations before and after sludge was applied is given in Table 6. Even at the highest sludge application rate (184 t/ha), the trace metal concentrations in the surface spoil (0 to 15 cm) were only slightly increased. In general, the trace metal concentrations in the spoil were all extremely low in comparison to published normal ranges for soils.

Concentrations of extractable trace metals at the 0-15 cm depth increased over the 5-year period (Table 6), with the greatest increase in the fifth year, probably as a result of the decrease in pH. All values are well within the normal ranges of these elements in US soils (8).

Table 6. Changes in the concentration (mg/kg) of extractable trace metals from spoil collected at the 0-15 cm depth.

Sampling date	Cu	Zn	Cr	Pb	Co	Cd	Ni
May 1977[1]	2.5	2.9	0.2	0.5	0.7	0.02	1.1
Sept. 1977	10.8	7.7	0.4	3.5	1.3	0.04	0.9
1978	8.8	7.7	0.2	2.3	1.2	0.02	1.2
1979	58.7	56.9	1.7	13.0	1.2	0.27	1.5
1981	87.3	74.6	3.5	22.7	1.6	0.95	2.8
Normal range for US soils (8)	2-100	10-300	5-3,000	2-200	1-40	0.01-7.00	5-500

[1] May 1977 values represent pre-treatment conditions.

A comparison between Tables 1, 2 and 6 can be made if concentration values in Table 6 are converted to kg/ha. Conversions for 1981 indicate that there are 175 kg Cu/ha, 149 kg Zn/ha, 45 kg Pb/ha, 7 kg Cr/ha, 5.6 kg Ni/ha and 1.3 kg Cd/ha in the 0-15 cm spoil depth. All of these trace metal amounts, except Cu, are within the EPA and PDER recommendations. The amount of Cu is within the EPA guidelines, but is higher than that allowed under the Pennsylvania guidelines.

3.3 Spoil percolate water quality

Results of the analyses of soil percolate water at the 90 cm depth for the highest sludge application and the control plot are given in Table 7. Average monthly concentrations of NO_3-N in the percolate during the summer months in the first year (1977) on the plots treated with the highest applications of dewatered sludge were only slightly above potable water standards (10 mg/l). The highest monthly average was 13.0 mg/l for the month of August. Percolate NO_3-N concentrations were surprisingly low during May and June, immediately following the sludge application. This was probably due to the fact that rainfall during this period was below normal. As a result, there was little opportunity for leaching of nitrogen from the sludge to occur. By October, with the development of a complete vegetative cover, the concentrations of NO_3-N in the percolate decreased to levels below 10 mg/l. Concentrations of NO_3-N in the percolate remained at low levels throughout 1978 to 1981.

Results of the analyses for dissolved trace metals at the 90 cm depth for the highest sludge application together with the control plot are also given in Table 7. Results indicate that percolate water quality met EPA drinking water standards with only a few exceptions. During the first 3 months in the first year following sludge application, the concentrations of Zn and Ni significantly increased and exceeded drinking water standards. Concentrations of Cr and Pb slightly exceeded drinking water standards on both the control and sludge-treated plots. During the second (1978) and third (1979) years, only concentrations of Pb exceeded drinking water standards. These concentration increases were minimal and posed no threat to human nor animal health. It should be noted that the average monthly concentrations of Pb on the control plot also exceeded potable water standards during the study period (1977 to 1981).

Table 7. Results of analyses for trace metals and nitrate-nitrogen for soil percolate at the 90-cm depth (mg/l).

Sludge application rate (t/ha)	Year[1]	Cu	Zn	Cr	Pb	Co	Cd	Ni	NO_3-N
0	1977	0.63	2.75	0.23	0.07	0.67	0.005	1.37	1.8
	1978	0.14	1.20	0.05	0.10	0.22	0.002	0.33	0.7
	1979	0.10	0.68	0.05	0.05	0.12	<0.001	0.26	0.7
	1980	0.08	0.90	0.06	0.07	0.10	0.001	0.22	0.8
	1981	0.05	0.36	0.03	0.03	0.07	0.002	0.12	0.9
184	1977	0.24	5.91	0.04	0.05	1.50	0.011	2.82	7.3
	1978	0.04	1.16	<0.01	0.08	0.19	0.002	0.26	0.5
	1979	0.07	0.87	0.02	0.05	0.20	0.001	0.34	<0.5
	1980	0.02	0.51	0.01	0.03	0.06	0.001	0.11	0.6
	1981	0.03	0.36	0.02	0.02	0.07	0.001	0.07	0.7
EPA drinking water standard		1.00	5.00	0.05	0.05	-	0.010	-	10.0

[1] Values represent the mean of all samples collected from the plot for the year.

Total and faecal coliform analyses were conducted on all soil percolate water samples collected during the period from May 1977 to December 1981. No faecal coliform colonies were observed for any sample.

3.4 Groundwater quality

Results of the analyses of groundwater well samples are given in Table 8. The values for Well 1 (control) reflect the quality of the groundwater for the disturbed mine site. Well 2 reflects the effects of dewatered sludge on water quality. The results indicate that the sludge applications have not had any significant effect on groundwater pH nor on concentrations of NO_3-N, which were below 10 mg/l (maximum concentration for potable water) for all months sampled during the 5-year period. The highest monthly values were 3.0 mg/l for the control well and 2.4 mg/l for Well 2.

Table 8. Groundwater analyses for the period 1977-1981 (mg/l).

	Year[1]	pH	NO_3-N	Cu	Zn	Cr	Pb	Co	Cd	Ni
Well 1	1977	4.4	1.4	0.22	4.13	0.02	0.14	3.19	0.006	3.67
(Control)	1978	4.3	<0.5	0.23	2.02	0.01	0.19	1.04	0.002	0.98
	1979	4.6	<0.5	0.17	1.48	0.03	0.13	0.57	0.001	0.50
	1980	5.5	0.6	0.05	0.89	0.05	0.09	0.60	0.001	0.50
	1981	5.7	0.7	0.06	0.83	0.03	0.04	0.45	0.003	0.31
Well 2	1977	4.6	1.1	0.10	3.39	0.03	0.09	2.12	0.001	2.67
(184 t/ha)	1978	4.5	<0.5	0.14	3.29	0.01	0.20	1.16	0.002	1.26
	1979	4.4	<0.5	0.18	1.49	0.03	0.13	1.92	0.001	0.97
	1980	5.7	0.6	0.05	1.05	0.04	0.11	0.87	0.001	0.76
	1981	6.0	0.6	0.05	0.57	0.02	0.05	0.42	0.001	0.31
EPA drinking water standard			10.0	1.0	5.0	0.05	0.05	5.0[2]	0.01	2.02[2]

[1] Values are annual means of monthly samples.
[2] Recommended values for irrigation water for agricultural use.

There appear to be no significant increases in any of the trace metal concentrations over the 5-year period in the groundwater samples from Well 2 compared to the control well. From 1977 to 1981, most of the monthly concentrations were within the US EPA drinking water standards. The only exception was Pb, which exceeded the limit of 0.05 mg/l for both the control well and Well 2, probably resulting from increased release of the element from the spoil material due to the increased acidity (lower pH) resulting from the mining. The highest monthly Pb values were 0.28 mg/l in the control well and 0.33 mg/l in Well 2 in 1978, and the mean annual Pb concentrations were 0.19 and 0.20 mg/l for the control well and Well 2, respectively. By 1981, however, the mean annual Pb concentrations had decreased to 0.04 and 0.05 mg/l for the two wells.

Table 9. Site descriptions.

Site	Age years	Amendment	Application rate tds/ha	Date of application	Lime application t/ha	pH
1	1	Sludge	120	9/84	18	6.9
2	2	Sludge	128	6/83	18	7.0
3	3	Sludge	128	5/82	12	6.8
4	4	Sludge	134	7/81	18	6.7
5	5	Sludge	134	7/80	11	7.3
Fertiliser-amended	5	Fertiliser (23.24.24)	0.5	8/80	11	6.3

Total and faecal coliform analyses were conducted on monthly groundwater samples collected during the period May 1977 to December 1981. No faecal coliform colonies were observed for any sample.

4. EFFECTS OF MICROBIAL POPULATIONS

In 1985, a field study of five coal surface mine sites reclaimed with sewage sludge and one site reclaimed by conventional methods (chemical fertiliser) was conducted to assess the effects of sludge amendment and time on populations of bacteria, fungi, and actinomycetes, and on microbial respiration and organic matter decomposition. Although the immediate goal of reclamation is to establish a vegetative cover that will prevent soil erosion, the long-term goal is soil ecosystem development and stability. Microbial processes such as humification, soil aggregation and nitrogen cycling are essential in establishing productivity in mine spoils, and productivity should be evaluated not only on above-ground biomass, but also on the degree of development of functional microbial populations resembling those of an undisturbed soil. Five mine sites were selected that had been reclaimed using the same sludge at approximately the same application rate. A five-year-old conventional fertiliser-amended site was also used as a control for comparison. The sludge used was anaerobically digested and dewatered and composted with woodchips at a 1:2 ratio. The compost was then mixed with an equal amount of dewatered sludge cake. The mean and range of the concentrations of constituents in the sludge samples collected at the time of application on the five sites are given in Table 9. The mixture, called Philadelphia mine mix, was applied to the sites as shown in Table 10.

Table 10. Mean and range of the concentrations of constituents in sludge samples collected at the time of application on the five sites (dry weight basis).

Constituent	Mean	Range
Solids (%)	48	47 - 49
pH	7.3	6.6 - 8.0
	concentration (mg/kg)	
NO_3-N	260	89 - 493
NH_4-N	645	193 - 1,316
Organic-N	4,670	3,667 - 6,575
Total N	5,755	4,456 - 7,980
Total P	11,531	8,408 - 14,366
K	1,032	789 - 1,319
Ca	14,415	13,049 - 15,920
Mg	8,277	6,961 - 10,041
Na	460	367 - 613
Fe	19,272	18,010 - 20,017
Al	15,946	13,085 - 18,058
Mn	1,036	920 - 1,093
Zn	1,921	1,494 - 2,236
Cu	753	600 - 809
Cr	506	395 - 580
Pb	542	454 - 588
Ni	157	106 - 198
Co	24	7 - 38
Cd	6	4 - 8
Hg	0.5	0.5 - 0.6

Six soil samples (5 cm diameter by 5 cm high) were collected in July 1985 for microbiological analyses by standard methods (9). Details of this study have previously been reported by Seaker and Sopper (10, 11).

Table 11. Microbial populations on stripmine sites one to five years following sludge application, and on the fertiliser-amended site (means of 6 samples with standard error).

Site	Aerobic heterotrophic bacteria	Fungi	Actinomycetes
	10^6/g	10^5/g	10^4/g
1	63.67 ± 16.93 a	18.14 ± 5.45 a	1.48 ± 1.04 b
2	7.07 ± 1.32 b	5.80 ± 2.35 b	9.75 ± 5.48 b
3	4.09 ± 0.77 b	5.54 ± 1.32 b	56.21 ± 26.71 b
4	11.37 ± 3.64 b	3.98 ± 0.46 b	140.23 ± 59.57 a
5	13.74 ± 3.58 b	4.03 ± 1.05 b	40.89 ± 22.68 b
F-value	***	**	*
	10^0/g	10^0/g	10^0/g
Fertiliser-amended	3.06 ± 1.17	0.16 ± 0.04	6.94 ± 4.01

Means followed by different letters are significantly different at the 0.05 level of probability by the Waller-Duncan k-ratio t-test.
*, ** and *** = significant effect at $p<0.05$, <0.01, and <0.001, respectively.

4.1 Aerobic heterotrophic bacteria

Bacterial populations on the sludge-amended sites ranged from 4 to 63 x 10^6/g (Table 11). Bacterial counts were 5 to 15 times higher on Site 1 than on the older sites, and were dramatically increased on all sludge-amended sites compared to the fertiliser-amended site. The first-year peak and subsequent stabilisation of bacterial populations is a typical response following organic matter additions to mine spoils (12). Considering the extremely low initial pH of the mine spoils in this study, commonly ranging from 3.0 to 5.0 prior to lime additions, the microbial populations achieved with lime and sludge amendments after only one year, are remarkably high. They compare favourably with estimates of 1 to 34 x 10^6/g reported for undisturbed soils (11, 12, 13, 14, 15).

Table 12. Concentrations and amounts of trace metals (dry weight) applied to the site compared to the PDER interim guidelines for maximum trace element loading for land reclamation.

	Metal concentration (mg/kg)	Amount applied at 134 t/ha (kg/ha)	PDER Maximum loading rate land reclamation (kg/ha)
Cu	666	50	140
Zn	2,298	175	280
Cr	411	31	560
Pb	688	53	560
Cd	20	2	5.6
Ni	85	7	56

4.2 Fungi

Sludge application resulted in fungal populations in the range of 4 to 18 x 10^5/g (Table 11.) These compare favourably with fungal populations in undisturbed soils which have been reported to range from 0.05 to 9 x 10^5/g (13, 14, 15, 16, 17). Fungal numbers were 3 to 4 times higher on Site 1 than on the older sites, and were greatly increased on all sludge-amended sites compared to the fertiliser-amended site.

4.3 Actinomycetes

Sludge applications resulted in actinomycete populations ranging from 1.48 to 140.23 x 10^4/g (Table 11), compared to actinomycete populations for undisturbed soils, reported in the range of 1 to 436 x 10^4/g (13, 14, 15, 16, 17). Actinomycetes exhibited a different pattern of development to the bacteria and fungi. These microbes are less competitive than the other groups and their populations were significantly lower on Sites 1 and 2 than on the older sites. The pattern follows that described by Alexander (17), whereby the bacteria and fungi proliferate initially upon the addition of organic matter to the soil, and the actinomycete responses do not occur until the later stages of decay, when competition has decreased. Actinomycete populations on Sites 3, 4, and 5 were considerably higher than on the fertiliser-amended site.

5. EFFECTS ON WILDLIFE

Many studies have been conducted on domestic animals that graze sludge-amended land or that have been fed sludge-grown maize or hay. The results of most of those studies have shown little or no adverse effect on the health of the animals. Fewer studies have been made on free-ranging wildlife species. In 1983, a study was conducted to determine the effects of sludge applications on the health of cottontail rabbits (*Sylvilagus floridanus*) and meadow voles (*Microtus pennsylvanicus*).

Rabbits and voles were trapped on a 121 ha mined area that had been amended with 134 t/ha of Philadelphia mine mix and seeded with tall fescue, orchardgrass and birdsfoot trefoil in June 1981. A nearby 81 ha mine site reclaimed with fertiliser and lime and seeded with similar species, also in 1981, was used as a control site. The concentrations and amounts of trace metals applied in the sludge are given in Table 12.

The trapped animals were necropsied and samples of liver, kidney, muscle, and bone tissue were analysed for trace metals by atomic absorption spectrophotometer methods. Six vegetation samples of each species were also collected and similarly analysed.

Results of the vegetation foliar analyses are given in Table 13. Concentrations of Zn in all plant species and Cd and Cu in three of the four species sampled were significantly higher ($p<0.05$) on the treated site compared to the control (Table 13).

Levels of most metals in the cottontail rabbits collected were not different ($p<0.05$) between males and females and therefore the data were combined (Table 14). Only Zn concentrations in the femur were significantly higher in the rabbits from the sludge-amended site. Concentrations of all other trace metals were not significantly different. The mean levels of Cd in rabbit liver from the sludge-treated site were above mean Cd amounts for selected foods evaluated by the US Food and Drug Administration, e.g. raw beef liver mean - 0.183 mg/kg, but within the range of Cd values observed for several foods, such as ground beef, breakfast cereal, and sugar (18). Occasional consumption of cottontail rabbit muscle tissue from the sludge-treated site should pose no threat to predators nor human health. A detailed report on this study has been previously published (19).

Results of tissue analyses for the meadow voles are given in Table 15. Concentrations of Cu, Zn, Co, Cd and Ni were not significantly different between control and sludge-treated sites. However, Cr concentrations in kidney and bone, and Pb concentrations in liver and bone were higher ($p<0.05$) on the control site than on the sludge-treated site. It appears that the sludge application has not had an adverse effect on the health of meadow voles. A detailed report on this study has been previously published (20).

Table 13. Trace metal concentrations (mg/kg dry weight) in vegetation foliar samples.

Element	Control		Treatment	
	Average	Median	Average	Median
Bromus sp.				
Cu	5.8	5.5**	8.4	7.4**
Zn	21.9	21.5**	41.8	37.6**
Cr	0.0	0.0	0.6	0.0
Pb	2.7	2.9	2.0	1.9
Co	0.3	0.1	<0.1	0.0
Cd	0.02	0.01	0.06	0.03
Ni	0.5	0.5	1.6	0.6
Orchardgrass				
Cu	9.4	9.4	10.5	10.4
Zn	23.7	23.5**	43.9	45.5**
Cr	0.0	0.0**	8.6	1.3**
Pb	4.2	4.0	3.2	3.3
Co	0.5	0.5	0.2	0.1
Cd	0.02	<0.01**	0.2	0.2**
Ni	1.0	1.0*	4.7	2.3*
Tall fescue				
Cu	7.0	7.0*	8.0	7.9*
Zn	21.7	21.8**	41.4	37.6**
Cr	0.1	0.0	4.2	0.0
Pb	2.8	2.9	2.3	2.1
Co	0.1	0.0	0.3	0.0
Cd	0.02	0.02**	0.1	0.12**
Ni	1.4	1.4	4.4	1.1
Trefoil				
Cu	10.7	10.8*	11.6	11.5*
Zn	43.0	43.0*	60.0	59.9*
Cr	0.3	0.0*	20.5	25.1*
Pb	4.3	4.5	3.6	3.7
Co	1.8	1.8*	0.5	0.4*
Cd	0.02	0.03*	0.2	0.16*
Ni	13.8	13.5	17.0	18.4

*, ** Significant at $p<0.05$ and $p<0.01$, respectively.

Table 14. Heavy metal concentrations (mg/kg dry weight) of male and female cottontail tissues[1].

Element	Control Average	Median	Treatment Average	Median
Femur				
Cu	2.9	3.0	2.8	2.8
Zn	126	125*	148	147*
Cr	1.4	1.5	1.3	1.3
Pb	12.2	12.0	12.9	12.8
Co	5.3	5.5	5.4	5.4
Cd	0.01	0.01	0.01	0.01
Ni	6.8	6.8	6.5	6.3
Kidney				
Cu	10.6	11.0	10.9	11.3
Zn	81.6	77.0	87.5	84.3
Cr	0.0	0.0	0.0	0.0
Pb	1.5	1.5	2.3	1.5
Co	0.0	0.0	0.1	0.0
Cd	9.6	5.3	17.0	12.3
Ni	0.5	0.0	0.8	0.0
Liver				
Cu	11.0	10.8	11.8	11.4
Zn	110	120	115	114
Cr	0.0	0.0	0.0	0.0
Pb	0.2	0.0	0.8	0.9
Co	0.0	0.0	0.2	0.0
Cd	1.60	1.63	2.40	2.25
Ni	0.1	0.0	0.2	0.3
Muscle				
Cu	4.3	4.0	4.0	4.3
Zn	41.1	41.8	44.1	43.6
Cr	0.1	0.0	0.5	0.0
Pb	0.3	0.0	0.1	0.0
Co	0.1	0.0	0.0	0.0
Cd	0.01	0.01	0.01	0.01
Ni	0.5	0.3	0.9	0.1

*Significant at $p<0.05$.
[1] Control values based on 11 replicates; treatment values based on 10 replicates (replicate - one sample from one rabbit).

Table 15. Trace metal concentrations (mg/kg dry weight) in meadow vole tissues.

Tissue	Metal	Control Mean	SE	Treated Mean	SE
Kidney	Cu	11.70±	0.15	12.29±	0.33
	Zn	55.07±	1.24	59.32±	0.31
	Cr*	0.68±	0.08	0.51±	0.07
	Pb	1.06±	0.26	2.27±	0.32
	Co	0.72±	0.07	0.77±	0.06
	Cd	0.48±	0.02	1.41±	0.22
	Ni	1.01±	0.10	0.59±	0.11
Liver	Cu	13.36±	0.20	13.60±	0.13
	Zn	81.26±	1.18	83.27±	0.88
	Cr	0.43±	0.30	0.37±	0.25
	Pb*	2.50±	0.20	0.41±	0.11
	Co	0.49±	0.04	0.40±	0.03
	Cd	0.23±	0.03	0.27±	0.04
	Ni	0.30±	0.14	0.21±	0.15
Muscle	Cu	7.50±	0.28	7.03±	0.10
	Zn	43.35±	0.77	49.16±	0.92
	Cr	0.43±	0.39	0.00±	0.00
	Pb	3.29±	0.13	3.00±	0.07
	Co	0.86±	0.06	0.91±	0.04
	Cd	0.38±	0.03	0.30±	0.01
	Ni	2.49±	0.35	2.38±	0.06
Bone	Cu	4.26±	0.53	3.09±	0.18
	Zn	179.28±	3.09	157.65±	2.72
	Cr*	4.53±	1.26	0.35±	0.16
	Pb**	12.90±	0.40	11.53±	0.42
	Co	5.77±	0.19	5.15±	0.23
	Cd	1.93±	0.08	1.47±	0.09
	Ni	8.80±	0.64	8.27±	0.35

*, ** Significant at $p < 0.05$ and $p < 0.01$, respectively.

6. CONCLUSIONS

One of the best potential uses for municipal sludge is its utilisation for improvement of marginal land disturbed by mining activities. One of the greatest obstacles to the concept of recycling sludge on the land is the lack of knowledge on the part of the general public and public officials. During the past decade, a considerable amount of research has been conducted and the gap between available technical information and public understanding is slowly closing. Generally, the results of research, such as the studies described in this paper, show that stabilised municipal sludges can be used to reclaim barren mine land in an environmentally safe manner with no adverse effects on vegetation, soil, or groundwater quality and with little risk to animal or human health.

This technology helped the City of Philadelphia to become the first east coast city to cease ocean dumping. Since 1979 Philadelphia sludge has been applied at 134 t/ha to strip-mined sites in Pennsylvania for revegetation purposes on over 1600 ha. This programme is continuing and is a good example of how municipal sludge can be used beneficially on a large scale.

REFERENCES

(1) United States Environmental Protection Agency (1977). Municipal sludge management: Environmental factors. Tech. Bull. EPA 430/9-76-004, MCD-28.
(2) Pennsylvania Department of Environmental Resources (1988). Guidelines for sewage sludge use for land reclamation. In: The rules and regulations of the Department of Environmental Resources. Commonwealth of Pennsylvania, Chapter 275.
(3) Ellis, B. G (1976). Analyses and their interpretation for wastewater application on agricultural land. North Central Regional Research Publication 235-Sec. 6.
(4) United States Environmental Protection Agency (1979). Methods for chemical analysis of water and wastes. Washington D.C.
(5) American Public Health Association (1981). Standard methods for the examination of water and wastewater. Washington D.C.
(6) Jackson, M. L. (1958). Soil chemical analysis. Prentice-Hall, Englewood Cliffs, N.J.
(7) Melsted, S. W. (1973). Soil-plant relationships. In: Recycling municipal sludges and effluents on land, pp. 121-128. National Assoc. of State Universities and Land Grant Colleges, Washington D.C.

(8) Allaway, W. H. (1968). Agronomic controls over the environmental cycling of trace metals. Advances in Agronomy, 20, 235-271.
(9) Page, A. L., R. H. Miller and D. R. Kenney (1982). Methods of soil analysis, Part 2, Chemical and microbiological properties. American Soc. of Agronomy, Madison, WI.
(10) Seaker, E. M. and W. E. Sopper (1988). Municipal sludge for minespoil reclamation: I. Effects on microbial populations and activity. J. Environ. Qual. 17 (4), 591-597.
(11) Seaker, E. M. and W. E. Sopper (1988). Municipal sludge for minespoil reclamation: II. Effects on organic matter. J. Environ. Qual., 17 (4), 598-602.
(12) Fresquez, P. R. and E. F. Aldon (1986). Microbial reestablishment and the diversity of fungal genera in reclaimed coal mine spoils and soils. Reclam. Reveg. Res., 4, 245-258.
(13) Wilson, H. A. (1965). The microbiology of strip mine spoil. West Virginia Univ. Agric. Exp. Stn. Bull. 506T.
(14) Segal, N. and R. L. Mancinelli (1987). Extent of regeneration of the microbial community in reclaimed spent oil shale land. J. Environ. Qual., 16, 44-48.
(15) Visser, S. (1985). Management of microbial processes in surface mined land reclamation in western Canada. In: Soil reclamation processes (R. L. Tate and D. A. Klein, Eds.), pp. 203-341. Marcel Dekker, New York.
(16) Miller, R. M. and S. W. May (1979). Stauton 1 reclamation demonstration project, Progress Report II, Argonne National Lab Rep. ANL/LRP-4. Nat'l Tech. Information Serv., V.S. Dep. of Commerce, Springfield, VA.
(17) Alexander, M. (1977). Introduction to soil microbiology. John Wiley and Sons, New York.
(18) Sharma, R. P. (1981). Soil-plant-animal distribution of cadmium in the environment. In: Cadmium in the environment, Part I. Ecological cycling, (J. O. Nriagu, Ed.), pp. 587-605. John Wiley and Sons, New York.
(19) Dressler, R. L., G. L. Storm, W. M. Tzilkowski and W. E. Sopper (1986). Heavy metals in cottontail rabbits on mined lands treated with sewage sludge. J. Environ. Qual., 15, 278-281.
(20) Alberici, T. M., W. E. Sopper, G. L. Storm and R. H. Yahner (1989). Trace metals in soil, vegetation, and voles from mine land treated with sewage sludge. J. Environ. Qual., 18, 115-120.

Sewage sludge as an amendment for reclaimed colliery spoil

I. D. PULFORD

Agricultural, Food and Environmental Chemistry,
University of Glasgow, Glasgow, G12 8QQ, UK.

SUMMARY

Colliery spoil is crushed rock brought to the surface by mining and dumped in tips. To reclaim and revegetate this material it is essential to improve its chemical, biological and physical properties. The reclaimer is essentially trying to create "soil conditions" in a relatively short period. Two necessary improvements are: the inhibition of acid production in pyritic spoil, and the initiation of nutrient cycling. The production of acid due to oxidation of pyrite (FeS_2) is a difficult problem. The rate limiting step in this process is the oxidation of Fe(II) to Fe(III), and treatments which interfere with this reaction can inhibit the rate of acid production. Manures are very effective at doing this by complexing iron. The problem of nutrient supply to plants growing on colliery spoil is not simply one of providing sufficient for establishment, but nutrient cycling over the long term. This is particularly relevant to the supply of nitrogen. Even small additions of sewage sludge to colliery spoil can enhance microbial transformations of nitrogen six years after reclamation. This effect seems to persist, although the increased plant yields obtained initially by addition of manures do not. The value of sewage sludge as an amendment to achieve these aims is discussed here.

1. INTRODUCTION

The materials known collectively as colliery spoil are the rocks brought to the surface during coal mining operations and then separated from the marketable coal. These rocks comprise mainly shales, with lesser amounts of sandstones, ironstones, mudstones and limestones. They are tipped to form mounds of varying size, which may contain spoil from a number of seams, and in many cases from more than one colliery. Thus a spoil tip may contain a mixture of materials with different properties. A tip initially comprises rock which has been physically broken down to some degree. With time other processes act on the spoil, resulting in further physical breakdown and, depending on spoil composition, chemical attack and weathering.

When a spoil tip is reclaimed the material that it contains represents the initial state of a soil, but one which often requires considerable modification in order to become a suitable medium for plant growth. In many tips the factor which distinguishes the spoil from a soil is the lack of an organic matter component. Inputs of organic matter are minimal until a sustainable vegetation cover is established, and even then there is poor incorporation of the plant debris into the spoil. This is a consequence of the low level of biological activity, both microbial and mesofaunal, in most spoils. So in the period immediately following reclamation, soil properties which depend on this activity are absent or only poorly developed. These include transformations of nutrients, especially nitrogen, which help to build up storage pools and to release nutrients from them. Another consequence of the lack of organic matter is the poor physical nature of many spoils. This manifests itself as instability, leading to erosion, and poor waterholding properties in the spoil.

Conditions in some spoils lead to particular chemical factors which have to be overcome in order to establish a good vegetation cover. A common example of this is the presence of pyrite (FeS_2) in some colliery spoils. When pyrite oxidises it produces large quantities of acid, which can result in spoil pH values of less than 3. The acid must be neutralised, or its formation prevented, for vegetation to be established.

A reclamation programme for colliery spoil must therefore attempt to combat these adverse factors and create more suitable conditions for the development of a vegetation cover. As more land is reclaimed, however, the costs of management increase, and become a greater part of the total budget. So in recent years attention has turned to treatments which can reduce the amount of aftercare, particularly in terms of additional fertiliser and lime applications. Sewage sludge is a low-cost amendment which can have a number of beneficial effects on colliery spoil. In this discussion its role in inhibiting acid production and in promoting nutrient cycling will be considered.

2. INHIBITION OF ACID PRODUCTION

A major problem encountered on many colliery spoil tips is the production of acid by the oxidation of pyrite (FeS_2), which can commonly result in values below pH 3. Pyrite is a diagenetic mineral formed in the shales and coals as a result of the reduced environment under which they were deposited. It is commonly found on sites in central Scotland as discrete nodules, or as veins in shale particles. Pyritic spoil is also found in the coal fields of northern England, but not in south Wales nor in eastern Scotland.

Oxidation of pyrite occurs when pyritic spoil is exposed to the action of air and water. It can follow one of two pathways, depending on conditions within the spoil. If the pH is maintained above 4, the pathway is one in which oxygen is the oxidizing agent.

$$4FeS_2 + 15O_2 + 14H_2O \rightarrow 4Fe(OH)_3 + 8H_2SO_4 \tag{1}$$

This reaction is slow as it is controlled by the rate of diffusion of oxygen into the spoil. So if the spoil pH is held above 4 by natural buffering systems or by the addition of lime, then the rate of acid production is slow and can usually be dealt with by leaching and natural dispersion. If, on the other hand, the pH falls below 4 a different oxidation pathway is followed in which Fe^{3+} ions act as the oxidizing agent.

$$FeS_2 + 14Fe^{3+} + 8H_2O \rightarrow 15Fe^{2+} + 2SO_4^{2-} + 16H^+ \tag{2}$$

In this case the rate of reaction is controlled by the oxidation of ferrous to ferric ions, which is catalysed by the bacterium *Thiobacillus ferrooxidans*. It has been estimated that the rate of oxidation of pyrite can be increased by a factor of 10^6 by the bacterial action (1). So if the pH of the spoil falls sufficiently to allow this pathway to be followed, very large amounts of acid are produced which can neutralise any natural buffering capacity of the spoil or any added lime.

When reclaiming a pyritic spoil, the normal practice has been to add very high amounts of a liming material, so that the pH is held at a level high enough to preclude oxidation by reaction 2. Attempts have been made to measure the pyrite content of spoil and to calculate the maximum potential acidity which would be produced if all the pyrite oxidised (2). This gives a liming rate of 40 tonnes $CaCO_3$ per hectare per 1% FeS_2, assuming an incorporation depth of 15 cm and a spoil bulk density of 1.2 t/m^3. This approach assumes that pyrite is uniformly distributed through the spoil and over a tip, which is not the case. On one site in Scotland, pyrite contents varying from 0.1% to 6.6% were measured in an area of 0.25 m^2 (3), which would give a lime addition of between 4 and 264 tonnes $CaCO_3$ per hectare. In practice, therefore, it is unlikely that any estimate of pyrite content would allow for maximum acid production. So any lime addition made on this basis is inevitably a compromise figure which is too high in some places, but too low in others. The latter case tends to occur around nodules of pyrite or pieces of highly pyritic shale. The result is a number of small acidic spots, which spread out to form large acid patches.

The crucial step in the pyrite oxidation pathway is the conversion of ferrous to ferric ions. It is rate-determining in high pyrite systems, whereas in low pyrite systems the oxidation of pyrite by ferric ions is rate determining (4). In either case, any factor which inhibits the ferrous to ferric oxidation could result in a decrease in acid production.

Backes *et al.* showed that acid and iron production in incubated suspensions of pyritic spoil could be inhibited by treatments which specifically interfered with the ferrous-ferric oxidation (4). Initially this was done using a bactericide to kill *Thiobacillus ferrooxidans* or 1:10 phenanthroline, which is a chelating agent specific for ferrous ions. They showed that treatments maintained iron in the ferrous form, while in an untreated control Fe^{2+} ions were oxidized to Fe^{3+}. Subsequent studies showed that acid production could be inhibited by phosphate and silicate, which precipitated iron out of solution, and citrate, which complexed the iron (5). Organic wastes were also extremely effective at inhibiting acid production in these incubations. Chicken manure and wood waste were used in these studies, and these results are considered here. Sewage sludge would be expected to behave in a manner similar to chicken manure.

Table 1. Titratable acidity and soluble iron and sulphate concentrations in solution in incubated suspensions of pyritic spoil.

Treatment	Time of incubation (days)				
	12	25	39	67	88
	mg H^+/l				
Control	21.9	40.5	72.0	136.0	260.0
Chicken manure	22.1	4.0	8.5	2.0	0
Wood waste	22.1	16.7	29.0	45.0	55.0
	mg Fe/l				
Control	313	900	725	1,300	3,800
Chicken manure	318	130	100	23	3
Wood waste	300	293	330	405	775
	g SO_4^{2-}/l				
Control	0.57	2.47	4.37	6.96	9.00
Chicken manure	0.60	2.10	1.60	0.94	0.11
Wood waste	0.55	1.66	2.15	2.68	3.35

Table 1 shows the effect of chicken manure and wood waste on titratable acidity, soluble iron and sulphate in suspensions of pyritic spoil incubated at 24°C for 88 days (3). At the end of this period the manure had completely inhibited production of acid, iron and sulphate while the wood waste maintained concentrations which were approximately one-third to one-fifth those in the controls. These figures show that the manure was more effective than the wood waste at inhibiting pyrite oxidation. Backes showed that the manure removed from solution 45 mg Fe/g on an oven dry basis, while the wood waste removed 5 mg Fe/g (3). It was suggested that iron was complexed onto a solid phase by the manure, but by a soluble component of the wood waste. Infra-red spectra of the wastes after equilibration with iron solutions confirmed this (3).

If organic wastes such as sewage sludge, chicken manure or wood waste are to be used as colliery spoil amendments to inhibit acid production, then they must be shown to be effective in more realistic systems than incubated suspensions. As their role is to remove iron from solution by complexation, they must be in intimate contact with the grains of pyrite. As a first step, organic wastes were mixed with pyritic spoil in pots. These mixtures were kept in a moist condition and periodically leached. The acidity and iron contents of the leachates showed that the organic wastes were again very effective at inhibiting the oxidation of pyrite (Figure 1).

The value of organic amendments such as those studied is that they are available at relatively low cost, they provide an organic component in the spoil which can complex iron during the establishment phase and they promote vegetation growth by their nutrient content. This last point may be important over the long term as the manure is mineralised and replaced by plant debris.

3. NUTRIENT SUPPLY AND CYCLING

Sewage sludge has commonly been used as an amendment for a variety of mining wastes. In most cases it has been used primarily as a nutrient supply and attention has been paid to the initial response in plant growth. There has generally been less concern with the effect of the sewage sludge on biological activity in the spoil and the cycling of nutrients in the long term. Many studies have reported yield responses resulting from the addition of sewage sludge to colliery spoil or to open cast mine spoil (6,7). As there is a possibility of high heavy metal contents in sludges from industrial areas, attention has been paid to metal uptake by plants and metal leaching from spoil tips (7).

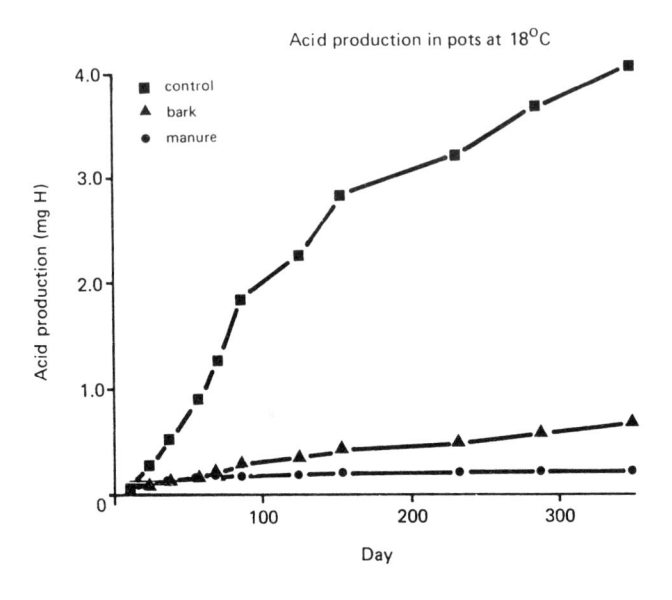

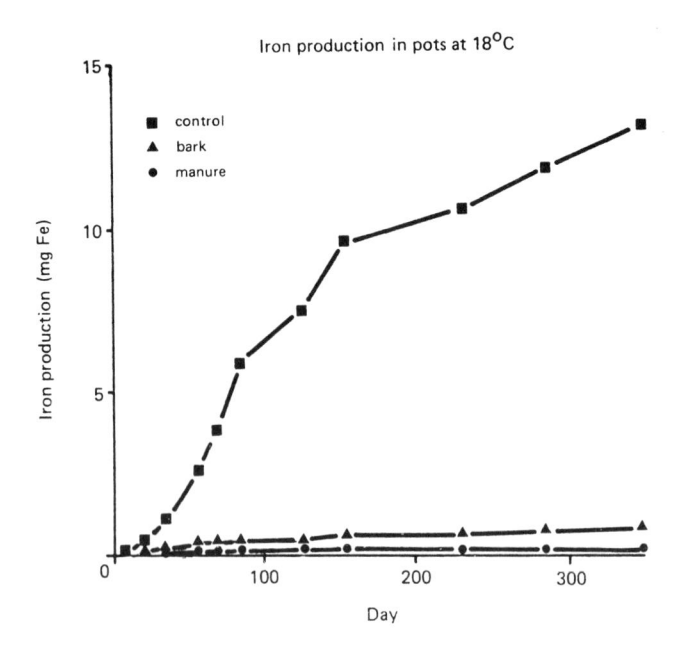

Fig. 1. Effects of manure and wood bark treatments on the production of acid and iron in pyritic colliery spoil.

Sewage sludge as an amendment for reclaimed colliery spoil

A field trial was established in 1980 on a reclaimed colliery spoil tip at Baads, West Lothian, in order to compare various organic amendments. Sewage sludge (40% moisture content) and peat (60% moisture content) were added at a rate of 20 t/ha; chicken manure (14% moisture content) was added at 4 t/ha; and a seaweed-based soil conditioner, Alginure, was also used. Lime was added at a rate of 25 t/ha and all plots received 300 kg/ha of a 15:10:10 NPK compound fertiliser. Control plots, having no organic amendment, with and without lime addition, were also included. Table 2 shows the inputs of N, P and K provided by the amendment and fertiliser applications.

As would be expected, the higher nutrient input by the manure treatments resulted in a higher vegetation yield in those plots compared to the peat, Alginure and no organic amendment treatments. This is shown in Table 3, where vegetation yields in the treated plots are expressed as a percentage of the control plots (no organic amendment with lime), for a period of six years following reclamation. By the sixth year the enhanced yield due to sewage sludge or chicken manure is no longer evident.

Table 2. Inputs of N, P and K to experimental plots by organic amendments plus fertiliser (kg/ha).

Treatment	N	P	K
Chicken manure	195	97	118
Sewage sludge	153	44	44
Peat	63	18	46
Alginure and no organic amendment	45	13	25

As a result of the higher yields on the manure treated plots, much of the nitrogen in those systems is found in the vegetation and dead litter (8). Table 4 shows the distribution of nitrogen in the plots at Baads for the 1984 harvest. The dead litter is a major reservoir for nitrogen and so the higher yield resulting from sewage sludge and chicken manure treatments are an important factor in the long term nitrogen supply.

While manure treatments have a role in the initial supply of nutrients following reclamation, a more important consequence of their use may be the enhancement of biological activity in the spoil. This is particularly relevant to nitrogen supply, where it is essential to initiate the various transformations which ensure the continuing availability of nitrogen to

Table 3. Vegetation yield for plots treated with an organic amendment, expressed as a percentage of the yield for the no organic amendment with lime control.

Treatment	1980	1982	1983	1984	1986
Chicken manure	722	499	310	226	85
Sewage sludge	226	288	250	227	116
Peat	62	105	83	112	99
Alginure	34	88	106	157	150

Table 4. Distribution of nitrogen between spoil, litter and vegetation 4 years after reclamation.

Treatment	N in the three compartments kg N/ha	
Chicken manure	Spoil	3.9
	Litter	98
	Vegetation	14
Sewage sludge	Spoil	5.2
	Litter	92
	Vegetation	15
Peat	Spoil	7.5
	Litter 53	
	Vegetation	8.8
Alginure	Spoil	6.4
	Litter	63
	Vegetation	11
No organic amendment with lime	Spoil	6.7
	Litter	50
	Vegetation	8.4
No organic amendment without lime	Spoil	28
	Litter	-
	Vegetation	-

plants. Figure 2 shows the respiration rate in the Baads spoils in 1983, three years after reclamation (8). All four organic amendment treatments resulted in higher amounts of carbon dioxide evolved compared with the plots which had received lime alone. Spoil which had had no organic amendment or lime added had a very low respiration rate. In all cases, the rate of carbon dioxide evolution was linear from week 4 to 10 of the incubation. These values are given in Table 5 for 1983 and 1984.

Table 5. Carbon dioxide evolution rates in reclaimed colliery spoil.

Treatment	Carbon dioxide evolution rate mg CO_2-C/kg spoil per week	
	1983	1984
Chicken manure	144	146
Sewage sludge	113	127
Peat	131	143
Alginure	120	101
No organic amendment with lime	84	102
No organic amendment without lime	11	16

Figure 3 shows the effect of a manure treatment on the mineralisation of nitrogen in colliery spoil. Although this example is for chicken manure treatment, sewage sludge behaves in the same way. When no organic amendment is added, little nitrogen is mineralised, and that which is is mostly in the ammonium form. Addition of a manure stimulates nitrogen mineralisation and nitrification.

Table 6 shows data from an incubation study, carried out in 1986 using spoil which had received sewage sludge, chicken manure or no organic amendment in 1980 (9). Nitrogen was added to these spoils prior to incubation at a rate of 100 mg N/kg as ammonium sulphate, urea or chicken manure. Comparison of the nitrogen mineralisation rates in the control spoils shows that the manure treated spoil mineralised more nitrogen than the unamended spoil. Addition of nitrogen as ammonium sulphate or urea resulted in no change in the nitrogen mineralisation rate in the sewage

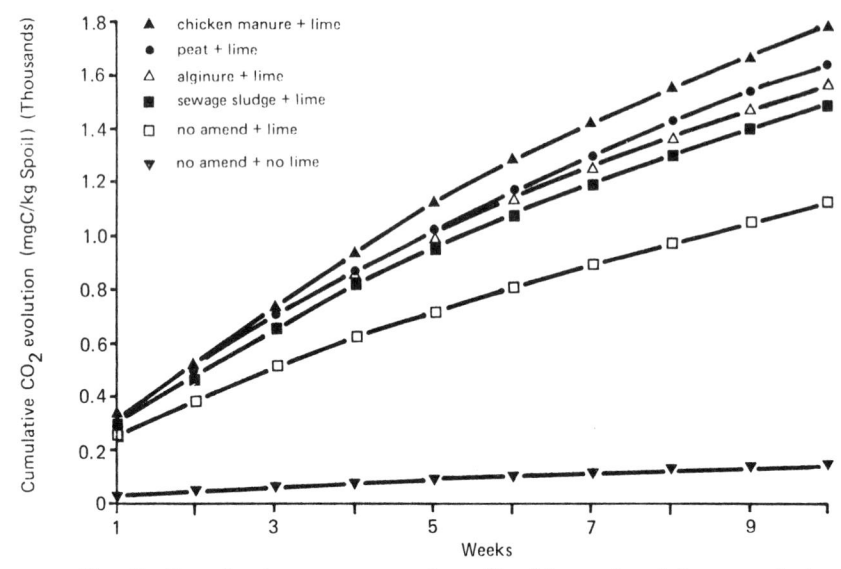

Fig. 2. Respiration rate, as carbon dioxide evolved, in amended colliery spoil, 3 years after treatment at reclamation.

Table 6. Nitrogen mineralisation rates in colliery spoils following treatment.

	Chicken manure	Sewage sludge	No organic amendment With lime	No lime
pH	6.2	7.2	7.0	3.8
N addition	------- N mineralisation rate (mg N/kg/week) ----------			
Control (no N)	2.59	2.73	0.14	0.77
Ammonium sulphate	3.99	2.31	0.21	0.07
Urea	4.27	2.31	0.00	-0.14
Chicken manure	8.82	4.62	3.50	5.04

sludge treated or unamended spoil. There was, however, an increase in nitrogen mineralisation rate when these forms of nitrogen were added to the chicken manure treated spoil. Addition of nitrogen to spoil as chicken

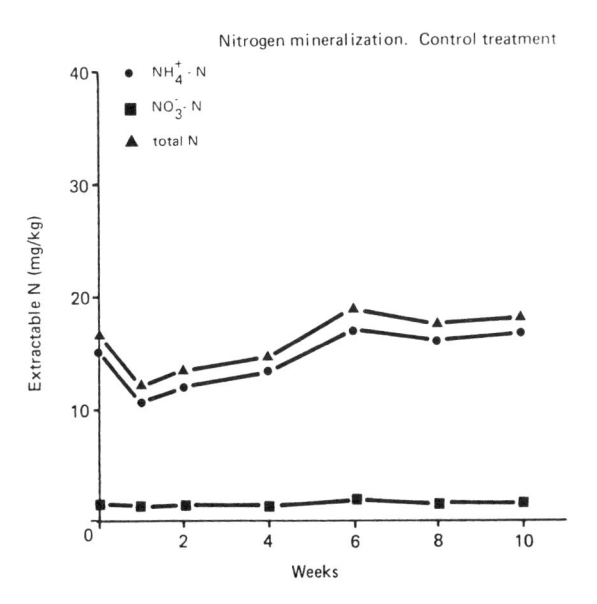

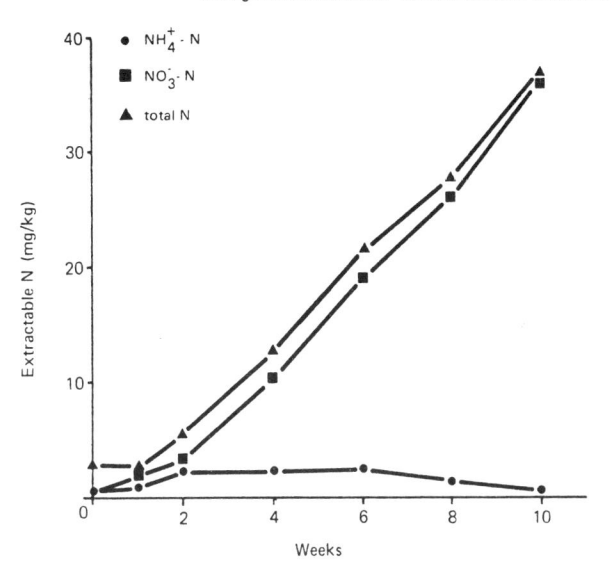

Fig. 3. Effect of manure treatment on nitrogen mineralisation in colliery spoil.

manure resulted in an increase in the mineralisation rate in all spoils. So six years after reclamation the original manure treatments still have an effect on the rate of nitrogen mineralisation in the spoil, whereas their effect on vegetation yield has not persisted (Table 3). Additions of a manure can further increase the mineralisation rate. Sewage sludge treated spoil had a high nitrification rate (Table 7). Addition of chicken manure to this spoil and to the untreated spoils resulted in an increase in the nitrification rate. The spoil treated with chicken manure in 1980 had a low nitrification rate which was not stimulated by further addition of nitrogen.

Table 7. Nitrification rates in colliery spoils following treatment.

	Chicken manure	Sewage sludge	No organic amendment With lime	No lime
pH	6.2	7.2	7.0	3.8
N addition	\multicolumn{4}{c}{Nitrification rate (mg N/kg/week)}	
Ammonium sulphate	12.7	41.4	27.8	0
Urea	14.3	44.0	28.1	0
Chicken manure	12.8	50.8	43.5	7.9

4. CONCLUSIONS

Sewage sludge provides a low-cost, readily available amendment for improving the soil environment in reclaimed colliery spoil. When added to pyritic spoil in particular, it may inhibit the production of acid by pyrite oxidation. On these sites, this is the factor most likely to destroy any established vegetation cover. More generally, sewage sludge provides an input of nutrients and organic matter. As well as improving conditions for plant growth, these inputs help to initiate biological activity which is necessary for the long-term supply of nutrients to the vegetation.

Studies on various types of spoil in different locations and under a range of climatic conditions have suggested that any application of sewage sludge will be beneficial to plant growth and soil conditions (6, 8, 9, 10, 11, 12, 13, 14, 15). There has, however, been no systematic study of application rates and ensuing benefits, nor on the value of repeat applications. This last point raises the problems of cost and practicality.

From the point of view of the disposal of sewage sludge onto land, colliery spoil tips would be suitable receptor sites. Few sites are used for crop production, although some are used for grazing. On such sites metal uptake by plants or animals should be monitored. This will not be necessary on amenity areas. There may be a problem of metal leaching into streams and rivers, especially from acidic spoil. Of more concern may be the amount of colliery spoil land available and the quantity of sewage sludge which can be disposed on it.

ACKNOWLEDGEMENTS

Much of the work described in this paper has been carried out with the support of the Scottish Development Agency and the Landscape Development Unit of Lothian Regional Council. The author is grateful to C Backes, S Shah and B Walker for the use of some as yet unpublished data.

REFERENCES

(1) Singer, P. C. and W. Stumm (1970). Acidic mine drainage: the rate-determining step. Science, *167*, 1121-1123.
(2) Dacey, P. W. and P. Colbourn (1979). An assessment of methods for the determination of iron pyrites in coal mine spoil. Reclamation and Revegetation Research, *2*, 113-121.
(3) Backes, C. A. (1984). The oxidation of pyrite and its environmental consequences. Ph.D. thesis, University of Glasgow. p. 281.
(4) Backes, C. A., I. D. Pulford and H. J. Duncan (1986). Studies on the oxidation of pyrite in colliery spoil. I. The oxidation pathway and inhibition of the ferrous-ferric oxidation. Reclamation and Revegetation Research, *4*, 279-291.
(5) Backes, C. A., I. D. Pulford and H .J. Duncan (1987). Studies on the oxidation of pyrite in colliery spoil. II. Inhibition of the oxidation by amendment treatments. Reclamation and Revegetation Research, *6*, 1-11.
(6) Topper, K. F. and B. R. Sabey (1986). Sewage sludge as a coal mine spoil amendment for revegetation in Colorado. J. Environmental Quality, *15*, 44-49.
(7) Sopper, W. E. and E. M. Seaker (1984). Use of municipal sewage sludge to reclaim mined land. CRC Critical Reviews in Environmental Control, *13*, 227-271.

(8) Walker, T. A. B. (1988). The use of organic amendments in the reclamation of acidic coal mine waste. Ph.D. thesis, University of Glasgow. p. 224.

(9) Shah, S. S. H. (1988). Transformations of nitrogen and its availability to plants in coal mine soils. Ph.D. thesis, University of Glasgow. p. 269.

(10) Fresquez, P. R. and W. C. Lindemann (1983). Greenhouse and laboratory evaluations of amended coal mine spoils. Reclamation and Revegetation Research, 2, 205-215.

(11) Hill, R. D., K. R. Hinkle and M. L. Apel (1982). Reclamation of pyritic waste. In: 1982 Symposium on surface mining hydrology, sedimentology and reclamation (D. H. Graves, Ed.), pp. 687-697. University of Kentucky, OES Publications.

(12) Joost, R. E., F. J. Olsen and J. H. Jones (1987). Revegetation and mine soil development of coal refuse amended with sewage sludge and limestone. J. Environmental Quality, *16*, 65-68.

(13) Coker, E. G., R. D. Davis, J. E. Hall and C. H. Carlton-Smith (1982). Field experiments on the use of consolidated sewage sludge for land reclamation: Effects on crop yield and composition and soil conditions, 1976-1981. Technical Report TR183. Water Research Centre, Medmenham, UK.

(14) Hall, J. E., A. P. Daw and C. D. Bayes (1986). The use of sewage sludge in land reclamation. Report ER 1346-M. Water Research Centre, Medmenham, UK.

(15) Metcalfe, B. (1984). The use of consolidated sewage sludge as a soil substitute in colliery spoil reclamation. J. Water Pollution Control, *83*, 289-299.

Reclamation of acidic soils treated with industrial sludge

D. DAUDIN, J. F. DEVAUX, D. FULCHIRON,
M. C. LARRE-LARROUY and P. LORTHIOS

Chambre d'Agriculture de Vaucluse, Cantarel BP 734,
84034 Avignon Cédex, France.

SUMMARY

The use of industrial sewage sludge on acidic, infertile soils from Vaucluse (France) required prior knowledge of the interactions between sludge, soil and plants. In order to evaluate the effects of sludge applications on crop production and soil chemical composition, growth medium tests as well as field and pot experiments were conducted over five years. A set of analyses for sludge, soils and sludge-treated soils are presented.

1. INTRODUCTION

In 1980, because of the high cost of the sludge disposal method adopted until then from a gelatin processing factory (thermal dewatering of the sludge which was pelletised to be used as a fertiliser), another disposal system was proposed by the departmental Chamber of Agriculture in Vaucluse (France); the suggested method was to apply the sludge to agricultural land without a preliminary thermal treatment.

The first trials carried out (sludge used as a growth medium), although unsuccessful, showed that the sludge could be used as a lime amendment. A first prospective study identified an area of approximately 2,000 ha of acidic and quasi-infertile soils located in the Plateau de Sault (Vaucluse). The farmers there were particularly interested in alternative solutions that would help them to save on mineral fertiliser and lime amendment costs, mainly because their farms were economically fragile.

In order to evaluate this project successfully, growth medium tests and field and pot experiments were conducted.

They were designed:
a) To study the behaviour of soil and a cereal crop after spreading the sludge at different rates;
b) To determine the optimum rate of sludge application, according to the soil type;

c) To specify the possible problems that could arise from chlorides and nitrates contained in sludge;
d) To evaluate land application as a means of sludge disposal and crop fertilisation.

The results of all these experiments are presented in this paper. They finally developed into a large-scale work which is still operating in this region of Vaucluse.

2. MATERIALS AND METHODS

2.1 Sludge characterisation

The sludge comes from a gelatin processing factory. After thickening, it is treated with lime and ferric chloride and then dewatered by filter-pressing. Typical chemical analyses of the sludge are given in Tables 1 and 2.

Table 1. Mean analysis of the sludge used in the experiment.

Analysis	Mean value
Dry matter (%)	38.5
pH_{H_2O}	11.6
Conductivity (mS)	18
Organic-C (% of dry wt)	16.5
Total-N (% of dry wt)	2.2
P_2O_5 Truog (% of dry wt)	2.5
K_2O (% of dry wt)	0.08
MgO (% of dry wt)	0.68
CaO (% of dry wt)	37.6
NO_3 (mg/kg dry wt)	90
NH_4 (mg/kg dry wt)	11
Chloride (%)	0.91

2.2 Growth medium tests

The sludge was examined as a growth medium in admixture with fresh ground pine bark. The sludge/ bark ratio was studied using the following sludge contents: 0, 5, 15, 25 and 50% of dry weight. Five plants were tested: *Rosmarinus officinalis*, *Prunus laurocerasus* "Caucasica", *Cupressus sempervirens* L., *Pinus halepensis* Mill, *and Nerium oleander* "Mont Blanc". The experiments were carried out in 500 pots (5 species x 5 mixtures x 10 plants with two replicates). Survival counts and height growth measurements were made periodically, together with other assessments (diseases, deficiencies etc.).

Table 2. Mean heavy metal content of the sludge used in the experiment (mg/kg dry matter).

Element	Mean value	Recommended maximum not to be exceeded (French directive AFNOR V 44-041)
Cd	3.8	20
Co	16	-
Cr	24	1000
Cu	23	1000
Hg	0.46	10
Mn	164	-
Ni	42	200
Pb	74	800
Zn	531	3000

2.3 Field trials

During 1981 and 1982, field experiments were undertaken to investigate the effects of the sludge on three soils representative of the region of Plateau de Sault (the results relying upon crop yield and soil physico-chemical changes). Two plots, which were acidic soils to be corrected for organic matter deficiency and by liming, were studied in particular (see Table 3). After spreading the sludge, winter wheat was grown as a test crop in the 1981-1982 and 1982-1983 seasons.

1981-1982: It was decided to apply the sludge on three plots, at five rates of addition including a control without any sludge (0, 15, 25, 32 and 50 t/ha). However, one plot could not be harvested. Sludge was applied during the spring prior to planting in the autumn. All plots were fertilised with 150 kg/ha K_2O, except for the control plot which received 100 kg/ha nitrogen and 100 kg/ha P_2O_5. Crops were examined for wheat grain yield, number of plants/m², rate of tillering and number of grain/head. Soil samples were analysed for pH, C, N, P, CEC, Ca, Mg, K and Na.

Table 3. Some physico-chemical properties of the soils studied.

Soils	St-Christol (S_1)	St-Trinit (S_2)
Sand (%)	29	24
Silt (%)	48	50
Clay (%)	23	26
pH_{H2O}	4.5	6.4
Organic-C (%)	1.48	1.98
Total-N (%)	0.079	0.087
CEC (meq/100g)	11.81	16.41
P Truog	21	37
Ca (meq/100g)	2.71	13.76
Mg (meq/100g)	0.14	0.75
K (meq/100g)	0.13	0.39
Na (meq/100g)	0.09	0.09

1982-1983: One plot was subdivided into 4 sub-plots for spreading with three different sludge applications: 20, 40 and 60 t/ha. One plot without any sludge was left as a control. Because of the low potassium content of the sludge, a potassium supplement equivalent to 150 kg/ha K_2O was applied to the plot.

As in the 1981-1982 season, the effects of sludge application on crop production were studied (yield, height, number of plants/m², rate of tillering, etc). Soil chemical levels were measured after harvesting the crop.

2.4 Pot trials

The lay-out of the pot experiments set up in 1983 for two years was as follows:
a) Pots: 324 containers, 12 litre capacity
(1 plant x 3 soils x 9 rates x 4 pots x 3 replications)
b) Test plant: Winter wheat (Talent).
c) Soils: 3 well-defined regional soils:
Grande Partie, pH 5.5
Garusse, pH 6.5
Teyssonnières, pH 7.5.
d) Fertilisation: 0.85 g/container K_2O (as K_2SO_4).
Fertilised control 6g/container of 0.14.14 fertiliser.

e) Sludge: 0, 10, 20, 30, 40, 50, 100, 300 t/ha (0, 65, 130, 195, 260, 325, 650, and 1,950g/container). First year: all the containers (except the two control ones, the fertilised control and the non-fertilised control) received sludge. Second year half the containers receive sludge.

f) Analysis: Plant yield and height. Soil (after the trial) pH, N, P, CEC, Ca, Mg and K.

3. RESULTS AND DISCUSSION

3.1 Sludge characterisation

The lime-treated sludge (20,000- 25,000 tonnes produced per year) must be considered primarily as a lime amendment, since the sludge contains about 40% CaO on a dry matter basis. This high level of CaO influences the availability of phosphorus. Therefore, it could be assumed that the "P-effect" would not be noticeable unless sludge was spread on acidic soils. As for its organic matter level, the sludge is less interesting since it only contains 15-20% total-C. Moreover, the organic matter, due to its animal origin, has no bulking effect on soils. The sludge is also characterised by a low heavy metal content and, due to its high pH value, contains no pathogens.

3.2 Growth medium tests

The effects of sludge when used as a growth medium on plant growth of five species are shown in Table 4. All of the plants tolerated a sludge content up to 25% (w/w), above which growth was depressed and even inhibited for the most sensitive species (*Rosmarinus off.*, *Prunus laur.*, *Cupressus semp.*). No such negative effect was observed for *Nerium oleander* and *Pinus halep.*

Rosmarinus officinalis: growth increased with a sludge content up to 25% (w/w). However, mortality occurred at 5% and became more important with increasing sludge content, affecting all the plants grown on sludge alone.

Prunus laurocerasus: the results were similar to those obtained for *Rosmarinus off.* However, there was no significant effect of sludge content on the rate of mortality up to 25% (w/w). Above 50% sludge, the bark/sludge mixtures had a negative growth effect on almost all the pots.

Cupressus sempervirens: growth increased up to 50% (w/w) of sludge. Above this content, the negative growth effect was more important in the second replication of the test.

Pinus halepensis: bark/sludge mixtures enhanced the growth of Pinus plants. For a sludge content of 50% (w/w), mortality affected half the pots.

Nerium oleander: growth increased with increasing sludge content, without a significant rate of mortality at higher amounts of sludge. *Nerium oleander* appears to be the species that grows best in bark/sludge mixtures.

Table 4. Rate of mortality for the species grown on bark/sludge mixtures (%).

Species	Sludge content (%)					
	0	5	15	25	50	100
Rosmarinus off.	0	6	16	36	100	-
Prunus laur.	0	0	0	3	98	100
Cupressus semp.	6	3	0	0	12	-
Pinus halep.	0	3	3	0	44	-
Nerium oleander	0	3	3	3	6	-

The rate of mortality is related to the high conductivity values of the bark/sludge mixtures (see Table 5). The conductivity value of the bark itself is high (2.34 mS) due to the excessive addition of a N-fertiliser. Moreover, the heaps made of the different bark/sludge mixtures had been stored indoors and thus were kept dry. As soon as mortality occurred, an intensive watering of the containers was carried out, which then reduced this phenomenon.

Another reason for the high conductivity value is the presence of chlorides, since the sludge had been conditioned with ferric chloride. The data suggest that the levels of chlorides were phytotoxic.

It should be also noticed that the pH value of the sludge was high (pH_{H2O} = 11.6). Consequently, the pH values of the bark/sludge mixtures were also very high (close to 8).

The main conclusions for practical use from this experiment is that, despite an obvious economic interest, bark/sludge mixtures should not be recommended as growth media due to their high conductivity and pH values.

Table 5. Chemical characteristics of the bark/sludge mixtures used in the experiment.

Analysis	Sludge content (%)					
	0	5	15	25	50	100
Conductivity (mS)	2.34	2.34	3.62	6.2	7.6	11.9
pH_{H20}	7.5	8.2	8.1	7.9	7.9	12.7
Chloride (mg/kg)	5	1,200	3,615	8,440	12,000	22,100

3.3 Field experiments

3.3.1 1981-1982: Crop responses The different crop responses to sludge applications are shown in Table 6.

Table 6. Field experiments (1981-1982). Crop responses to sludge applications.

Soils	St-Christol (S_1)			St-Trinit (S_2)		
Rate of sludge application (t/ha)	0	25	50	0	32	50
No. of plants/m² (A)	182	182	182	217	217	217
No. of heads/m²	454	610	658	468	485	607
Rate of tillering (B)	2.50	3.32	3.61	2.15	2.23	2.79
No. of grain/head (C)	18.6	18.2	23.1	27.5	22.3	22.9
1000 grain wt (D)	36	31	31	35	42	34.5
Yield (t/ha) Y=A x B x C x D	3.04	3.45	4.72	4.5	4.56	4.81
Yield x 100 / control yield	100	113	155	100	101	107

The effect of sludge applications on seedling emergence and number of plants/m² was not significant. At this stage, no phytotoxicity was observed. Rate of tillering on the two plots increased with increasing sludge applications, especially on the most acidic soil (S_1). The higher rate of tillering was probably due to the significant supply of mineralised nitrogen from sludge at the end of the winter period. The number of grain per head is generally a function of fertilisation (nitrogen) up to ear emergence. On the most acidic soil (S_1), maximum yield of grain was obtained with the highest sludge rate (50 t/ha). S_2 exhibited opposite results. The decrease in specific weight probably resulted from a serious scorching when sludge was applied. Yield increased with increasing sludge applications and showed higher values for the most acidic soil (S_1). It was affected most by the rate of tillering.

Soil analyses Soil analytical data are summarised in Table 7. The highest rate of sludge application (50 t/ha) modified the soil characteristics, especially those of the most acidic soil (S_1). The results suggest that the clay-humus complex was saturated at this rate of application (see Table 8).

Table 7. Field experiments (1981-1982). Soil physico-chemical properties, eight months after spreading the sludge.

Soils	St-Christol (S_1)		St-Trinit (S_2)	
Sludge application rate (t/ha)	0	50	0	50
pH_{H2O}	5.1	5.6	5.4	5.3
C (%)	1.87	2.87	2.00	2.26
N (%)	3.22	4.93	3.45	3.89
P Truog	35	35	19	19
CEC (meq/100g)	13.29	18.71	13.08	14.04
Ca (meq/100g)	7.15	16.15	8.99	15.30
Mg (meq/100g)	1.17	2.34	1.72	1.81
K (meq/100g)	0.17	0.22	0.29	0.30
Na (meq/100g)	0.10	0.13	0.11	0.13
NaCl (%)	0.45	0.45	0.25	0.35

During this first year experiment, a 0.5 unit increase in pH occurred in the same soil (S_1).

In conclusion, the results of this experiment showed significant effects of the sludge, particularly when sludge was spread on the most acidic soil resulting in increases in pH and CEC. After sludge application, plant growth and crop yield were increased overall, particularly in highly acidic soils.

Table 8. Field experiments (1981-1982). CEC and exchangeable Ca, Mg, K and Na levels, eight months after spreading the sludge on St-Christol soil (S_1, pH 4.5).

		Rate of sludge application (t/ha)		
		0	50	Optimum
CEC	(meq/100g)	13.3	18.7	-
Ca	(% CEC)	53	86	80
Mg	(% CEC)	8.8	12.5	10
K	(% CEC)	1.3	1.2	5
Na	(% CEC)	0.75	0.7	-

3.3.2 1982-1983: Crop responses The parameters that influenced yields most (see Table 9) are in the order: the rate of tillering, the specific weight and the number of grain per head.

Table 9. Field experiments (1982-1983). Crop responses to sludge applications.

	Sludge application rate (t/ha)			
	0	20	40	60
No. of plants/m^2 (A)	250	250	250	250
No. of heads/m^2	255	282	343	414
Rate of tillering (B)	1.02	1.13	1.37	1.65
No. of grain /head (C)	16.55	16.94	17.53	17.96
1000 grain wt (D)	31.5	31.6	36.8	36.4
Height of plants	34.3	55.7	66.0	77.4
Yield (t/ha) $Y = A \times B \times C \times D$	1.33	1.51	2.22	2.71
Yield x 100 / control yield	100	113	134	163

It can be noticed that the yields were essentially influenced by the number of heads/m^2 owing to the rate of tillering which increased with increasing sludge rates. This fact may be explained by the amount of mineral nitrogen released from the applied sludge in the early spring, which enhanced the rate of tillering. The mineralisation of nitrogen takes place until ear emergence. This is reflected by the height of the plants, which increased with increasing rates of sludge applications. Sludge phosphate influenced the specific weight to a lesser extent.

The effects of sludge nitrogen and probably those of phosphate on crop yields were demonstrated in this experiment. However, it was difficult to evaluate the influence of sludge lime, except that wheat is known to grow better in a neutral soil.

Soil analyses The data presented in Table 10 show how sludge affected soil characteristics, thereby improving crop yields.

Table 10. Field experiments (1982-1983). Soil physico-chemical properties as affected by sludge applications.

	Sludge application rate (t/ha)			
	0	20	40	60
pH_{H2O}	6.5	6.8	7.3	7.5
pH_{KCl}	5.7	6.0	6.6	6.9
Conductivity (mS)	0.17	0.11	0.20	0.25
Organic-C (%)	1.73	1.59	1.74	1.41
Total-N (%)	0.20	0.18	0.20	0.16
C/N	8.65	8.83	8.7	8.81
NO_3-N (mg/kg)	23	15	21	22
P Truog (mg/kg)	24	46	72	83
CEC (meq/100g)	14.1	12.6	13.6	13.9
Ca (meq/100g)	10.5	10.7	15.2	15.0
Mg (meq/100g)	0.66	1.0	1.35	2.03
K (meq/100g)	0.42	0.33	0.35	0.33

Sludge was effective in:
a) Increasing pH values (this is confirmed by an increase in the amount of exchangeable calcium);
b) Increasing phosphorus levels significantly;
c) Re-balancing cations.

Exchangeable calcium becomes saturated between 20 and 40 t/ha of sludge. The magnesium level is even too high. As for potassium, its level did not increase because of the low K content of the sludge and because of the small amounts of K_2O used for fertilisation (which were below the soil requirements).

However, no sludge effects were observed on:
a) Carbon level and thus, organic matter;
b) Total nitrogen level (which did not increase significantly) nor

NO_3-N (because of the use of a nitrogen fertiliser, ammonium nitrate, on the control sub-plot);
c) CEC, since neither the organic matter level nor the clay level varied;
d) Conductivity, which actually increased slightly, but this was not related to chloride levels.

3.4 Pot experiments

3.4.1 Crop responses The yields obtained after different sludge applications in 1983 are shown in Table 11.

Table 11. Pot experiments. Wheat grain yield as affected by fertiliser and sludge applications (t/ha).

Soil	FC	UFC	\multicolumn{7}{c}{Sludge application rate (t/ha)}
			10	20	30	40	50	100	300
Grande Partie	2.26	1.23	1.74	2.21	2.99	2.94	3.58	3.53	1.55
Garusse	3.38	0.9	1.57	2.42	2.76	2.49	3.08	4.35	5.25
Teyssonnières	2.47	1.36	2.11	2.73	2.5	2.74	2.98	2.48	3.34

FC: fertilised control (140 kg/ha P_2O_5 + 140 k/ha K_2O).
UFC: unfertilised control (140 kg/ha K_2O).

In an attempt to determine if crop yields were related to sludge rates, several regression equations were developed. Utilising the data from Table 11, the following equations ($y = ax + b$) were generated:

Grande Partic (highly acidic pH) $y = 0.46 x + 13.0$ ($r = 0.98$)
Garusse (acidic pH) $y = 0.40 x + 12.0$ ($r = 0.92$)
Teyssonnières (alkaline pH) $y = 0.28 x + 17.0$ ($r = 0.89$)

The first conclusion to be drawn is that the effects of sludge applications on yields were highly significant up to a sludge rate equivalent to 50 t/ha. The best results were obtained when sludge was spread on acidic soils. The effect on alkaline soil, although positive, could be related to the effect of the fertiliser.

Plant height appeared to be a good index of the sludge effects (see Table 12). The plant response to sludge was primarily a function of pH and fertility of the soil before sludge application.

Accordingly, the highly acidic soil (Grande Partie) which is characterised by the lowest CEC level, exhibited higher yields (for the lowest and the highest rates of sludge application) than did the alkaline soil. The soil from

Table 12. Pot experiments. Wheat plant height as affected by fertiliser and sludge applications (cm).

Soil	FC	UFC	Sludge application rate (t/ha)						
			10	20	30	40	50	100	300
Grande Partie	40.3	32.0	36.9	43.7	45.1	49.5	51.5	48.5	51.6
Garusse	48.7	32.2	42.7	47.6	46.9	53.8	52.1	53.2	50.5
Teyssonnières	47.2	39.3	42.7	41.9	45.1	45.9	46.1	47.6	45.8

FC: fertilized control (140 kg/ha P_2O_5 + 140 kg/ha K_2O).
UFC: unfertilized control (140 kg/ha K_2O).

the Garusse plot, which was acidic but showed a higher CEC level, reacted differently and only showed significant yield differences (equivalent to that for the fertilised control) at the highest rates of sludge applications. It can be suggested that this soil took advantage of the sludge nitrogen in a more rapid and better way than the highly acidic soil.

During the growing season of the wheat, no phytotoxicity was observed except when 300 t/ha sludge were spread on the highly acidic soil.

3.4.2 Soil analyses The changes in soil pH values are shown in Table 13. After two years the effect of the sludge on soil pH had not stabilised and the pH kept on increasing even in 1984. The pH increased significantly after the additional application of a similar amount of sludge in 1983 and 1984, and at the highest rates of sludge application, the pH appeared to reach a maximum.

In the highly acidic soil, the effects of two years of sludge application on P levels were significant (see Table 14). However, the differences were not significant whether sludge was applied or not during the second year. In the Garusse soil (intermediary pH), the increase in P was greatest during the first year but decreased significantly in the second year, whether sludge was applied or not. In the alkaline soil, the effect of sludge on soil P was not significant. Overall, there appeared to be a reduction in extractable P concentrations, mainly under alkaline soil conditions.

The effects of sludge applications on CEC levels were not significant (see Table 15). However, exchangeable Ca levels increased with increasing sludge application rates and CEC became rapidly saturated. The K/CEC ratio was not greatly modified whatever the rate of sludge application (note the low K content of the sludge). The increasing amounts of Mg supplied by increasing sludge applications were adsorbed on the clay-humus com-

Table 13. Pot experiments. Soil acidity (pH $_{KCl}$) as affected by sludge applications.

Soil	Sludge application rate (t/ha)								
	UFC	$S_{20}+S_0$	$S_{20}+S_{20}$	$S_{40}+S_0$	$S_{40}+S_{40}$	$S_{100}+S_0$	$S_{100}+S_{100}$	$S_{300}+S_0$	$S_{300}+S_{300}$
Grande Partie									
1983	5.4	5.9	-	6.0	-	6.7	-	-	-
1984	5.4	6.4	7.5	7.5	7.8	7.6	7.9	7.8	7.9
Garusse									
1983	6.0	6.1	-	7.0	-	7.0	-	-	-
1984	6.0	6.5	7.5	7.4	7.7	7.7	7.8	-	-
Teyssonnières									
1983	6.9	6.8	-	6.8	-	7.3	-	-	-
1984	7.0	7.4	7.7	7.7	7.8	-	-	-	-

UFC: unfertilized control (140 kg/ha K_2O).
$S_{20}+S_0$: 20 t/ha sludge the first year + no sludge the second year.
$S_{100}+S_{100}$: 100 t/ha sludge the first year + 100 t/ha sludge the second year.

Table 14. Pot experiments. Soil extractable P[1] as affected by sludge applications (mg/kg).

Sludge application rate (t/ha)	Grande Partie		Garusse		Teyssonnières	
	1983	1984	1983	1984	1983	1984
0	-	22	-	69	-	24
$S_{20} + S_0$	31	45	84	73	34	36
$S_{20} + S_{20}$	31	40	84	95	34	49
$S_{40} + S_0$	37	89	156	127	34	38
$S_{40} + S_{40}$	37	75	156	138	34	60
$S_{100} + S_0$	64	84	248	189	78	-
$S_{100} + S_{100}$	64	60	248	152	78	-
$S_{300} + S_0$	-	87	-	-	-	-
$S_{300} + S_{300}$	-	181	-	-	-	-

[1] P evaluated by the Truog method.
$S_{20} + S_0$: 20 t/ha sludge the first year + no sludge the second year.
$S_{100} + S_{100}$: 100 t/ha the first year + 100 t/ha sludge the second year.

plex but were not sufficient enough to provide a suitable Mg/CEC ratio, unless high rates of sludge were applied (40 to 100 t/ha, according to the soil).

The total-N levels increased with increasing rates of sludge applied (see Table 16). Half of the organic-N is likely to mineralise during the first year following sludge spreading.

Table 15. Pot experiments. CEC and exchangeable Ca, Mg and K as affected by sludge applications.

		Sludge application rate (t/ha)								
1983	0	20	20	40	40	100	100	300	300
1984		0	20	0	40	0	100	0	300
Grande Partie									
CEC (meq/100g)	8.8	10.7	10.7	10.6	10.2	8.9	8.8	8.1	8.1
Ca (meq/100g)	5.7	11.7	22.4	33.0	40.7	37.2	43.1	50.6	57.5
Mg (% CEC)	3	3.6	5.0	4.6	8.0	6.2	11.8	13.3	35.5
K (% CEC)	7	5.0	4.4	4.3	4.4	4.6	4.2	4.0	3.8
Mg/K	0.4	0.7	1.1	1.0	1.8	1.4	2.8	3.3	9.0
Garusse									
CEC (meq/100g)	15.4	14.3	15.2	14.6	14.9	13.2	13.2	-	-
Ca (meq/100g)	17.7	14.3	23.5	22.7	25.0	38.7	49.0	-	-
Mg (% CEC)	-	7.2	2.7	7.4	8.7	8.4	12.2	-	-
K (% CEC)	-	4.1	4.0	3.6	3.4	3.6	3.4	-	-
Mg/K	-	1.7	2.0	2.0	2.5	2.3	3.6	-	-
Teyssonnières									
CEC (meq/100g)	18.6	17.1	19.1	17.0	17.0	-	-	-	-
Ca (meq/100g)	21.6	25.5	35.0	32.2	41.6	-	-	-	-
Mg (% CEC)	5.5	5.7	5.9	5.9	8.0	-	-	-	-
K (% CEC)	3.4	3.9	3.7	3.4	3.8	-	-	-	-
Mg/K	1.6	1.5	1.6	1.7	2.1	-	-	-	-

Table 16. Pot experiments. Soil total nitrogen as affected by sludge applications (%).

	Sludge application rate (t/ha)								
1983	0	20	20	40	40	100	100	300	300
1984	0	0	20	0	40	0	100	0	300
Grande Partie	1.37	1.54	1.88	1.81	1.82	1.55	2.05	2.35	3.80
Garusse	-	2.58	2.66	2.11	2.69	2.23	2.97	-	-
Teyssonnières	-	1.43	1.47	1.82	2.32	-	-	-	-

4. PRACTICAL IMPLICATIONS

On the basis of this preliminary five-year evaluation of crop growth and soil responses, it could be concluded that sludge at appropriate application rates (10 to 20 t/ha every two years) was successful in improving acidic soils. Therefore, large-scale demonstration projects supervised by the Chamber of Agriculture of Vaucluse were conducted to promote land application among farmers of Plateau de Sault, not only as a safe and economical method but also as a way of making soils return to productive agriculture. Farmers took note of the fertiliser value of the sludge and were satisfied with the results. Today, land application is still going on very successfully in this region with all the sludge production being used by the farmers, who even grouped themselves in a cooperative system in order to buy and share tractors and manure spreaders used in sludge application.

Experiences of the usage of heavy amounts of sewage sludge for reclaiming opencast mining areas and amelioration of very steep and stony vineyards

W. WERNER, H. W. SCHERER and F. REINARTZ

Rhein. Friedr.-Wilhelms-University,
Institute of Agricultural Chemistry, 5300 Bonn, Fed. Rep. of Germany.

SUMMARY

Trials were set up to examine whether pure or composted sewage sludge can be used for reclaiming opencast mining areas and for the amelioration of steep and stony vineyards. The results have shown that the application of sewage sludge or compost of sewage sludge with garbage in reclamation has a beneficial effect on humus content and soil fertility. The suitability of sewage sludge composts - even in combination with bark to get a wide C/N ratio - for storage accumulation of organic matter in steep and stony vineyards, however, is limited by the high mineralisation rate followed by extremely high leaching losses of NO_3^-.

1. INTRODUCTION

With regard to recycling and supply of organic matter and plant nutrients, mainly nitrogen and phosphate, the use of sewage sludge as soil amelioration material and fertiliser should be promoted in agriculture. According to Kick (1), in the FRG about 35% of the total amount of 34×10^6 metric tonnes of sewage sludge/year was used agriculturally about 10 years ago. Since then the readiness of farmers to use sewage sludge has decreased because of the possible contents of hazardous substances (heavy metals and organic compounds). According to a law concerning soil protection, 5 t/ha of sewage sludge dry matter may be supplied to the soil in the course of 3 years on the basis of maintaining tolerable heavy metal contents. This amount can be increased to 10 t/ha when the time between the applications is increased to 5 years.

As potential alternatives in the usage of sewage sludge, the application of heavy amounts in the reclamation of opencast mining areas and the amelioration of very steep and skeletal vineyards could be taken into consideration.

The aim of this paper, therefore, is to describe the principle objectives of these alternative uses of sewage sludge and to present some results of experiments on these subjects.

2. OBJECTIVES

Since 1948 in the "Rhenish brown-coal district", about 6,300 ha of land have been reclaimed after coal extraction for agricultural use and 6,100 ha for forestry. In the soil which is used for reclamation, mainly consisting of C-horizon loess material, the content of nutrients as well as the humus content are extremely low. To promote the crumb formation and to improve soil structure in this soil with its high silt content the enrichment with humus is very important. That means that huge amounts of organic material, besides plant nutrients, are necessary to reclaim opencast mining areas. Besides farmyard manure, sewage sludge provides organic matter, which is beneficial as a soil conditioner (2).

Another alternative use is the amelioration of very steep and especially skeletal vineyards, where bark or compost from sewage sludge and garbage are recommended as humus storage fertilisers (3). Side-effects of these materials are the reduction of soil erosion (4) and the increase of the sorption of plant nutrients, as well as water (5).

However, experiments have shown that the application of high amounts of compost from sewage sludge and garbage results in an increased leaching of NO_3^- (6,7). For this reason, in other experiments sewage sludge was composted together with bark to increase the C/N-ratio, in order to induce a temporary biological immobilisation of the sewage sludge nitrogen.

The aim of this paper is to demonstrate the influence of the application of high amounts of sewage sludge over a long period of time in opencast mining areas on the development of the humus content and on the N dynamics of the soil; and to discuss the application of composts from bark plus sewage sludge for the amelioration of steep and stony vineyards.

3. LAY OUT OF THE EXPERIMENTS

The first experiment has been running since 1969 on an opencast mining area in the "Rhenish brown-coal district", which was reclaimed with raw loess (17% clay, 72% silt, 11% sand) in the upper 90 cm soil layer. The mean application rate of organic matter amounts to 4.7 t/ha per year, except in the treatment with the higher dose of compost from garbage and sewage sludge, where the application rate was doubled.

The experiment has nine different treatments and the five following will be discussed:
a) Control (contr);
b) Farmyard manure (fym);
c) Sewage sludge (ss);
d) Compost from garbage and sewage sludge (gssc1);
e) Compost from garbage and sewage sludge (gssc2).

Based on the amount of mineral N in the soil layer from 0 to 90 cm in early spring, the treatments are adjusted with mineral fertiliser to the level which is usual for the different crops.

The second experiment with storage humus application was conducted on a stony loamy soil site with a gradient of slope of about 30° in the Ahr valley. The stone content ranged between 30 and 70% (mean value 50%). The fine earth had the following composition: 20% clay, 40% silt, 40% sand.

The experiment had the following treatments:
a) 50 t/ha organic substance as peat (contr);
b) 60 t/ha organic substance as compost from bark and sewage sludge (bssc1);
c) 120 t/ha organic substance as compost from bark and sewage sludge (bssc2);
d) 60 t/ha organic substance as compost from bark (bc1);
e) 120 t/ha organic substance as compost from bark (bc2).

These high amounts, which were chosen to cover the humus demand for about 15 years, were incorporated in the soil layer from 0-60 cm in winter 1984/85 before establishing the vineyard.

4. RESULTS

In the reclamation experiment the C content of the soil increased with time in all treatments (Figure 1). The increase was highest in the treatment gssc2, where the application rate of the organic material was doubled. In the other treatments the efficiency of ss and gssc1 was the same as of fym.

The total N content of the soil layers 0-30 cm and 30-60 cm was increased significantly by the application of all the organic materials (Figure 2). There was no difference between the treatments supplied with organic materials except with the higher dose of gssc in the deeper soil layer. The increase of total N was caused mainly by hydrolysable organic N compounds in the treatments ss and fym (Figure 3). With gssc also the non-hydrolysable organic N compounds were related to this increase (Figure 4).

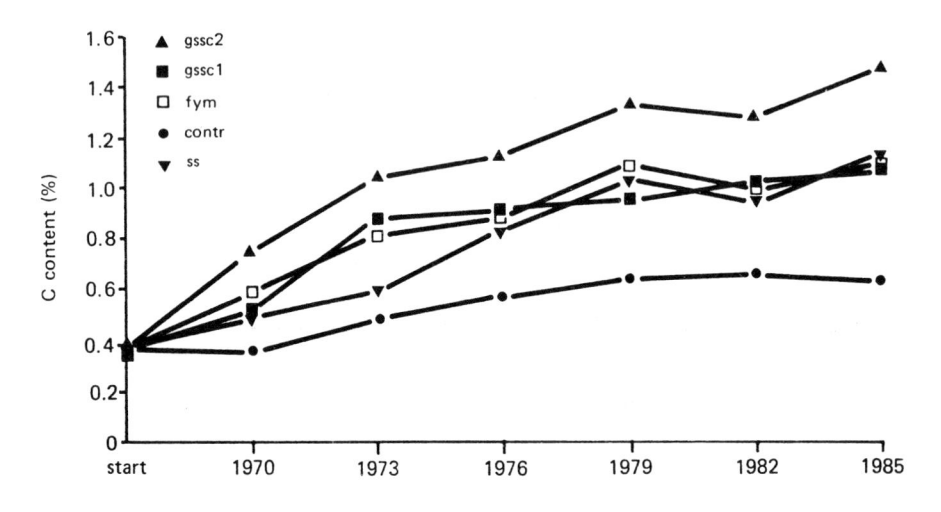

Fig. 1. Development of the total C contents of the soil with time by application of organic materials.

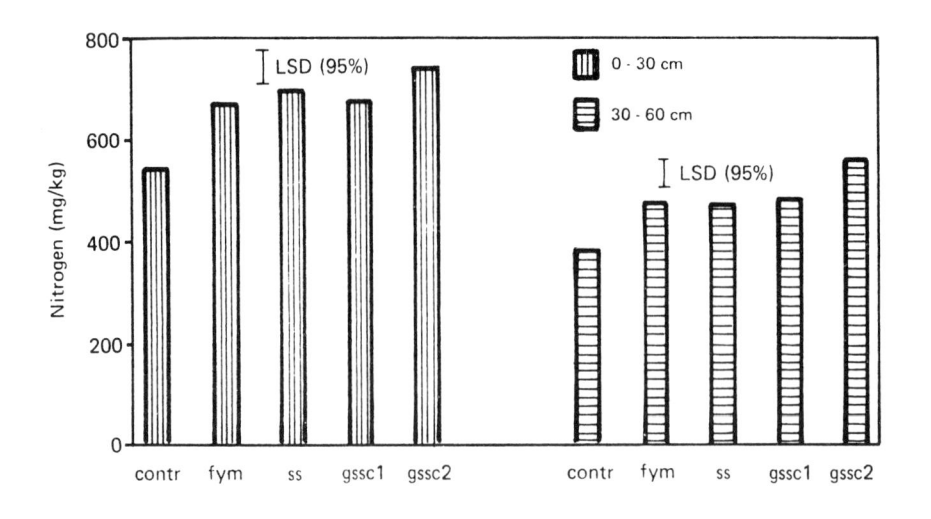

Fig. 2. Total N contents in the soil layers 0 to 30 cm and 30 to 60 cm after long-term application of organic materials.

Experiences of the usage of heavy amounts of sewage sludge for reclaiming opencast mining areas and amelioration of very steep and stony vineyards

The organic N fraction, extracted by means of electro-ultrafiltration (EUF) and with 0.1M $CaCl_2$, was higher in the treatments supplied with organic material. The effect of ss, gssc and fym on this N fraction was similar (Table 1).

Table 1. Contents of organic N (mg/kg) extracted by EUF and $CaCl_2$ in the soil layers 0-30 cm and 30-60 cm after the application of organic materials.

Treatment	EUF organic N 0 - 30 cm	30 - 60 cm	$CaCl_2$ organic N 0 - 30 cm	30 - 60 cm
contr	8.60	5.05	4.80	2.08
fym	13.68	9.33	6.88	4.08
ss	15.00	9.13	7.43	3.23
gssc1	13.95	10.40	7.00	3.98
gssc2	17.55	9.15	8.18	5.20
LSD (95%)	2.78	2.18	1.02	1.45

Table 2. Contents of NO_3-N (kg/ha) in spring in the soil layer 0-90 cm after application of organic material.

Treatment	1984	1985	1986	1987	1988
contr	45	27	57	28	20
fym	47	30	78	41	27
ss	126	50	166	77	54
gssc1	42	32	88	34	25
gssc2	44	28	86	31	23

The NO_3-N content of the soil in the layer 0-90 cm in spring was always highest in the treatment where sewage sludge has been applied, while there were no differences between the control and the other treatments (Table 2).

In the vineyard amelioration experiment, about two years after the incorporation of the organic materials the total C contents of the soil were significantly higher in the treatments with compost from bark and sewage sludge and from bark as compared with the start (Figure 5).

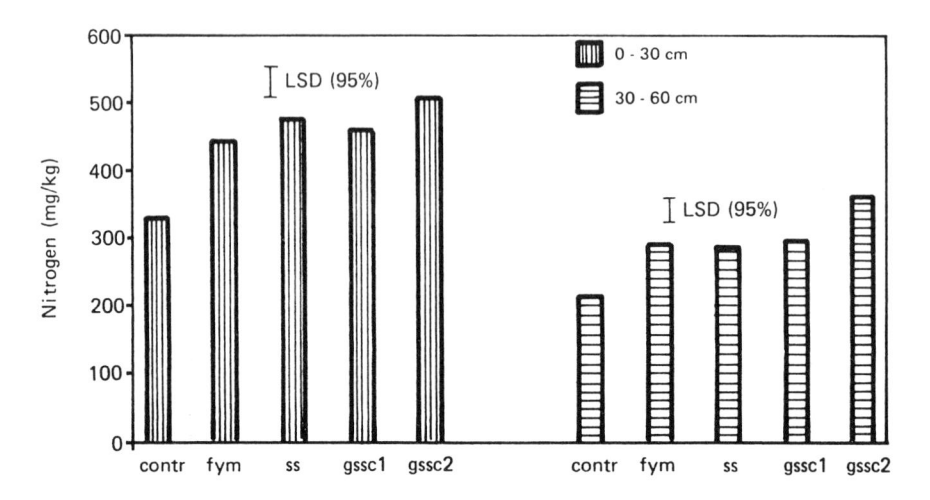

Fig. 3. Contents of hydrolysable organic N compounds in the soil layers 0 to 30 cm and 30 to 60 cm after long-term application of organic materials.

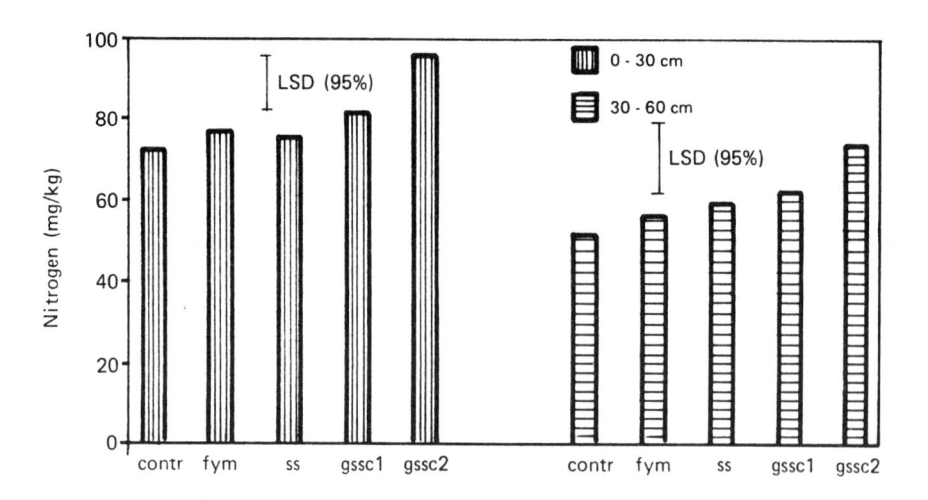

Fig. 4. Contents of non-hydrolysable organic N compounds in the soil layers 0 to 30 cm and 30 to 60 cm after long-term application of organic materials.

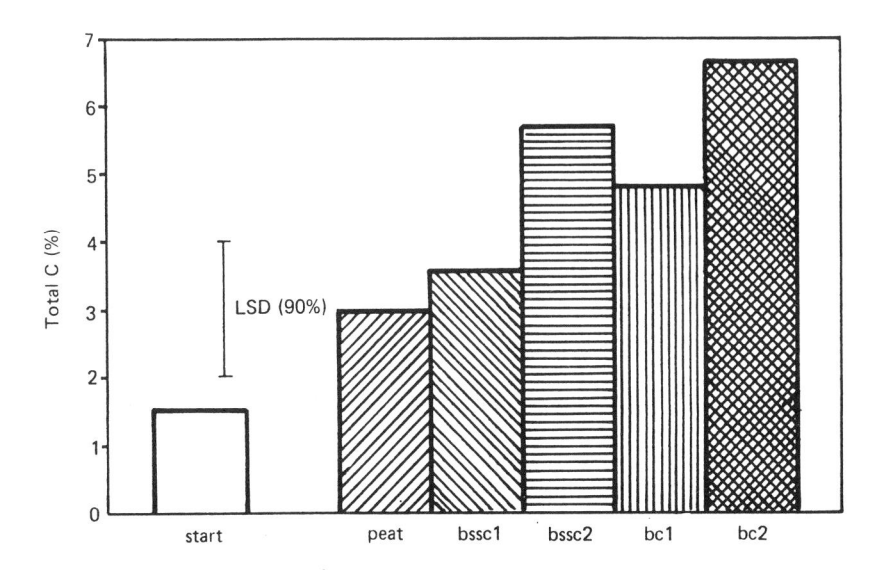

Experiences of the usage of heavy amounts of sewage sludge for reclaiming opencast mining areas and amelioration of very steep and stony vineyards

Fig. 5. Total C contents of the soil after storage application of organic materials.

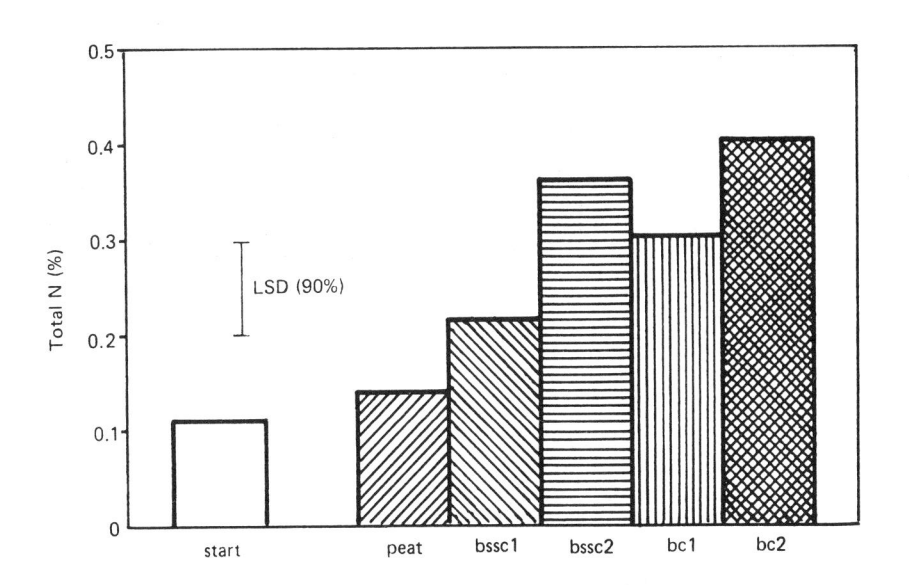

Fig. 6. Total N contents of the soil after storage application of organic materials.

Furthermore, the application of the composts resulted in higher total N contents of the soil (Figure 6) with the highest contents in the treatments where the high amounts of compost were supplied.

Between June 1985 and February 1987 up to 1,000 kg NO_3-N were leached (Figure 7). The amounts were highest in the treatments where compost from bark and the sewage sludge was supplied.

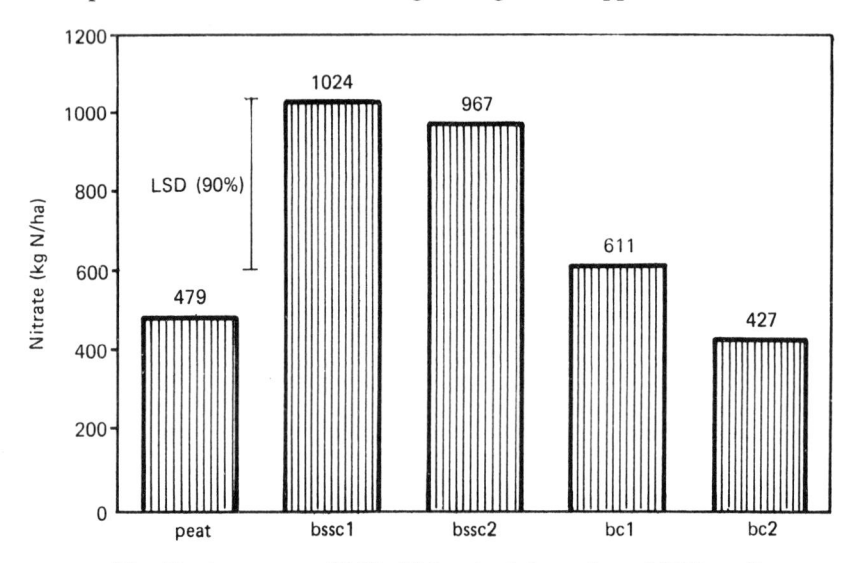

Fig. 7. Amounts of NO_3-N leached from June 1985 until February 1987.

5. DISCUSSION

Besides farmyard manure, sewage sludge and compost from garbage and sewage sludge are potential materials to increase the humus content of the soil. The organic matter derived from sewage sludge mainly consists of microbial biomass and is easily degradable (8). With regard to this property sewage sludge is an appropriate material to stimulate the microbial activity of raw loess material, which is used to reclaim opencast mining areas. Also sewage sludge has a positive effect on the physical and chemical properties of the soil that means improvement to soil structure and increase of the exchange capacity (9). In this reclamation experiment the increase of the humus content (total C) in the treatment with sewage sludge was the same as in the treatments with compost from garbage and sewage sludge and farmyard manure (Figure 1). According to results of Walisade (10) and Werner et al. (11), compost from garbage and sewage sludge is more effective in increasing the humus content. This apparent contradiction between the results is caused by higher amounts of compost compared with sewage sludge supplied in the experiments of these authors.

Liquid sewage sludge contains about 30 to 40% of total N as ammonium. The efficiency of nitrogen in sewage sludge is comparable to farmyard manure (1). In this experiment the application of all the organic materials resulted in higher total N contents in both soil layers investigated (Figure 2), which confirms results of Specht (12). Scherer *et al.* (13) demonstrated that the accumulation of total N could be detected in the soil layer 60 to 90cm. This enrichment, which was mainly caused by hydrolysable organic N compounds (Figure 3), has the advantage that especially in a dry summer when the upper soil layer is dried out the crop can recover N from the deeper soil layers.

To investigate whether the accumulation of organic N compounds in the soil is reflected by a higher N mineralisation potential, soil samples were extracted by means of electro-ultrafiltration (EUF) and with 0.1M $CaCl_2$. The organic N fraction in both extracts is claimed to represent the easily mineralisable soil nitrogen (13, 14, 15). As expected, the application of all the organic materials resulted in significantly higher organic N contents in both soil layers investigated (Table 1). The effect of sewage sludge was similar to the other materials. Interestingly, the content of mineral N in the soil profile in spring 1988 (about six months after taking soil samples for the analysis of the other N fractions) was highest in the treatment with sewage sludge (Table 2). It amounted to about 54 kg/ha in the soil layer down to 90cm depth, while the other treatments, including the control, ranged between 20 and 25 kg N/ha.

The yield was highest in the treatment where sewage sludge was applied. This must have been the result of the additional supply of N due to mineralisation of organic N compounds from sewage sludge during the growing season, because early in spring the different treatments were adjusted to the same level of mineral N in the soil with fertiliser.

From these results it can be concluded that sewage sludge can be used to reclaim the raw loess in opencast mining areas as a result of the supply of organic matter and with regard to recycling plant nutrients, provided levels of heavy metals are tolerable.

In steep and stony vineyards the retention of plant nutrients as well as water is a major problem. Additionally, soil losses caused by erosion are problematical. Under these conditions high contents of soil organic matter are likely to reduce the percolation of water and leaching of nutrients and to prevent soil erosion as well. However, on these sites the yearly application of organic material is very difficult and very time consuming. To overcome these problems, large amounts should be incorporated into the soil as a storage application before the vineyard is established, to cover the humus demand for the period of productivity of the vine plants (about 15 years). The degradation rate of humus in these vineyards is about 8 t/ha per year (16, 17), so huge amounts of organic materials must be supplied, but because of the shortage of farmyard manure on these special vineyards other organic materials are needed. In this case waste from garbage and sewage sludge are potential materials. However, as already mentioned in experiments with higher doses of compost from garbage and sewage sludge, large amounts of NO_3^- were leached. According to Späte (7), the NO_3^- concentration of the leaching water was about 200 ppm, when 1,200 t/ha compost from garbage and sewage sludge had been applied in one application. The reason for the leaching is the high mineralisation potential of the well-aerated stony slopes and the narrow C/N ratio of this compost. Therefore in this ameliorisation experiment sewage sludge was composted with bark to produce a material with a wider C/N ratio as compared with the compost of sewage sludge with garbage.

Thirty months after the application of the organic materials the C content was highest in the treatments with bark compost (Figure 5). With the lower C/N ratio of the compost from bark and sewage sludge it can be assumed that the mineralisation rate of this material was higher, and this can be shown by the fact that the NO_3-N content of the soil at different sampling dates was higher in these treatments.

Compared with the N demand of vines the mineralisation rate of all the organic materials, even those with a low C/N ratio, was still too high. The NO_3^- which is not taken up by plant, is prone to leaching. Most NO_3^- was leached in the treatment with compost from bark and sewage sludge, which may be the reason for the lower total N contents in these treatments.

Although huge amounts of compost were applied, the heavy metal contents of the soil did not reach the threshold values of the German government. Furthermore, no accumulation of heavy metals could be detected in vine plants.

It can be concluded that the composts applied in this experiment are valuable materials, particularly with regard to the enrichment of the soil with humus, but on the other hand they can create problems as far as the leaching of NO_3^- is concerned, when applied in storage amounts.

ACKNOWLEDGEMENTS

The authors wish to thank W König (LÖLF, Nordrhein-Westfalen, Düsseldorf) for the soil samples from the reclamation experiment and for unpublished data.

REFERENCES

(1) Kick, H. (1986). Agricultural applications of sludges and waste waters. Biotechnology, *8*, 307-334.
(2) Kick, H. and P. Weber (1974). Der Einsatz hoher Müllkompost- und Klärschlammengen auf Rekultivierungsfächen des Braunkohletagebaues. Landwirtsch. Forsch, *36*, 201-207.
(3) Keilen, K. (1977). Rinde ein billiger und zugleich wertvoller Humusspender. Der Deutsche Weinbau, *29*, 1260-1261.
(4) Zöttl, H. W. Die Wirkung von Rindenmulch im Weinbau. Der Deutsche Weinbau, *7*, 299-301.
(5) Walter, B. (1987). Einfluß von Rindenmulch auf Wasserhaushalt, Sickerwasserqualität, Mineralisierung und Nitrifikation in skelettreichen Weinbergsböden. VDLUFA-Schriftenreihe, *23*, 477-488.
(6) Höfer, T. (1986). Die Auswirkungen einmaliger, hoher Müllklärschlammkompostgaben vor der Wiederbestockung eines Weinberges im Steilhang insbesondere zur Bodenverbesserung, im Rahmen einer Flurbereinigung. PhD thesis, University of Bonn.
(7) Späte, A. (1988). Die Düngung von Müllklärschlamm-Kompost als Vorratsgabe im Weinbau und ihre Auswirkungen auf Ertragsentwicklung, Nitratauswaschung und Schwermetallbelastung des Systems Boden, Rebe und Sickerwasser. PhD thesis, University of Bonn.
(8) Beck, Th. and A. Süß (1979). Der Einfluß von Klärschlamm auf die mikrobielle Tätigkeit im Boden. Z. Pflanzenernähr. Bodenkd., *142*, 456-475.
(9) Kick, H. (1963). Zur Kennzeichnung der organischen Masse in Städtekomposten und Klärschlämmen sowie daraus hergestellten Humusdüngern. Landwirtsch. Forsch, *16*, 229-237.
(10) Walisade, D. (1978). Der Einfluß von Müllklärschlammkompost auf die Ertragsbildung, Inhaltsstoffe, der Pflanzen sowie auf den Boden und das Bodenwasser. PhD thesis, University of Gießen.
(11) Werner, W., H. W. Scherer and H.W. Olfs. (1988). Influence of long-term application of sewage sludge and compost from garbage with sewage sludge on soil fertility criteria. J. Agronomy and Crop Science, *160*, 173-179.
(12) Specht, G. (1970). Untersuchungen zur organischen Düngung. PhD thesis, University of Berlin.
(13) Scherer, H. W., W. Werner and H. Kick (1985). N-Fraktionen und N-Nachlieferungsvermögen einer Löß-Parabraunerde nach langjähriger Zufuhr von Stroh, Stallmist und Faulschlamm. VDLUFA-Schriftenreihe, *16*, 325-333.

(14) Nemeth, K. and H. Recke (1985). EUF-N-fractions in different soils during a vegetation period in field and pot experiments. Plant and Soil, *86*, 249-256.
(15) Kohl, A. and W. Werner (1986). Untersuchungen zur saisonalen Veränderung der EUF-N-Fraktionen und zur Charakterisierung des leicht mobilisierbaren Bodenstickstoffs durch Elektro-Ultrafiltration (EUF). VDLUFA-Schriftenreihe, *20*, 333-341.
(16) Platz, R. Qualitätsweinbau. Ratgeber für die Landwirtschaft Heft Nr. 2, 3. Auflage.

Consolidated sewage sludge as soil substitute in colliery spoil reclamation

B. METCALFE[1] and J. C. LAVIN[2]

[1] Mitchell Laithes Sewage Treatment Works, Yorkshire Water, Clough Lane, Dewsbury, West Yorkshire, WF12 8LQ, UK.
[2] West Yorkshire Ecological Advisory Service, Cliffe Castle, Keighley, West Yorkshire, BD20 6LH, UK.

SUMMARY

A joint project between Yorkshire Water and the West Yorkshire Ecological Advisory Service commenced in 1981 to investigate whether consolidated sewage sludges could provide a suitable alternative to expensive top soil in the reclamation of colliery spoil tips. The project consisted of laboratory/greenhouse trials and progressed, with the cooperation of Kirklees Metropolitan District Council, to field-scale reclamation. Two sites reclaimed by the method developed have been continually monitored for substrate pH and metal uptake into the sward for approximately five years. The technique has been shown to be successful, providing a seed bed resisting acid regression, high sward productivity and acceptable metals in the grass for grazing livestock, whilst having the advantage of considerable cost savings for the reclaiming authorities.

1. INTRODUCTION

In November 1980, the West Yorkshire Ecological Advisory Service was consulted regarding the regression of a grass sward following reclamation of a former colliery spoil and waste tip. It was apparent from tests carried out on site material that the major problem inhibiting site growth was sulphuric acid production, resulting from oxidation of pyrites present in the spoil. In some parts of the site the effect was toxic to established grass. It was assumed that recorded pH values of 2.9 - 3.5 would also have toxic effects on developing micro-fauna and flora, thereby retarding soil formation.

These problems have been well documented (1) but no conclusive solution could be found in the literature to overcome them, other than blinding the site with imported top soil. This option is expensive, especially when dealing with large sites, and also depends on availability. It was concluded that a method was required that would accelerate the soil forming process, whilst buffering the acid production. It followed that such a method would necessitate a considerable increase in the organic matter content of a colliery spoil tip.

Literature searches revealed that some attempts had been made in this direction, for example the incorporation of domestic wastes, composts etc. (1). None of these trials, however, had increased the organic matter significantly.

Further problems arose when initially investigating zootoxic levels for potentially toxic elements (PTEs). All publications agreed that it was difficult, if not impossible, to establish maximum dietary levels for certain elements (2). For example, the capacity for copper storage varies greatly among animal species and differences among species in tolerance to high copper intakes are also large (2). There are also metabolic interactions with many other elements, particularly zinc, iron, cadmium and molybdenum. Zinc and iron compete with copper for hepatic protein binding sites; consequently, theoretically a potentially toxic concentration of copper could be absorbed by a grazing animal (i.e. >30 mg/kg), but because the same animal would absorb simultaneously high concentrations of iron and zinc, the resultant effect could be of copper deficiency in some species rather than toxicity.

There was also a lack of data on naturally occurring concentrations of PTEs in soils, untreated with sewage sludges. Therefore, meaningful comparisons could not be made between sludges and soils.

It must be pointed out that standard chemical analysis of the sludges and spoils can be highly misleading. The extraction techniques used to determine 'available' PTEs does not mean that they will be available to plants on site. Site conditions which make PTEs available are very different and analyses of available metals extracted by acidified water are probably more realistic. Absorption of metal elements by plants is governed by a number of factors, such as pH and organic matter content of the substrate and the genotype of the grass cultivars, independent of the size of the plant-available reservoir of PTEs (3). Plant tissue concentrations were found in this study far below those expected from the perusal of standard chemical analyses.

As the complexity of the problems became clearer, a decision was made to confine the experimental work to the physical and chemical properties of particular sludges and particular colliery spoils and wastes from chosen sites, due to the inherent variability of both materials. To date, this work has included a feasibility study, greenhouse pot trials and the full scale reclamation of two sites, coupled with regular on-site monitoring of the herbage at monthly intervals for the succeeding five years.

In a previous paper (4), Metcalfe reported the results from one of the laboratory/greenhouse trials and the initial results from the first field-scale investigation. More comprehensive results are now available and additional data from the second field-scale trial have been obtained.

2. LABORATORY/GREENHOUSE TRIALS

The experiments conducted in these trials consisted of a series of 18 cm pots containing various mixtures of a site-specific spoil and sludge which were measured against the performance of spoil/soil controls mixed in the same ratio and a pure soil control. Each test was replicated 10 or 15 times and set up in a random block pattern in a cool greenhouse.

The pots were sown with specified amenity (Mommersteeg 22) and agricultural (Nickersons Red Circle) grass mixtures at a rate equivalent to 20 g/m^2. Every 28 days from germination, the pots were harvested, cutting to 1.5 cm above "soil" level, and the grasses dried and weighed to enable production to be determined. The dried grass tissues were then milled to pass a 1 mm mesh screen using a Tecator cyclotec mill, and analysed for their PTE content by atomic absorption spectrophotometry at Yorkshire Water's divisional laboratory in Bradford.

At the end of each major experiment, the pots containing the mixtures most closely resembling actual field practice were retained, cropping and analysis continuing in the manner described. Tables 1 and 2 show the production and average metal results relating to the East Bierley trial, which utilised lagoon dried sludge from Mitchell Laithes sewage treatment works. It is clearly evident that herbage production is greatly increased when sludge is included in the formation of the seed bed, indicating the nutrient value of the sludge as opposed to soil. Moreover, this increase in grass production is maintained well into the fifth year post application, without any additional nutrient being supplied, representing cost savings in mineral fertiliser. The pH remained stable throughout, representing further cost benefits as periodic liming is unnecessary.

With regard to the PTE uptake into the grasses, it is also evident that the only metals showing levels elevated above those of the soil control are nickel and zinc, both of which remain well below the phytotoxic limit (P) at which the yield begins to reduce, and copper which marginally exceeds the limit. Levels of 20-25 mg Cu/kg DM are quite common in pasture grasses grown on unamended soils; indeed, the fields surrounding the East Bierley trials area itself produce levels up to 24.5 mg Cu/kg DM in the grasses. Moreover, the analytical evidence shows that PTE levels in the herbage decline with time, contrary to the popularly expressed fear that they would become increasingly available to the grass plants.

The Mill Bank trial, utilising lime conditioned press cake from North Bierley sewage treatment works, also exhibited greatly increased crop yields and a similar pattern of PTE uptake, declining with time. The only PTEs to show increased absorption above the soil control levels were chromium and copper; the higher mean copper level was produced by one result of the 12 exceeding the upper critical level (26.6 mg/kg DM); the remaining 11 results produced an average of 14.9 mg Cu/kg DM, which was very similar to the control soil.

The pot trial results tend to predict the worst possible situation that could arise in site reclamation because of several factors. Leaching of any soluble PTEs is actively prevented by standing the pots in saucers, the plant roots are confined and the greatly increased herbage production under near-ideal conditions means that the rate of uptake from the substrate will be higher than under normal field conditions.

From the results of the first year it was decided to progress to field-scale trials with the reclamation of the East Bierley and Mill Bank sites.

Table 1. East Bierley. Annual average dry matter production (t DM/ha - Mommersteeg 22) and PTE concentrations (mg/kg DM).

PTE	Toxic limit (mg/kg DM)	Soil control	Spoil/ soil	1:1 Spoil/ sludge Year 1	1:1 Spoil/ sludge Year 2	1:1 Spoil/ sludge Year 3	1:1 Spoil/ sludge Year 4	1:1 Spoil/ sludge Year 5
		\multicolumn{7}{c}{Production (t DM/ha per year)}						
		3.836	3.183	17.125	8.288	7.906	6.053	6.574
		\multicolumn{7}{c}{PTE concentration (mg/kg DM)}						
Chromium	10.0 P	2.92	2.87	2.16	2.28	1.05	0.79	<1.76
Nickel	14.0 P	1.93	2.58	4.85	3.73	2.18	2.57	2.39
Copper	21.0 P	20.7	17.53	23.52	15.91	13.34	15.09	11.20
Zinc	221.0 P	40.1	32.0	153.90	91.42	60.75	66.31	89.20
Cadmium	3.0 Z	>0.50	>0.50	>0.50	>0.50	<0.50	<0.50	<0.50
Lead	3.0 Z sheep 10.0 Z cattle	2.33	2.19	2.03	2.85	1.33	<0.91	<1.55

P = phytotoxic, Z = zootoxic limits.

Table 2. Mill Bank. Annual average dry matter production (t DM/ha of Nickersons Red Circle) and PTE concentrations (mg/kg DM).

PTE	Toxic limit (mg/kg DM)	Soil Control	1:1 Spoil/ soil + NPK	1:1 Spoil/ sludge Year 1	1:1 Spoil/ sludge Year 2	1:1 Spoil/ sludge Year 3	1:1 Spoil/ sludge Year 4
		\multicolumn{6}{Production (t DM/ha per year)}					
		5.154	9.109	12.626	6.206	5.440	4.624
			PTE concentration (mg/kg DM)				
Chromium	10.0 P	1.41	1.28	2.05	1.57	1.17	<1.70
Nickel	14.0 P	8.04	2.75	3.43	<1.26	1.76	<1.39
Copper	21.0 P	14.00	10.42	15.87	12.54	12.48	9.09
Zinc	221.0 P	82.94	48.51	53.13	33.67	47.24	28.93
Cadmium	3.0 Z	<0.50	<0.50	<0.50	<0.50	<0.50	<0.50
Lead	3.0 Z sheep 10.0 Z cattle	2.71	1.62	1.79	<1.91	1.57	<2.74

P = phytotoxic, Z = zootoxic limits.

3. FIELD SCALE TRIALS

3.1 Site preparation

Prior to sludge application both tips were contoured and commercially acceptable coal was extracted from Mill Bank by washing the spoil. The spoil was then regraded such that a layer of finer material was deposited to form the surface "skin" of the tip. When the surface had been adequately prepared, a layer of consolidated sewage sludge was laid down and immediately cultivated into its own depth of spoil, thus creating a seed bed composed of a 1:1 (wet volume) mixture of spoil and sludge, equivalent to a 2:1 mixture by dry weight.

East Bierley was reclaimed using lagoon dried sludge from Mitchell Laithes sewage treatment works. 4,710 tonnes of wet sludge were spread in a 100 mm layer over 3.1 ha and cultivated in to a total depth of 200 mm in May 1982. At 45% ds this was equivalent to an application rate of 692 tds/ha (1).

Mill Bank was reclaimed using lime conditioned press cake from North Bierley sewage treatment works, 4,851 tonnes of wet sludge being spread in an 80 mm layer over 4.4 ha and cultivated in to a total depth of 160 mm in July 1983. At 59.5% ds this was equivalent to an application rate of 656 tds/ha.

Samples of spoils and sludges were taken and analysed, the results and metal and nutrient applications being shown in Table 3.

The sites were then rolled to compress the seed bed and left to settle. East Bierley was seeded with the amenity grass mixture, Mommersteeg 22, at a rate of 20 g/m^2 in August 1982. This germinated readily to form a good sward.

Mill Bank was seeded with the agricultural grass/clover mixture, Nickersons Red Circle, also at a sowing rate of 20 g/m^2, in September 1983, with similar results to East Bierley.

Table 3. Composition of colliery spoil and sludge at East Bierley and Mill Bank.

	East Bierley			Mill Bank		
	Colliery spoil	Sludge	Addition (kg/ha)	Colliery spoil	Sludge	Addition (kg/ha)
pH	7.6	6.1		7.2	7.3	
Total metals						
Cr (mg/kg)	66.7	1,750	1,211	29	1,060	695
Ni (mg/kg)	63.0	72.0	50	33	38	25
Cu (mg/kg)	45.0	365	253	78	930	610
Zn (mg/kg)	180	1,491	1,035	146	740	485
Cd (mg/kg)	1.5	1.0	0.7	0.5	1.7	1.1
Pb (mg/kg)	30	536	372	58	510	335
EDTA extr. metals						
Ni (mg/kg)	2.0			1.6		
Cu (mg/kg)	2.0			29		
Zn (mg/kg)	14.0			24.8		
Nutrients						
N (%)		2.2	15.6		0.98	6.4
P$_2$O$_5$ (%)		1.98	13.7		1.76	11.5
K$_2$O (%)		0.3	2.1		0.88	5.8

3.2 Survey methods

Twenty four herbage samples were taken every 28 days, following standard 'W' sampling patterns as shown in Figures 1 and 2. Soil pH was recorded at each sampling point. It should be noted that these sampling points were not fixed so that some randomisation was introduced. Further randomisation of sampling Mill Bank was obtained by rotating the sampling pathways through 90° each month.

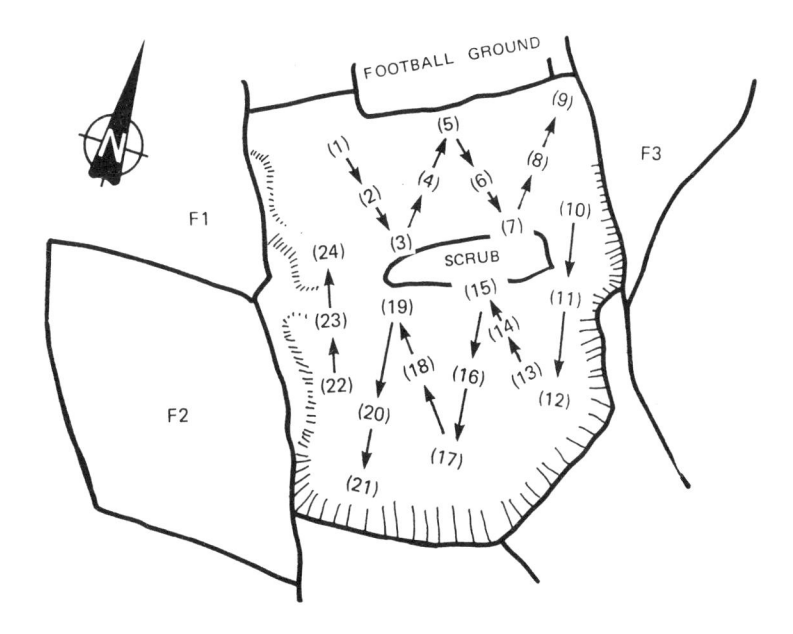

Fig. 1. East Bierley site plan.

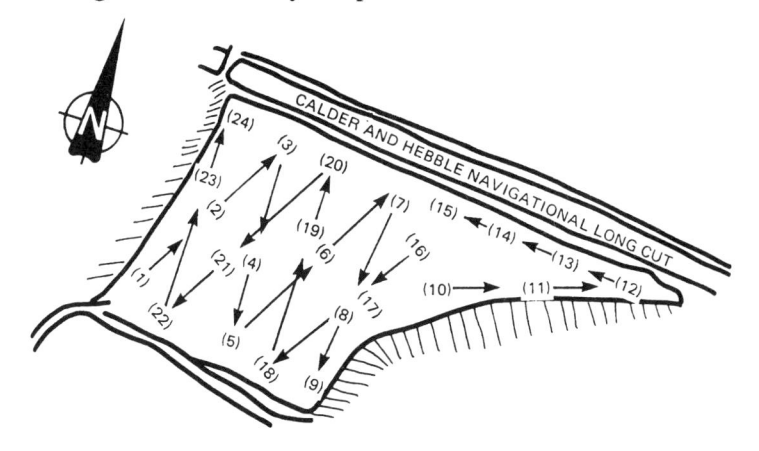

Fig. 2. Mill Bank site plan.

The grass samples collected were washed, dried in a moisture extraction oven at 100° C for 18 hours, milled, and analysed for their PTE content.

When considering PTE uptake it is important to define certain concentration criteria and much research has been devoted to establishing normally occurring background and toxic levels in herbage. The total concentrations of PTEs typically found in plants quoted by Allaway (3) are shown below in Table 4, which also includes background and upper critical levels specifically determined for ryegrass by Beckett and Davis (5). The upper critical level is defined by Beckett and Davis (5) as the level of the element in the tissue at which yield begins to reduce, not the level at which plant tissue death would occur.

Table 4. Ranges, background and upper critical levels of PTEs in plant tissue (mg/kg DM).

Metal	Range in plants (4)	Background	Ryegrass upper critical level (3)
Cr	-	1.0	10.0
Ni	1-10	2.0	14.0
Cu	4-15	11.0	21.0
Zn	15-200	50.0	221.0
Cd	0.2-0.8	<1.0	10.0
Pb	0.1-10.0	3.0	35.0

4. EAST BIERLEY

4.1 Assessment of results

The average PTE concentrations in the herbage at East Bierley for the years 1982-86 are shown in Figure 3. From this it is evident that, although tissue concentrations of Ni, Cu and Zn exceeded the normally occurring background levels, particularly during the first year's growth, they did not tend to exceed the upper critical concentrations. The only metal to have done so was copper at 22.8 and 33.44 mg/kg DM. Interestingly, the herbage samples taken from the surrounding fields in April 1983 also proved to contain copper levels in excess of the upper critical concentration, ranging from 22.0 to 24.5 mg/ kg DM, despite these grasses growing on normal pasture soils uncontaminated by sewage sludge.

The analytical evidence, moreover, shows that the copper levels in the grass tissues diminished to near background levels after 12 months and,

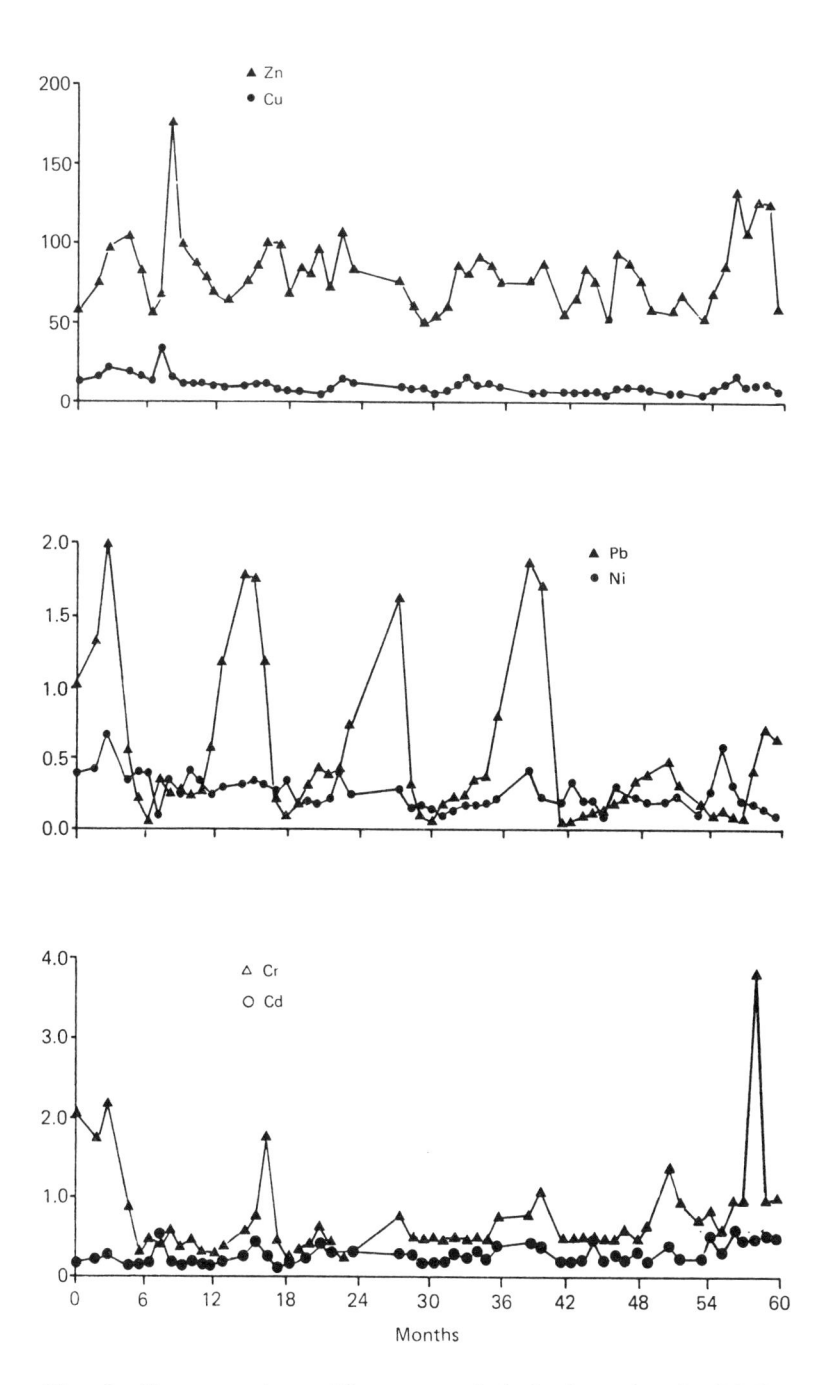

Fig. 3. Concentrations of heavy metals in herbage (mg/kg DM) at East Bierley. Monitoring started on 1 December 1982.

indeed, the majority of results from Year 2 onwards were below the 10.0 mg/kg DM levels recommended by MAFF as the dietary requirement for grazing cattle (2).

The relationship between the results, the upper critical and the background concentrations clearly demonstrates that, despite the high rate of sludge application necessary to form the seed bed, the PTEs are not taken up into the herbage in sufficient quantity to even reduce the rate of grass production.

Cadmium and lead are potentially deleterious to the health of grazing livestock in much lower quantities than would be necessary to prove phytotoxic, but neither element presents a problem in this respect since both are well below their critical levels for livestock (Table 1) (6).

The lead levels at first appear to present an interesting problem, the marked seasonal fluctuation in the results showing the concentration in the tissues to rise above the critical level for livestock in the winter months. In fact this could not represent the metal being taken up from the substrate; in six years' continuous monthly monitoring of the pot trial, lead concentrations in the herbage have never risen above 5.0 mg/kg DM. The most likely explanation for the field result is that the increased lead content arose as a consequence of seasonally increased aerial deposition. The site is bounded on two sides by motorways (M606 and M62) and the prevailing wind is from the direction of these major road systems. During the colder winter months the exhaust gases do not rise to the same extent and the lead in them is precipitated closer to the motorways by the higher rainfall. The phenomenon of increased lead deposition has been noted by other research workers, notably Cawse (7). The most exposed of the fields surrounding the trials area also exhibits higher lead concentrations in the grass tissues, 13.5 mg/kg DM, during the winter period, (2).

4.2 Mill Bank

The average PTE concentrations in the herbage at Mill Bank are shown in Figure 4. This indicates that PTEs, whilst present in the sludge, are unlikely to cause any deleterious effects to the grass as phytotoxins, or to grazing livestock.

Nickel and zinc tended to remain at, or near, normally occurring background levels. Copper exhibited a similar uptake pattern to that obtained at the East Bierley trials area, namely greater than background but not exceeding upper critical levels during the first twelve months and thereafter steadily declining to <10.0 mg/kg DM.

Chromium and cadmium also tended to remain at, or near, normally occurring background levels, whilst lead exhibited the same seasonal fluc-

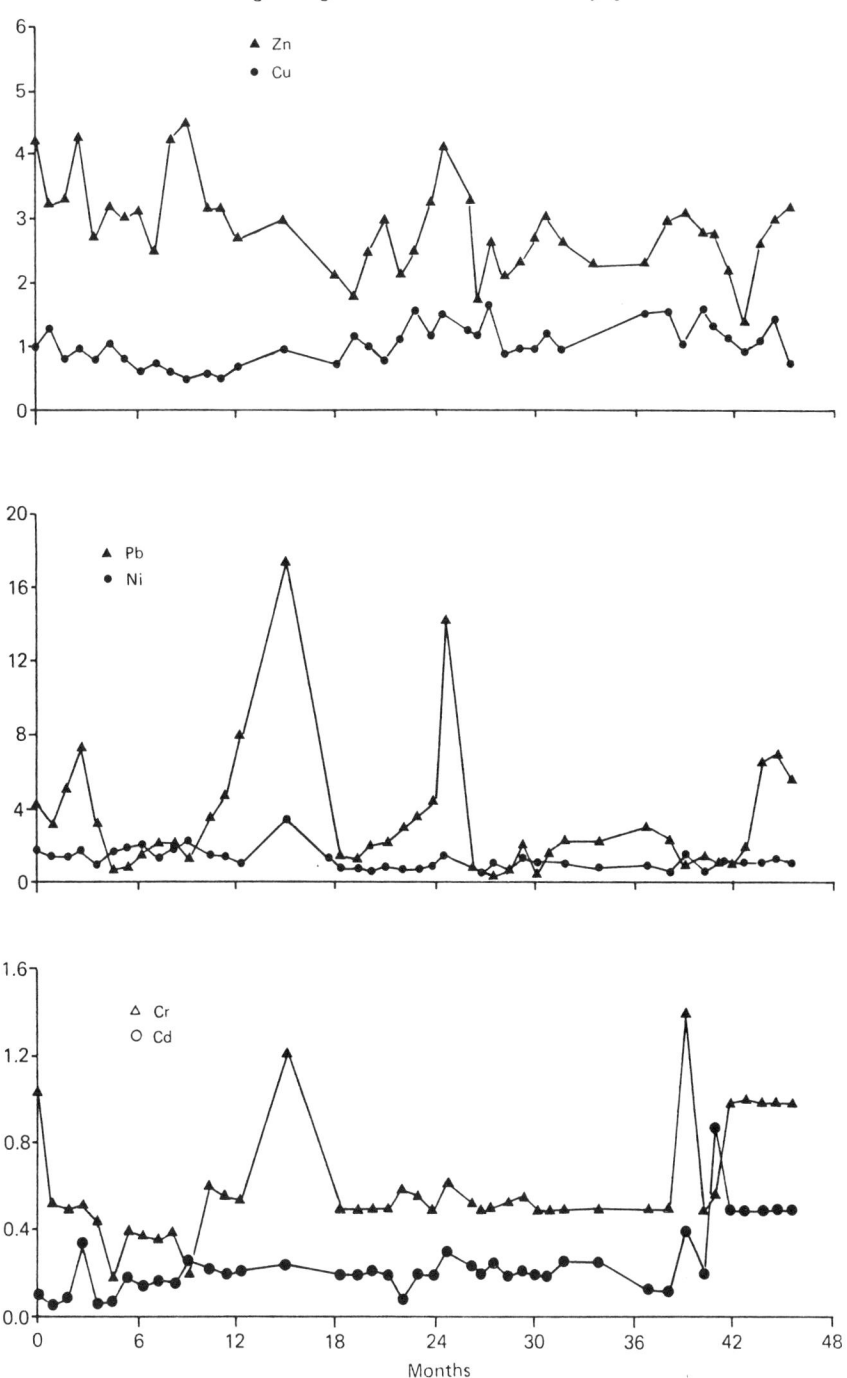

Fig. 4. Concentrations of heavy metals in herbage (mg/kg DM) at Mill Bank. Monitoring started on 14 December 1983.

tuating pattern, though to a lesser degree, as the site is considerably more sheltered with regard to possible atmospheric deposition from a nearby source of lead emission.

5. DISCUSSION AND CONCLUSIONS

Despite the seasonal fluctuations in metal levels in the herbage it is evident that uptake from the substrate at both sites is insufficient to cause toxicity to the grass sward. These tissue concentrations are also unlikely to cause any deleterious effects to grazing livestock. Underwood (6) quoted a dietary intake of 500 mg Zn/kg as having no effect upon cattle, toxicity symptoms appearing only at approximately 900 mg/kg total diet. For copper, nickel and cadmium, the same author quoted figures of 50.0, 50.0 and 3.0 mg/kg respectively as being critical levels in herbage.

The only problem that could arise with regard to the continued success of the field operation is the possibility of acid regression occurring. Normal reclamation practice on colliery spoil sites treated with top-soil is to lime the spoil heavily and subsequently apply lime as necessary to stabilise pH. Provided that the practice of post-reclamation liming is adhered to on sludge-treated sites, where necessary, acid regression should not occur. Indeed, one striking feature of the trials areas is the relative stability of the substrate pH. Certainly, the spoil at East Bierley does have the capacity to become extremely acidic as the pH of 50 kg samples of spoil, which had a pH of 7.6 on collection, dropped to pH 3.1 following drying in the laboratory in a matter of weeks.

Evidence collected in the USA (8) appears to suggest that sludge itself exercises a stabilising influence upon spoil pH and that PTE levels in the herbage tend to fall over a period of years if the spoil is treated with one large application of sewage sludge rather than repeated small applications.

The reclamation of East Bierley and Mill Bank colliery sites far exceeded expectations from a Local Authority viewpoint. The costs of reclamation using sludge as compared to topsoiling were reduced by approximately £2,000 per hectare. The general performance of the two sites has been very good; lush, even swards were developed within two months of seeding and no fertiliser applications have been required since establishment to maintain productivity, representing a further cost-benefit. The inclusion of sludge to form the seed bed promotes a very rapid growth of the sward and, once reclaimed, it is recommended that provisions should be made for two cuts of the sward during the first growing season. If the grass is allowed to grow unchecked, it can be beaten down by high winds or heavy rain, forming a dense mat which excludes light, thus hindering the development of the young growth and damaging the sward. Cuttings should be removed for the same reason.

In order to achieve comparable results with this technique, it is of paramount importance to use consolidated sludge, i.e. lagoon dried or pressed, with a dry matter content of not less than 40%, which enables normal contractual machinery to be used on site. Moreover, it is essential to investigate the chemical and physical properties of each colliery spoil site to be reclaimed, as well as identifying the most suitable sludge by pre-reclamation greenhouse pot trials.

It is also important that site practice should conform closely to experimental procedure, namely that predetermined mixing ratios are used and that the experimental grass seed be used in the field. Site monitoring, following sludge-based reclamation, is also to be recommended.

Greenhouse pot trials need to be continued after reclamation, and continuously monitored to detect problems that may arise and to isolate problems that may have been caused by reclamation techniques or heterogeneous conditions on site not detected during pre-investigations.

Given these types of control during the operation, consolidated sewage sludges can form substitutes for top soil in the reclamation of colliery wastes. The advantages to the Local Authority include an inexpensive alternative to soil and cost savings in both lime and fertiliser, whilst for the supplying Water Authority it is a convenient outlet for large quantities of sludge.

In conclusion, the success achieved at the sites treated with this technique is largely due to the detailed pre- and post- investigations carried out, and, equally importantly, the close working tri-partite relationship between the Local Authority, its scientists and the scientists of Yorkshire Water. This small task force was able to contribute its skills to the project, unfettered by unnecessary bureaucratic procedures, each member taking full responsibility for their aspect of the work.

REFERENCES

(1) Gemmel, R. P. (1977). Colonisation of industrial wasteland. Studies in biology No. 80, 24-26.
(2) Agricultural Research Council (1965). The nutrient requirements of farm livestock, No. 2. Ruminants. ARC.
(3) Allaway, W. H. (1968). Agronomic controls over environmental cycling of trace elements. Advanc. Agron., *20*, 235-274.

(4) Metcalfe, B. (1984). The use of consolidated sewage sludge as a soil substitute in colliery spoil reclamation. Wat. Pollut. Cont., *83*, 288-299.

(5) Davis, R. D. and P. H. T. Beckett (1978). Upper critical levels of toxic elements in plants II. Critical levels of copper in young barley, wheat, rape, lettuce and ryegrass, and of nickel and zinc in young barley and ryegrass. New Phytol., *80*, 23-32.

(6) Underwood, E. J. (1977). Trace elements in human and animal nutrition. Academic Press.

(7) Cawse, P. A. (1977). Deposition of trace elements from atmosphere. In: Inorganic pollution and agriculture, MAFF Ref. Book 326, 22-46.

(8) Sopper, W. E. and E. M. Seaker (1984). The use of municipal sewage sludge to reclaim mined land. Critical Reviews in Environmental Control, *13*, 227-271.

DISCUSSION SESSION 1

QUESTION: R. J. UNWIN, MINISTRY OF AGRICULTURE.
Is there no cutting management of the grass/legume mixture once established on the minespoil, and is it just allowed to die back each fall because, if so, it would seem surprising that the legumes assume dominance and that they are not shaded or killed out in the first two seasons?

ANSWER: W. E. SOPPER.
On most of these mine sites there is no management after establishment. On a few of the sites where some topsoil was saved and there were not very many large rocks, they have gone back into agriculture production, growing hay and row crops like corn. Then they are managed because now nutrients are removed. On the other sites it was simply to get vegetation and microbial activity established, and once that happens it is more or less permanent. The reason we put the legumes in that mixture is that the grasses may not be a long-term cover because they cannot fix nitrogen from the air whereas the legumes can. Our oldest site is now about 15 years old and it still has a 100% cover on it, with absolutely no indication of deterioration.

QUESTION: C. D. BAYES, WATER RESEARCH CENTRE.
The actual application rates of sewage sludge and chicken waste used by Dr Pulford were about 20-30 t/ha. What sort of extrapolation can he make of his results?

ANSWER: I. D. PULFORD.
If you can get responses of that type with 20 t/ha I would expect you would get greater responses the more you add, but I think there will be an upper limit. We have not had a concerted attempt to look at rates, methods of application and types of sewage sludge. We need to look at these in a more ordered manner.

QUESTION: J. A. GLEGG, GREENPEACE.
Did Prof. Sopper look for any organic contaminants like PCBs in the sludge and groundwater?

ANSWER: W. E. SOPPER.
Before any of the sludge goes on land it has to have a complete analysis for all the organics etc. and the concentrations have to meet certain maximum criteria that have been laid down. All of these sludges from the various cities and even the city of Philadelphia pass all those EPA toxicity tests, and so that is why in the research in the field we did not analyse ground water,

not only because it is very expensive but also if organics are not in the sludge, it is not a problem. PCB's are very very low, I think that under the current regulations they have to be <10 ppm and they are probably 3 ppm in Philadelphia sludge. The Authority producing the sludge would also have control over the industrial discharges into the sewers and would therefore be in a very good position to identify any potential contributions to the sludge.

QUESTION: J. WALKER, YORKSHIRE WATER.
Are there any odour problems with the actual application of sewage sludge, particularly in relatively urban areas?

ANSWER: W. E. SOPPER.
No, because almost all of the sludges that we have used have been treated. Despite this, we have tremendous problems with public acceptance, because the source is human waste they think there are going to be odour and pollution problems.

QUESTION: P. J. MATTHEWS, ANGLIAN WATER.
Dr Pulford suggested that there was a declining yield from a site with the different treatments and that one would have to revisit the site to keep the activity going, yet the previous papers suggested that one application did a good job for several years and that you did not have to revisit? Now there does seem to be a bit of a contrast between those two.

ANSWER: I. D. PULFORD.
One of the reasons for the declining yield was to do with cutting. These sites were not cut and so where we had very high vegetation yields due to the manures, we got a lot of grass produced which essentially fell over in the autumn and virtually thatched the surface and so growth in the subsequent year had to come up through that dead vegetation. Although that is a store of the nitrogen it is also a barrier to growth. So the point about cutting is an important one. If you go in for a low-cost sustainable system then you have ultimately got to have legumes in as part of the system. With high initial rates you may damage the legumes, either by shading or by just having too much nitrogen there.

ANSWER: K. BYROM.
It depends what your end use is. Obviously if you are looking to take a harvest off your land you are going to have to refertilise it but if you are looking for just a stable grass cover, low growth may actually be an advantage and you want something which is self-sustaining but not necessarily going to give you a huge sward that you have to cut every year.

QUESTION: P. HYDES, SOUTHERN WATER.
With the garbage sludge compost in Dr Scherer's experiments, was there a non-degradable plastic problem from the garbage and, if so, what did he do to sort this out?

ANSWER: H. W. SCHERER.
The garbage compost was sieved with 0.8cm mesh so only small particles of plastic were in it, the amount of which was negligible.

QUESTION: P. HILEY, YORKSHIRE WATER.
Could Dr Scherer tell us a bit more about the methods he used to measure the leaching rate of the nitrogen on the steep vineyards?

ANSWER: H. W. SCHERER.
The angle of these vineyards is about 30°. Down the hill we had lysimeters covered with plastic film so that running water from the top could not go down. This meant that we only collected the leachate in a depth of about 1m and the amount of water and its nitrogen content were determined. The water was collected every 10 or 12 days depending on the rainfall.

QUESTION: B. LEE-HARWOOD, FRIENDS OF THE EARTH.
Could Mr Lavin elaborate on his relationship with the Water Authority?

ANSWER: J. C. LAVIN.
The work of my colleagues in the Water Authority was absolutely first class; they did exactly what I wanted done, and the results they presented were excellent. It has been an exceedingly fruitful relationship, particularly when you consider you have three organisations working together: a large Water Authority, a large local government organisation and its scientific advisers.

COMMENT: R. J. UNWIN, MINISTRY OF AGRICULTURE.
The conclusions of Mr Lavin's paper are that there are no risks to grazing livestock based on analysis of the herbage tissue. On sites with elevated metal levels, regardless of the growing medium, the greatest risk to livestock is not through the ingestion of herbage but through the ingestion of soil adhering to that herbage. The risk to the stock therefore is of a complex interaction between the amount of soil-forming material ingested and the availability of the metals in that material. Those looking for an agricultural after-use of such reclaimed sites should consider the total metal level in these soil amendments and also their management because the establishment of a closely knit sward can do a lot to reduce the ingestion of soil by grazing livestock. The total amount ingested can also be limited by

the amount of access given to stock.

QUESTION: C G NICHOLS, ORANGE COUNTY, CALIF., USA.
Was there any analysis to learn whether the character of the grapes were altered because of the sludge applied?

ANSWER: H. W. SCHERER.
The contents of heavy metals in the leaves increased a little but not in the grapes. The nitrate content in the grapes increased but during processing of the wine this nitrate was reduced to levels similar to the control.

QUESTION: B. METCALFE, YORKSHIRE WATER.
In Mr Lavin's experience, what preparations need to be made before the sludge is put on to land and what do the local authorities want from the Water Authorities?

ANSWER: J. C. LAVIN.
Local governments are political institutions and therefore our political masters are extremely keen on one thing - votes - so it is vital not to upset the local population around a site. The biggest problems tend to be odour or the desire by the authority to get rid of unsuitable sludges. It is of paramount importance that the Water Authority cooperates with the local authority to provide sludges suitable for land reclamation and not to get rid of problematic sewages in large amounts.

COMMENT: P. J. MATTHEWS, ANGLIAN WATER.
The objective of the Water Companies is to maintain long-term outlets; there is no mileage in short-term short cuts. We want to dispose of sludge in such a way that we protect the environment and the public are encouraged by our activities. If we take measures which upset local politicians or local people, it is quite clearly our future operations that are going to be compromised. We are not seeking to dump sludge in any environmentally unfriendly way. It is very important that everyone should know and remember that.

QUESTION: J PERSSON, AGRIC. UNIVERSITY OF SWEDEN.
Has Mr Lavin any investigations to show if there are any affects of the very high amounts of sludge, 600 tds/ha, on soil processes such as nitrification, nitrogen fixation by legumes and so on?

ANSWER: J. C. LAVIN.
We have analysed the site every 28 days for the last seven years and , as yet, we have not detected such problems.

QUESTION: R. D. DAVIS, WATER RESEARCH CENTRE.
We have heard that sludge is a good material to use for land reclamation but nevertheless it remains a minor outlet for sludge. How can this outlet be expanded?

ANSWER: J. C. LAVIN.
I think it is a matter of organisation between local government and the Water Authorities. The organisation for land reclamation has been beset with all sorts of minor problems; one is that sewage works managers desire to get rid of sludge rather than store the thousands of tonnes required. Secondly, local government is beset with problems in the acquisition of sites. I think a lot of the problems could be resolved providing we both forward-planned.

COMMENT: G. A. FLEMING, JOHNSTOWN CASTLE, EIRE.
When we speak of metals in sludge we tend to take the negative view and speak of the harm that accrues which it does sometimes. I think we rarely look at possible positive effects from the application of sewage sludge.

COMMENT: J. C. LAVIN.
The absorption of metal into grass tissue is complex and we found that the cultivar was controlling to some extent the absorption of metal irrespective of the size of the available metal reservoir. It is not quite so straightforward but we found in general terms that, for instance, amenity strains of grasses absorbed considerably more metals than agricultural strains. In fact on one site we have completed recently we have found we now have a copper deficiency as laid down by ADAS.

COMMENT: P. J. MATTHEWS.
We rightly spend a lot of time looking at the potential disadvantages of using sludge and we should make sure that we do not do any harm to the environment, but we should also give every thought and consideration to looking at positive aspects of sludge and indeed many research institutions do this already.

SESSION 2
Forestry

The potential for utilising sewage sludge in forestry in Great Britain

C. M. A. TAYLOR[1] and A. J. MOFFAT[2]

[1]Forestry Commission Northern Research Station, Roslin, Midlothian, EH25 9SY, UK.
[2]Forestry Commission Research, Alice Holt Lodge, Farnham, Surrey, GU10 4LH, UK.

SUMMARY

In Great Britain, the concern over reductions in the amount of sewage sludge that can be disposed by current methods has initiated the search for new, more environmentally safe outlets. Recent experimental results have indicated that sludge can be successfully utilised as a forest fertiliser. This paper describes two surveys carried out in 1987 to assess the potential programme and summarises the results. It is estimated that up to 11% of the total sludge production of 1.2 million tonnes dry solids could be applied to forest land. When the transport distance is reduced from 32 km to 16 km the proportion decreases to 6%. Therefore, there is a substantial potential programme on forest land where considerable benefits could be derived by both the disposal authorities and the forest industry.

1. INTRODUCTION

In Great Britain, over 450,000 tonnes dry solids (tds) of sewage sludge is currently applied on agricultural land (40% of production) and a further 300,000 tonnes dry solids (30%) is disposed of at sea (1). However, there is increasing pressure against the use of these main outlets, with constraints being imposed on disposal on farmland by a recent directive from the European Community (2) and possible future restrictions on sea disposal (3). Alternative disposal methods are now being sought to secure environmentally safe outlets and alleviate these anticipated problems. One of the outlets being actively pursued is the use of sludge as a forest fertiliser, a common practice in other countries such as the United States (4).

Since the introduction of the Forestry Act in 1919 there has been a progressive programme of afforestation in Britain, which had been extensively cleared by man's agricultural and industrial activities since prehistoric times. The area of productive woodland now stands at a little over 2 million hectares (ha), or 10% of the land area (Table 1), of which nearly 58% is privately owned and 42% is state owned and managed by the Forestry Commission. The forest area is currently being expanded at the rate of 25,000 ha per annum, mainly by private investment. The Government has recently increased the target to 33,000 ha and augmented it by a further target of 12,000 ha per annum during 1989-91 on surplus agricultural land. The main reason for continued expansion is to reduce Britain's dependence on imports of timber, which currently stand at nearly 48 million m³ per annum (90% of requirements) and costs £6 billion.

Table 1. Area of productive woodlands, annual afforestation and restocking programmes in Great Britain (thousands of hectares).

	Productive woodlands	Annual afforestation programme (av. 1984-88)	Annual restocking programme (av. 1984-88)
Forestry Commission	888	5.6	7.2
Private woodlands	1,224	19.3	4.1
Total	2,112	24.9	11.3

Most of the forest area is on nutrient-poor soils, and recent planting has taken place increasingly on poor sites. Currently, inorganic fertilisers are applied to correct nutrient deficiencies in most forest crops, but these are costly and therefore used mainly on those sites where fertiliser is essential for the establishment of a satisfactory crop. Phosphate is the principal nutrient applied with potassium the next most important (Table 2), although the application of nitrogen is increasing and is a vital treatment on many sites (5). However, it would only be possible to substitute part of this programme with sludge application due to the high proportion of moorland sites (gleys and peats) planted in recent years where high water tables would mean an unacceptable risk of runoff. The main area of interest would be on the more freely-draining heathland sites where spruce in the establishment phase and pine in the pole stage would respond to applications of nitrogen and phosphate.

Table 2. The annual programme of fertiliser application to forestry in Britain over the period 1981-86 (19).

Fertiliser applied	Average annual area fertilised (ha)	Quantity of fertiliser applied (tonnes)
Unground rock phosphate (P)	54,000	24,300
Muriate of potash (K)	29,000	5,800
Urea (N)	2,000	700
Total	56,000[1]	30,800

[1] Muriate of potash is always applied in combination with unground rock phosphate - hence the total area fertilised is 29,000 ha less than the sum of the figures in this column.

Potentially, sludge has much to offer as a forest fertiliser for such sites. It contains 2-4% phosphate and 3-5% nitrogen, much of which is in organic forms withheld in the soil against leaching. Its use may therefore produce a longer response period of enhanced growth than the three years gained by the use of conventional inorganic fertilisers (6). Furthermore, the large quantities of organic matter in sludge can also improve soil structure, particularly important on degraded reclamation sites.

A programme of research has been conducted since 1982 to quantify the growth benefits from applying sewage sludge to forest crops and to monitor the effects on soil and the environment (7). This has provided sufficient encouragement to consider utilising sludge on a larger scale in British forestry. In 1987 surveys were made to assess the potential scale of this programme (8) (9), the results of which are summarised in this paper.

2. SURVEY

2.1 Methods

2.1.1 Crop selection Suitable site and crop types were defined as those where growth responses to the application of sludge with low risk of pollution from runoff could be predicted. Suitable soil types are principally podzols, ironpan soils, rendzina, ranker and littoral soils, which are freely-draining and capable of absorbing the large quantities of water present in sludge (95-98%), and low in available phosphate and nitrogen. Gleys and peats, soil types where sludge would be more difficult to apply due to high water tables, were also included in the survey for England and Wales (8) but have been excluded from this paper. Terrain type, as defined by the Forestry Commission (10), was also used in England and Wales so the sites that were considered too rough or steep could be excluded.

There are three stages in the forest rotation when access is possible to enable application of sludge:
a) After clearfelling and before replanting;
b) During the early establishment phase before the growth of the crop limits access ($c.$ 5 years after planting);
c) When stands have been thinned and extraction racks cut, allowing access to be gained under the tree canopy (normally after 25 years).

Both surveys were restricted to the principal conifers, i.e. spruces, pines and firs, although the survey in Scotland concentrated on areas of (a) Douglas fir (*Pseudotsuga menziesii* (Mirb.) Franco), Sitka spruce (*Picea sitchensis* (Bong.) Carr) and Norway spruce (*Picea abies* (L.) Karst) growing at less than 14 m^3/ha per year in the 1-5 year stage and (b) pine crops (*Pinus* spp.) growing at less than 8 m^3/ha per year in the older crops. These species produce the largest and most predictable responses on the soil types and crop stages chosen (5) (11). Larch (*Larix* spp.) was excluded from the survey in England and Wales due to concern over the deterioration of form that often accompanies the application of large amounts of nitrogen (12).

2.1.2 Site selection within existing forests The area of suitable site and crop types available within the state forests was assessed using the Forestry Commission subcompartment database for 1985 and 1986 (13). Unfortunately, detailed soil surveys were not available for all areas and estimations were made for areas without detailed soil coverage utilising either 1:250,000 scale soil maps (Soil Survey for England and Wales) or a system of sample forests (Scotland). The latter approach involved zoning the country according to similarities in climate, rainfall, soils and geology (9). Sample forests were then chosen within each zone where there was over 90% detailed soil survey coverage (Forestry Commission Site Survey maps, 1:10,000). The ratios of soil types indicated for the sample forests were then extrapolated to the remaining forests in the same zone. Crop type, as previously defined by species and age, was then quantified by soil type for each forest. In addition, the amount of clearfelled land awaiting replanting on suitable soil types was identified for each forest by the same method.

The estimation of the suitable land area available within the private sector was more difficult due to the lack of information on crops and soils. However, the Census of Woodlands and Trees (14) provided gross areas by species which were converted by using appropriate Forestry Commission forests as sample forests for extrapolation as described previously. This assumes that private woodlands occur on the same soil types and in the same proportions. Although it is unlikely to be strictly accurate, this method is believed to give a reasonable approximation of the potential area available for sludge application in private woodland.

2.1.3 Sludge application rates Suitable application rates were chosen to convert the area of forest available to the quantity of sludge that could be applied (Table 3). These rates were based on previous experience with the rates of inorganic fertiliser required (11) (15) (16) (17), the desire to limit heavy metal loading and experience gained from early experimental results (18). Cake sludge was chosen in preference to liquid sludge for restocking sites in England and Wales due to the higher application rates possible without risk of runoff. This option was not included in the Scottish study because cake sludge is less available in Scotland. The application rate for liquid sludge was increased for the post-thinning category (>25 years) in England and Wales as it was felt that the rate previously chosen (5 tds/ha) (8) was too conservative.

Table 3. Sludge application rates for the various crop stages.

Crop stage	Sludge type	Sludge application rate (tds/ha)
Restocking	Cake sludge (England/Wales)	37.5
	Liquid sludge (Scotland)	20.0
Crops <5 years	Liquid sludge	10.0
Crops >25 years	Liquid sludge	10.0
Afforestation	Liquid sludge	20.0
Reclamation	Cake sludge	50.0
	Liquid sludge	5.0

Note: 5% dry solids assumed for liquid sludge.

In the Scottish survey an attempt was made to match the potential forest area available for sludge application to sources of sludge production. These were defined as those sewage treatment works providing at least primary settlement or biological treatment of crude sewage and were selected from data supplied by the Scottish Development Department (19). Total sludge production per annum was calculated from population estimates using *per capita* sludge production figures for the various sludge processes. The sludge production was then compared to the available forest area within 16 or 32 km of each sewage treatment works to define further the potential sludge application programme.

2.1.4 Other potential forest outlets There are two further areas where there is potential to apply sludge, namely newly afforested land and areas where land is being reclaimed to forestry following opencast coal mining or mineral extraction.

Changes in land use policy to reduce agricultural surpluses within the European Community are likely to release better land types for afforestation. This applies to both the revised target of 33,000 ha for conventional afforestation and the additional 36,000 ha that the British Government has set as a target for farmers to plant trees on their land over the period 1989-1991. Although this land will be inherently more fertile than land previously available for forestry, a proportion will be on unimproved heathland or freely-draining soils where sludge application would be beneficial. Woodlands planted on this type of land will generally be closer to sludge supplies and will have better on-site access than existing forests. In addition, it is likely that tree crops such as *Populus* spp. on short, 10-15 year rotations, will be planted on improved land under this programme. These crops have larger nutrient demands than conventional tree crops, providing further opportunities for sludge disposal.

While agricultural surpluses exist, there is less need to restore opencast coal mines or mineral extraction sites to agricultural use and an increase in reclamation to forestry above the current modest programme is likely. However, the low nutrient status and lack of organic matter make many sites initially unattractive for forestry, an industry which cannot justify large, early investment in ameliorative treatments for crop establishment and maintenance. The use of sludge would be an ideal treatment for such sites (20) (21) and could offer a substantial outlet. Cake sludge could be applied in large quantities (e.g. 150 tds/ha) to try to restore the soil prior to planting. Alternatively, liquid sludge could be applied on a frequent basis for the first 10 years, for example 100 m^3 liquid sludge (5 tds)/ha per year.

2.2 Results

The potential forest area for treatment is indicated in Table 4, with an estimate of the annual programme and the quantity of sludge that could be applied. This is based on a single application to clearfelled sites or crops less than five years old and an application every five years to crops in the post-thinning category. The area available annually is 16,580 ha which, at application rates as described in Table 3, would allow 164,500 tds of sludge (14% of total production) to be applied to forest land.

This potential disposal programme has to be modified to take account of the distance from the sludge source to the forest outlet. This has been attempted in Scotland (described previously) which indicates that, from the available area of 5,284 ha of Forestry Commission land, only 1,334 ha could be treated if transport distance is limited to 16 km. If this is increased to 32 km the potential area increases to 3,303 ha, although this latter figure is a slight overestimate caused by the same sludge from certain sources being allocated to more than one forest outlet (9). However, these calculations indicate the major impact that transport distance has on the potential programme.

If a simple extrapolation is made from the Scottish study, where 25% of the available forest area was within 16 km of a suitable sludge source, then

Table 4. Summary of the total area available in existing forests and the associated potential programme of sludge application.

	Area of suitable land (ha)	Area available annually (ha)	Potential annual sludge application (tds)
Forestry Commission:			
England	21,122	4,884	43,693
Scotland	24,010	5,284	58,871
Wales	8,014	1,884	16,645
Total	53,146	12,052	119,209
Private woodlands:			
England	8,642	1,728	17,280
Scotland	13,420	2,684	26,840
Wales	586	117	1,170
Total	22,648	4,529	45,290
Grand total	75,794	16,581	164,499

this would indicate a potential annual programme of 4,063 ha for the whole of Britain. The amount of sludge that could be applied on this area would be 42,140 tds, or 4% of the total sludge production (Table 5). This may well be a conservative estimate for the potential programme in England and Wales, where the influence of higher densities of population close to the forest areas is currently being investigated. Conversely, the preference for recreational use of state woodlands near population centres may preclude the use of some areas for the application of sludge. An upper limit of 32 km increases the potential programme to approximately 102,000 tds, or 8% of the total sludge production (Table 5). This proportion can become much higher when comparisons are made on a regional level. For example, in the north-east of Scotland (Grampian Region) the potential annual programme is greater than the sludge supply.

It is not possible to predict accurately the potential programme on afforestation or reclamation sites. If it is assumed that there is a 70% achievement of the afforestation target of 45,000 ha, of which only 5% was suitable and near to sludge supplies, then this would provide an outlet for a further 31,500 tds of sludge (Table 5). An additional 2,000 tds could be disposed on reclamation sites, assuming a modest programme of 400 ha of afforestation per annum (Table 5).

Table 5. Summary of the potential annual programme accounting for the area annually available within existing forests, the quantity of sludge available and the transport distance, and an estimate of the potential in the afforestation programme of bare land and reclamation sites.

	Area available annually (ha)		Quantity of sludge (tds)	
	Pessimistic (<16 km)	Optimistic (>32 km)	Pessimistic (<16 km)	Optimistic (>32 km)
Existing forest:				
Forestry Commission	3,037	7,532	30,041	74,506
Private woodlands	1,141	2,831	11,413	28,306
Afforestation:				
Bare land	1,575	1,575	31,500	31,500
Reclamation	400	400	2,000	2,000
Total	6,153	12,338	74,954	136,312

3. CONCLUSIONS

The total area of productive forest in Britain is 2 million ha, of which it is estimated that over 75,000 ha are potentially suitable for utilising sewage sludge as a forest fertiliser.

Based on a five year return period for application to post-thinning crops and single applications to clearfelled sites and crops less than five years old, the estimated area available annually is over 16,000 ha (73% Forestry Commission, 27% private) which could accommodate over 160,000 tds of sludge. When limitations on transport distance are imposed, the area available in existing forests is reduced to approximately 10,000 ha (32 km limit) and 4,000 ha (16 km limit), with the amount of sludge able to be applied being 103,000 tds and 40,000 tds respectively.

The potential for using sludge on newly afforested land is difficult to assess but cautious estimates would suggest that there is an additional potential programme for 32,000 tds on bare land (1,600 ha) and 2,000 tds on reclamation sites (400 ha).

The total potential annual programme for sludge application would be 136,000 tds on 12,000 ha of forest land (optimistic case) or 75,000 tds on 6,000 ha of forest land (pessimistic case). This could utilise either 11% or 6% of the total sludge production in Britain. This scale of programme will not solve the predicted problems over sludge disposal through existing outlets but it could make a significant contribution. In certain locations the proportion would be much higher, particularly where there are rural sludge sources near to the main suitable forest areas. In return, the dependence on expensive imports of inorganic fertilisers in forestry could be dramatically reduced and the overall productivity of the forest area would be increased, thus reducing Britain's dependence on timber imports.

ACKNOWLEDGEMENTS

The authors would like to thank the Water Research Centre and the Scottish Development Department for funding the original surveys, and L. Bennet, C. Bayes and D. Bird for their contributions.

REFERENCES

(1) Department of the Environment (1983). Sewage sludge survey, 1980 data. Standing committee on the disposal of sewage sludge, Department of Environment, UK.

(2) Commission of the European Community (1986). Council directive on protection of the environment, and in particular of the soil, when sewage sludge is used in agriculture. Official Journal of the European Communities, L181/6-12.

(3) House of Commons (1987). Third report from the Environment Committee, session 1986-87. Pollution of rivers and estuaries in Britain. HMS0, London, UK.

(4) Bastion, R. K. (1986). Overview on sludge utilization. In: The forest alternative for the treatment and utilisation of municipal and industrial waste (D. W. Cole, C. L. Henry and W. L. Nutter, Eds.), pp. 7-25. Univ. of Washington Press, Seattle.

(5) Taylor, C. M. A. (1986). Forest fertilisation in Great Britain. Proceedings of the Fertiliser Society, *251*, London, UK.

(6) McIntosh, R. (1983). Nitrogen deficiency in establishment phase Sitka spruce in upland Britain. Scottish Forestry, *37*, 185-193.

(7) Bayes, C. D., C. M. A. Taylor and A. J. Moffat (1989). Sewage sludge utilisation in forestry: the UK research programme. Proceedings of the WRc international workshop on alternative uses for sewage sludge, York, UK. (This volume).

(8) Moffat, A. J. (1987). The use of sewage sludge in forestry: an assessment of the potential in England and Wales. Report PRU 1606-M. Water Research Centre, Medmenham, UK.

(9) Taylor, C. M. A., L. Bennet and C. D. Bayes (1987). The use of sewage sludge in forestry: an assessment of the potential in Scotland. Report PRU 1707-M. Water Research Centre, Medmenham, UK.

(10) Rowan, A. A. (1977). Terrain classification. Forestry Commission Forest Record 114, HMSO, London, UK.

(11) McIntosh, R. (1984). Fertiliser experiments in established conifer stands. Forestry Commission Forest Record 127, HMSO, London, UK.

(12) Moffat, A. J. and D. Bird (1989). The potential for using sewage sludge in forestry in England and Wales. Forestry, *62* (1), 1-17.

(13) Horne, A. I. D. and M. D. Whitlock (1984). The Forestry Commission subcompartment database. I. Description of the data held. Forestry Commission Research Information Note 94/84/FS, UK.

(14) Locke, G. M. L. (1987). Census of woodland and trees. Forestry Commission Bulletin 63, HMSO, London, UK.

(15) Everard, J. E. (1974). Fertilisers in the establishment of conifers in Wales and southern England. Forestry Commission Booklet 41, HMSO, London, UK.

(16) McIntosh, R. (1984). Phosphate fertilisers in upland forestry - types, application rates and placement methods. Forestry Commission Research Information Note 89/84/SILN, UK.
(17) Taylor, C. M. A. (1987). The effects of nitrogen fertiliser at different rates and times of application on the growth of Sitka spruce in upland Britain. Forestry, *60* (1), 87-99.
(18) Bayes, C. D., J. M. Davis and C. M. A. Taylor (1987). Sewage sludge as a forest fertiliser: experiences to date. J. Inst. Water Pollution Control, *86*, 158-171.
(19) Scottish Development Department (1987). Survey of the discharges of crude sewage and treated sewage effluent. Scottish Development Department RPS4, Edinburgh, UK.
(20) Hall, J. E., A. P. Daw and C. D. Bayes (1986). The use of sewage sludge in land reclamation. Report ER 1346-M. Water Research Centre, Medmenham, UK.
(21) Moffat, A. J. (1988). Sewage sludge as a fertiliser in amenity and reclamation plantings. Forestry Commission Arboricultural Research Note 76/88/SSS, UK.

Sewage sludge utilisation in forestry: the UK research programme

C. D. BAYES[1], C. M. A. TAYLOR[2] and A. J. MOFFAT[3]

[1] Water Research Centre, Unit 16 BETA Centre,
University of Stirling Innovation Park, Stirling, FK9 4NF, UK.
[2] Forestry Commission Northern Research Station,
Roslin, Midlothian, EH25 9SY, UK.
[3] Forestry Commission Research, Alice Holt Lodge,
Farnham, Surrey, GU10 4LH, UK.

SUMMARY

The UK research programme for utilising sewage sludge in forestry commenced in 1981 and has defined the potential in terms of tree species, crop stage and soil type. Tree growth responses are presented from the forest experiments, some of which show large benefits from sludge, and from these appropriate sludge application regimes are proposed. The results of the environmental monitoring for heavy metal additions, vegetation effects, health implications and water resource protection are described and used to determine the constraints. Methods of application are also reviewed. It is concluded that sludge can be used beneficially as a forest fertiliser and that there are operationally attractive phases in the forest rotation for developing this alternative outlet.

1. INTRODUCTION

Until the early 1980s there was little interest in the UK in utilising sewage sludge as a forest fertiliser. Yet, there are over 2 million hectares (ha) of productive woodland, which has been expanding at a rate in excess of 20,000 ha per annum during the decade and now covers 10% of the total land area in Great Britain (1). At the same time sewage treatment authorities have been experiencing increasing constraints on existing outlets for sludge.

Many of the afforested areas are on nutrient-poor soils and require the application of nitrogen (N), phosphorus (P) and/or potassium (K) to achieve satisfactory tree growth, particularly in the early years. Sewage sludge, which typically contains 3-5% N and 2-4% P_2O_5, is a potential alternative source of these nutrients for forest fertilisation and the slow release of nitrogen could be particularly beneficial in enhancing growth rates over a longer period than conventional inorganic fertilisers.

There are several attractions to using sewage sludge on forestry land, namely:
a) Forestry is a non-food chain outlet;
b) Sludge may be applied throughout the year unconstrained by the crop;
c) Coniferous trees are able to assimilate nutrients throughout the year;
d) Sites are generally more remote from habitation and avoid potential or perceived nuisances associated with sludge applications;
e) Forest road systems are built to a high standard for timber extraction and provide ready access for sludge vehicles.

However, there are constraints, namely:
a) Within the forest rotation, typically 50 years, there are periods when the stands are impenetrable for sludge applications;
b) The terrain is more difficult than on agricultural sites and the potential for runoff can be exacerbated by site cultivations;
c) There is public access to forest areas.

2. RESEARCH PROGRAMME

In 1981 a joint WRc/Forestry Commission research programme was initiated, with collaboration from both industries, to evaluate the use of sewage sludge and in particular:
a) Tree growth responses and appropriate sludge application rates;
b) Environmental implications:
   - soil
   - ground vegetation
   - health aspects
   - water resource protection;
c) Methods of application, equipment needs and costs.

The strategy was based on defining the species, crop stages and soil types where sludge would be of positive benefit to tree growth and where it is operationally practical to apply it. Figure 1 shows a typical forest cycle on heathland soil, with the associated fertiliser regimes. There are potentially four phases for utilising sludge, namely:
1) Preplanting on areas to be newly afforested.
2) Early establishment of spruce plantations (including Sitka spruce (*Picea sitchensis* (Bong.) Carr.), Norway spruce (*Picea abies* (L.) Karst.) and Douglas fir (*Pseudotsuga menziesii* (Mirb.) Franco.) from 1-5 years old (possibly extending to 8 years).
3) Pole stage pine plantations (25 years or older) which are thinned and thus

Sewage sludge utilisation in forestry: the UK research programme

the extraction racks at 12-20 m intervals provide convenient application avenues through the stands.
4) Clearfelled areas, similar to phase 1, awaiting replanting.

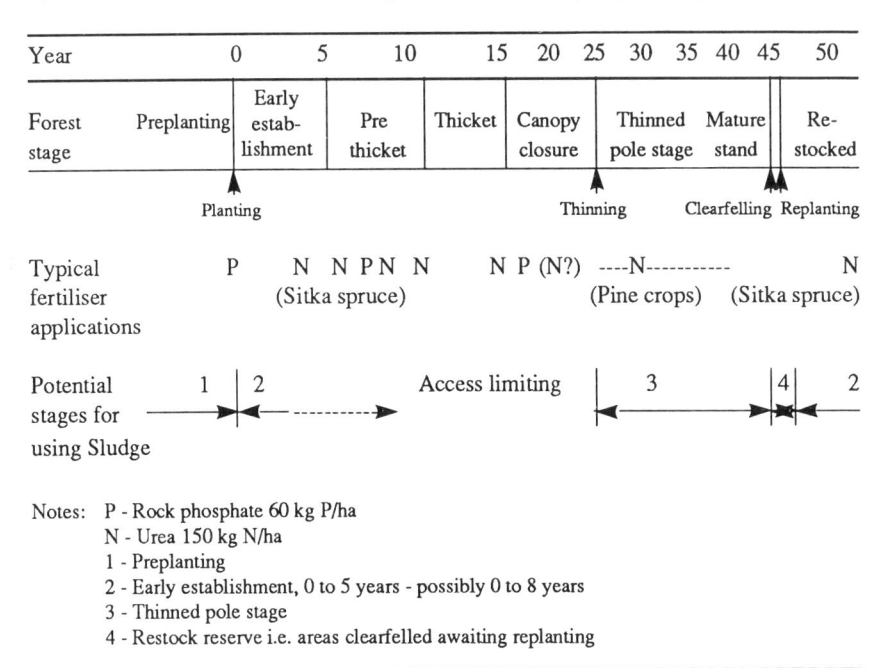

Year	0	5	10	15	20	25	30	35	40	45	50
Forest stage	Preplanting	Early establishment	Pre thicket	Thicket	Canopy closure		Thinned pole stage	Mature stand			Restocked
	▲ Planting							▲ Thinning		▲▲ Clearfelling Replanting	
Typical fertiliser applications	P	N (Sitka spruce)	N P N	N	N P (N?)		----N---- (Pine crops)			N (Sitka spruce)	
Potential stages for using Sludge	1	2		Access limiting			3			4	2

Notes: P - Rock phosphate 60 kg P/ha
        N - Urea 150 kg N/ha
        1 - Preplanting
        2 - Early establishment, 0 to 5 years - possibly 0 to 8 years
        3 - Thinned pole stage
        4 - Restock reserve i.e. areas clearfelled awaiting replanting

Fig.1. Typical forestry rotation for spruce and pines on heathland soils.

Access is not possible once the trees enter the pre-thicket stage until thinning, a period of approximately 20 years. However, within the 50 year rotation there may be 30 years when sludge can be applied, the majority of which is at the pole stage.

Soil types have been selected on a combination of their nutrient deficiency and free-draining characteristics, the latter being important to give good bearing capacity, allowing access throughout the year, and low risk of surface runoff. Appropriate soil types are podzolic, iron pan and littoral soils, and possibly disturbed soils.

Using the above criteria, eleven forest experiments have been established since 1982. These range from replicated randomised plot experiments, with up to seven different treatments, to unreplicated trials and they are summarised in Table 1.

3. GROWTH RESPONSES

Improvements in tree growth from fertiliser applications must prevail over a number of years if they are to be of tangible value to the crop over its long lifespan. For example, rock phosphate applications can have an effect for up to 6-10 years while urea-nitrogen may improve growth for three years. As a result, forest experiments need to be monitored for a long period to evaluate fully the benefits. Data are now available for several years from the earlier sites and they are considered for the various forest phases.

3.1 Preplanting

The forest experiment on replicated 0.1 ha plots at Ardross, Highland Region, was established in 1983 on a low mineralising heathland soil (2). Applications of phosphorus would normally be required on this type of site plus regular applications of nitrogen for Sitka spruce, the preferred commercial species. The sewage sludge applications were therefore designed to supply phosphorus and two high rates of nitrogen (S500 supplying 500 kgN/ha; S1000 supplying 1000 kgN/ha). The sludge treatments are being compared with a combination of fertiliser nitrogen and phosphate and herbicide treatments, the latter to remove heather (*Calluna vulgaris* (L.) Hull) which has an allelopathic effect on Sitka spruce. There is also an untreated demonstration control plot.

Liquid undigested sewage sludge was applied to the virgin heathland by slurry tanker at rates of 364 and 728 m^3/ha. The site was then ploughed before planting with Sitka spruce in spring 1984. The nutrient and potentially toxic element application rates are given in Table 2.

Tree height is the main measure of response in young trees and the results for the first five years are shown in Figure 2. Early growth of the trees on the sludge treated plots has been exceptional, with yearly increments relative to the inorganic fertiliser treatments of 185%, 189%, 121% and 91% for the S500 rate, and 199%, 197%, 135% and 111% for the S1000 rate. In the last two years (1987 and 1988) the site suffered from the green spruce aphid (*Elatobium abietinum*) which periodically defoliates spruce plantations in Scotland. It preferentially attacked the trees on the sludge treated plots and in 1989 had started to infest the others. Nevertheless, the annual growth rate in 1988 still matched the inorganic treatments in that year. The growth rate of the untreated control plot clearly highlights the problems of nutrient deficiency on these sites.

Table 1. Summary of forest experiments/trials 1982-1988.

Experiment/ trial	Forest stage/ species	Soil type/ vegetation	Sludge type	Treatments		Commenced
Ardross Forest, Highland Region	Preploughing/ planting of heathland site. Sitka spruce	Peaty podzol. Heather	Stored liquid undigested sludge 3.5% dry solids	Sludge Controls	- 364 m³/ha - 728 m³/ha - no treatment - P - NP - P + herbicide - NP + herbicide	1983
Teindland Forest, Grampian Region	Preploughing/ replanting of clearfelled. Sitka spruce	Podzol,podzolic ironpan/gley. Heather recolonisation	Liquid undigested sludge 2.5% dry solids	Sludge Controls	- 200 m³/ha - 400 m³/ha - 800 m³/ha - no treatment - NP	1987
Clydesdale Forest, Strathclyde Region	Preripping/ planting of former opencast coal site. Sitka spruce	Restored mineral soil. Rushes	Dewatered digested sludge 51% dry solids	Sludge Controls	-106 tds/ha - 194 tds/ha (- 200 m³/ha in yr. 3) - no treatment - NPK	1987
Gore Heath, Wareham, Dorset	Preripping/ planting of restocked site. Corsican pine	Podzol. Heather	Liquid digested sludge 1.9% dry solids	Sludge Control	- 285m³/ha - no treatment	1988
Clydesdale Forest, Strathclyde Region	During ripping/ preplanting of former opencast coal site. Sitka spruce	Restored mineral soil. Rushes	Liquid digested sludge 3.7% dry solids	Sludge Control	- 340m³/ha (injected) - ripped only	1987
Rogate, West Sussex	Postripping/ preplanting of restored oil exploration site. Japanese Larch	Sandy loam mineral soil (no topsoil). Sparse grass	Dewatered digested sludge 31% dry solids	Sludge Control	- 300 tds/ha - no treatment - NP	1987
Ringwood Numbers, Ringwood Forest, Dorset	Early establishment of replanted site. Corsican pine (5 yr.old)	Podzolic gley. Purplemoor grass/bracken	Liquid digested sludge 2.9% dry solids	Sludge	- 50 m³/ha - 100 m³/ha - 150 m³/ha - 200 m³/ha	1985
Clydesdale Forest, Strathclyde Region	Early establishment of former opencast coal site. Sitka spruce (3 yr. old)	Restored mineral soil. Open grass dominated sward	Liquid digested sludge 3.8% dry solids	Sludge Control	- 94 m³/ha - 195 m³/ha - no treatment	1986

... continued

C. D. Bayes, C. M. A. Taylor and A. J. Moffat

Table 1. Summary of forest experiments/trials 1982-1988.
... continued

Experiment/ trial	Forest stage/ species	Soil type/ vegetation	Sludge type	Treatments		Commenced
Tredeg, West Glamorgan	Early establishment of former opencast coal site. Japanese larch, common and red alder (7 yr. old)	Restored mineral soil (no topsoil). Sparse grass	Liquid digested sludge 3-10% dry solids	Sludge Control	- 75 m³/ha - 150 m³/ha - 250 m³/ha - no treatment	1988
Ringwood Uddens, Ringwood Forest, Dorset	Thinned pole stage. Corsican pine (30 yr.old)	Podzolic gley. Calcifuge grasses/bracken	Liquid digested sludge 2.3% dry solids	Sludge Controls	- 200 m³/ha - 400 m³/ha - 200 m³/ha/yr for 5 yr. - no treatment - P	1982
Montreathmont Angus Forest, Tayside Region	Thinned pole stage. Scots pine (40 yr. old)	Imperfectly drained podzol. Ferns and grasses	Stored liquid undigested sludge 4.2% dry solids	Sludge Control	- 239 m³/ha - 477 m³/ha - 716 m³/ha - no treatment	1982

Notes: Typical conventional fertiliser control treatments:
N - 350 kg urea/ha (150 kgN/ha)
P - 450 kg rock phosphate/ha (60 kgP/ha)
K - 200 kg muriate of potash/ha (100 kgK/ha)

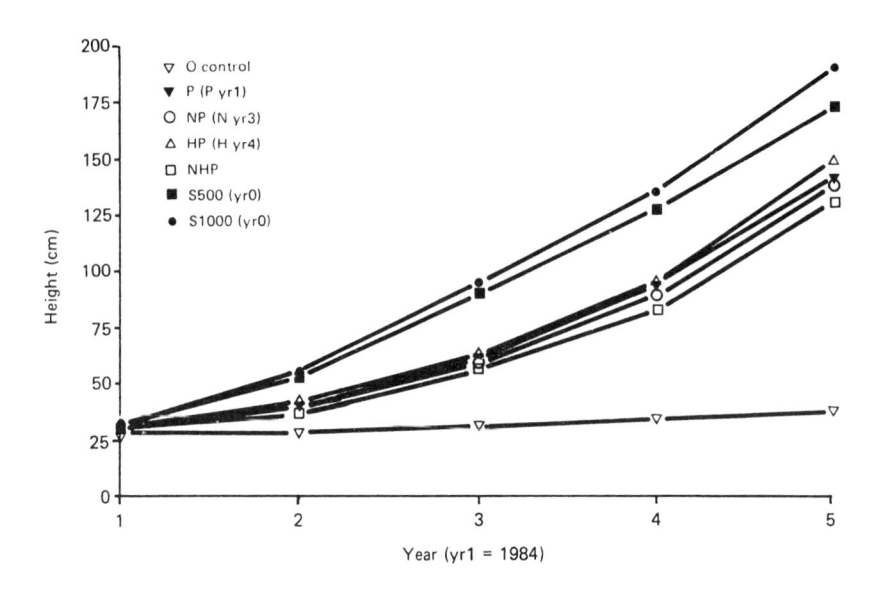

Fig. 2. Ardross Forest experiment tree growth responses.

Table 2. Sludge application rates, nutrients and heavy metals for selected forest experiments.

Experiment/ trial	Treatment	Rate m³/ha	Dry solids tds/ha	NH₄-N	Org-N	P	K	Cd	Cr	Cu	Pb	Ni	Zn
								kg/ha					
Ardross (1983)	S500	364	12.7	140	445	128	51	0.03	0.47	3.4	3.2	0.17	6.5
	S1000	728	25.5	280	893	256	103	0.07	0.94	6.8	6.4	0.35	13.0
Clydesdale (1986)	S1	94	3.6	60	114	46	17	0.014	0.24	1.14	1.48	0.11	3.6
	S2	195	7.4	123	234	95	34	0.030	0.50	2.35	3.05	0.22	7.5
Ringwood Uddens (1982)	S1	200	4.4	170	340	104	58	0.04	0.32	2.7	3.6	0.17	8.2
	S2	400	8.8	340	680	208	116	0.08	0.64	5.4	7.2	0.34	16.4 41.0
	SR⁽¹⁾	1000⁽¹⁾	22	850	1700	520	290	0.20	1.60	13.5	18.0	0.85	
Montreathmont (1982)	S1	239	10.1	133	299	87	25	0.02	0.46	6.5	3.0	0.19	5.7
	S2	477	20.2	266	498	174	48	0.03	0.92	13.0	6.0	0.38	11.4
	S3	716	30.3	399	897	261	72	0.05	1.38	19.5	9.0	0.57	17.1
EC Directive limits for agriculture. Annual average for 10 years								0.15	4.5⁽²⁾	12	15	3	30

⁽¹⁾ SR rate applied as 200 m³/ha per annum for 5 years.
⁽²⁾ Provisional

3.2 Early establishment

The Clydesdale site is a former opencast coal mine in Lanarkshire which was restored in the late 1970s. Sitka spruce were planted in 1983 and suffered from poor growth and yellow green foliage indicative of nitrogen deficiency. The ground vegetation similarly reflected the poor mineral soil with an open sparse grass sward.

Liquid digested sludge was applied in August 1986 to unreplicated trial plots by a tanker-fed static rain gun. The two treatment rates, S1 and S2, comprised 94 and 195 m³/ha and supplied 60 and 123 kg/ha of readily available ammoniacal nitrogen and 114 and 234 kg/ha of slow release organic nitrogen respectively; see Table 2 (3).

Within two weeks of the sludge application the foliage of the treated trees became a healthy dark grey/green colour indicating the uptake of nitrogen. Needle growth was also more vigorous and extended later into the season than on the control plot, which in addition suffered earlier frost damage. Foliar samples in autumn 1986 confirmed these observations, with needle weights increasing by 11% and 27%, and foliar nitrogen concentrations 64% and 84% higher for the two treatment rates relative to the control; see Table 3.

Tree growth in 1987 responded to the improved nutrient status and was maintained in 1988, exceeding the controls by 72% and 90% over the two year period. Monitoring is continuing to determine the duration of the response.

Table 3. Clydesdale Forest trial (1986). Foliar weights, nutrient concentrations and tree height assessments.

	Year	Control	S1 (94m³/ha)	S2 (195m³/ha)
Mean needle weight (mg)	1986	3.55	3.95	4.50
	1987	3.45	4.83	5.15
	1988	3.90	5.63	5.93
Nitrogen (% dry weight)	1986	1.17	1.92	2.13
	1987	1.17	1.92	2.21
	1988	0.97	1.44	1.68
Phosphorus (% dry weight)	1986	0.22	0.25	0.25
	1987	0.23	0.24	0.26
	1988	0.19	0.19	0.18
Tree height (cm)	1986	83.6	77.9	82.2
	1987	109.7	117.8	125.5
	1988	129.4	156.7	168.9
Annual height increment relative to control (%)	1986/7	-	153%	166%
	1987/8	-	197%	220%

3.3 Pole stage

Although pole stage stands are not normally fertilised in Britain, due to fertiliser needs on younger crops, it is known that they will benefit (4). In Scandinavia large areas of older pine crops are fertilised with nitrogen.

At the Ringwood-Uddens site Corsican pine (*Pinus nigra* var. maritima) had been planted on a podzolic gley in 1950 and line-thinned. Soil and herbage analysis showed phosphate deficiency and in 1982 liquid digested sludge was applied at rates of 200 m³/ha (S1) and 400 m³/ha (S2), and 200 m³/ha annually for five years (SR). Untreated and rock phosphate treated control plots were also established and all treatments were replicated six times. The sludge treatment rates are summarised in Table 2.

In 1983, one growing season after the applications, all the sludge treatments produced significant increases in needle weight and foliar phosphorus concentrations compared to the untreated and rock phosphate treated controls. The high sludge rate also produced a significant increase in foliar nitrogen (5). Further samples in 1987 showed the significant increases in needle weight and phosphorus content above deficiency levels to have prevailed (see Table 4), while the levels in the control plots remained deficient (6).

Tree growth responses have been monitored by the diameter at breast height (dbh) with a significant, but small, mean increase relative to the controls. There are indications that the 400 m³/ha application may have been detrimental to tree growth but annual applications of 200 m³/ha have significantly increased dbh beyond that obtained from the single application. The crop had not been silviculturally thinned so contained many trees which had begun to regress naturally. Analysis of the upper quartile results covering dominant and sub-dominant trees reveals greater differences (see Table 4), suggesting that the single sludge application has been effective in redressing the detrimental effect of nutrient deficiency on the larger trees.

Table 4. Ringwood-Uddens Forest experiment. Foliar weight and composition, and tree growth (1987).

	Control 0	P 50 kg/ha	S1 200m³/ha	S2 400m³/ha	SR 1000m³/ha
Needle weight (mg)	68.8	74.1	73.3	85.8	85.3
Nitrogen (% dry weight)	1.55	1.44	1.44	1.68	1.70
Phosphorus (% dry weight)	0.09	0.13	0.13	0.14	0.14
Potassium (% dry weight)	0.57	0.59	0.54	0.48	0.55
Diameter at breast height (mm)					
Mean 1983	124.3	130.6	131.5	130.0	127.3
Upper quartile 1983	153.0	166.5	163.2	163.3	168.2
Mean 1987	138.6	141.1	140.5	139.4	142.2
Upper quartile 1987	174.8	193.1	193.1	199.0	198.9

However, the replicated experiment at Montreathmont, Tayside Region, showed that there are constraints on sludge applications to pole stage stands. In March 1982 liquid undigested sludge was applied at three rates (239, 477 and 716 m^3/ha; S1, S2 and S3) to 0.1 ha plots of 40 year old Scots pine (*Pinus sylvestris* L.) using a retracting reel rain gun system; see Table 2 for the treatment rates (7). The stand was slow growing and it was anticipated that it would respond to applications of nitrogen.

Foliar nitrogen levels were significantly increased in the three years following the sludge treatments and there would normally have been an associated increase in tree growth, measured by basal area increment, in this type of crop (8). However, growth on the high rate treatment plots was significantly worse than the control and there has been a general trend for poorer growth following all the sludge treatments, with only the low rate being significantly better for one year (1983); see Figure 3. Tree damage due to the physical impaction of sludge from the irrigator jet during the treatment was investigated and discounted. Further research indicated that ammonia toxicity of the roots was not involved, but that prolonged anoxia was the most likely cause (9). This is associated with additions of large quantities of water and carbon and/or blinding of the soil surface, leading to a reduction in the number of active mycorrhizal feeding roots (10).

The Montreathmont site is an imperfectly drained podzol and the original sludge applications in March 1982 occurred when the soil was at field capacity. The sludge therefore remained on the soil surface without drying for two to three weeks allowing the soil to become anoxic.

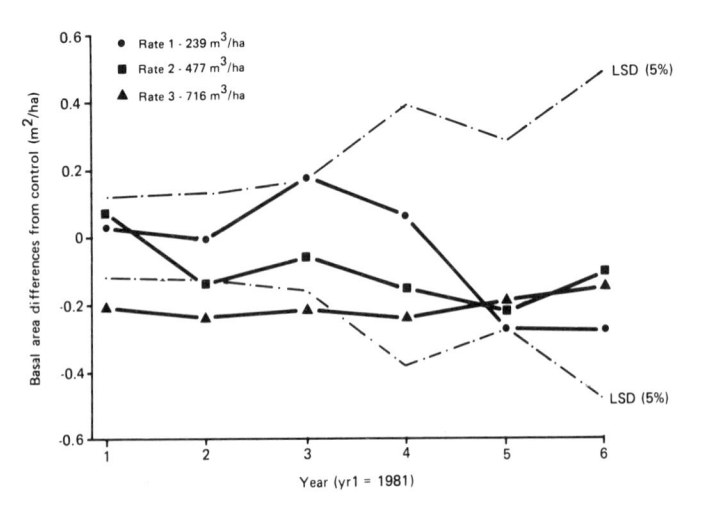

Fig.3. Montreathmont Forest experiment. Basal area changes for pole stage Scots pine relative to the control.

Thus on the more poorly drained sites sludge applications of 200 m^3/ha to growing trees should be limited to periods when there is a significant soil moisture deficit. At other times lower application rates must be used to avoid tree damage.

4. METAL ADDITIONS

There are obvious attractions in utilising sludge on non-food chain crops in respect of heavy metals and other potential toxic constituents such as trace organics. Trees are also less sensitive to metal additions than many agricultural crops and are grown on soils with generally low background levels; in some cases they are deficient, for example in copper. However, the applications must be below phytotoxic levels and avoid adverse effects on wildlife such as deer, grazing in the forest either on ground vegetation or tree foliage. Furthermore, forest soil pHs are generally lower than agricultural values and therefore heavy metals are potentially more mobile. Thus soil surveys and foliar monitoring have been undertaken at the forest experiments to investigate metal movement down the soil profile and tree uptake; see Table 5.

The treatment plots at Montreathmont were sampled at three horizons in 1988, six years after the sludge applications. Copper and zinc concentrations were found to be increased on the sludge treated plots in the top 0-75 mm organic layer, with lead levels also elevated although variable. The increases were significant for copper at the medium and high rate treatments and for zinc at the high rate and were predictable from the application rates (Table 2). However there was little movement detectable down the soil profile to the 75-150 and 150-250 mm layers, with only copper showing a significant but small increase in the 75-150 mm layer at the high rate treatment. Foliar samples taken annually have shown no significant uptake of metals except for copper at the medium rate treatment in autumn 1982 after the first growing season.

A similar soil survey at Ardross, five years after the sludge applications, revealed significant increases in the 0-75 mm layer in chromium, copper, zinc and lead relative to the phosphate controls. Again, lead levels were highly variable and only copper showed a significant but small increase in the 75-150 mm layer, with no differences in the 150-250 mm layer. For both Montreathmont and Ardross the levels of metals over the 0-250 mm depth are less than the EC Directive limit values for agriculture (11) and in the 0-75 mm layer they also comply with the UK code of practice for grassland (12).

The foliar samples at Ardross showed significant increases in copper and zinc levels. This suggested that young trees may take up metals more readily than mature ones, possibly enhanced by space/furrow ploughing of the site which incorporated the sludge into the ridges where the trees were planted.

Table 5. Soil and foliar metal concentrations for Montreathmont, Ardross and Ringwood-Uddens Forest experiments.

	pH	Cr	Ni	Cu	Zn	Pb	Cd
			.... mg/kg				

Montreathmont

Soil 0-75 mm horizon

	pH	Cr	Ni	Cu	Zn	Pb	Cd
0 (0 m³/ha)	3.57	48	12	15	59	90	0.39
S1 (239 m³/ha)	3.59	49	12	32	73	77	0.35
S2 (477 m³/ha)	3.65	49	13	62	72	135	0.41
S3 (716 m³/ha)	3.65	52	14	102	108	102	0.51

Soil 75-150 mm horizon

	pH	Cr	Ni	Cu	Zn	Pb	Cd
0	3.66	64	13	3	44	38	0.28
S1	3.65	58	11	4	37	41	0.24
S2	3.75	61	11	5	37	54	0.19
S3	3.75	63	13	7	45	38	0.24

Soil 150-250 mm horizon

	pH	Cr	Ni	Cu	Zn	Pb	Cd
0	4.21	73	20	4	57	16	0.21
S1	4.10	68	18	3	41	17	0.18
S2	4.19	74	18	5	44	18	0.16
S3	4.21	69	17	4	46	16	0.14

Foliage

		pH	Cr	Ni	Cu	Zn	Pb	Cd
1982	0	-	-	-	3.53	42.3	<2.5	<1
	S1	-	-	-	3.77	40.4	<2.5	<1
	S2	-	-	-	4.41	38.2	<2.5	<1
	S3	-	-	-	3.91	32.1	<2.5	<1
1985	0	-	-	-	3.34	41.3	<2.5	<1
	S1	-	-	-	3.47	43.0	<2.5	<1
	S2	-	-	-	3.14	42.0	<2.5	<1
	S3	-	-	-	3.33	44.5	<2.5	<1

Table 5. Soil and foliar metal concentrations for Montreathmont, Ardross and Ringwood-Uddens Forest experiments ... continued

	pH	Cr	Ni	Cu	Zn	Pb	Cd
				mg/kg			

Ardross

Soil 0-75 mm horizon

0 (Control)	3.60	14	6	10	34	32	0.40
P (Pyr. 1)	3.90	8	3	12	46	36	0.43
NP (Nyr. 3)	3.82	9	4	11	47	49	0.41
HP (Hyr, 4)	3.84	10	2	8	35	31	0.49
NHP (Hyr. 4)	3.87	9	3	10	47	95	0.46
S500 (364 m³/ha)	3.86	13	3	31	84	137	0.50
S1000 (728 m³/ha)	3.87	14	3	41	100	79	0.59

Soil 75-150 mm horizon

0	3.65	6	1	9	13	37	0.05
P	3.88	14	1	2	9	14	0.16
NP	3.85	9	1	4	13	22	0.13
HP	3.93	13	1	1	6	12	0.13
NHP	3.91	13	1	2	7	38	0.09
S500	3.89	9	1	6	17	29	0.15
S1000	3.95	12	2	4	16	40	0.16

Soil 150-250 mm horizon

0	4.11	16	1	1	1	1	0.05
P	4.10	22	1	1	2	6	0.05
NP	4.04	19	1	1	2	4	0.09
HP	4.12	18	1	1	2	6	0.10
NHP	4.10	24	1	1	4	6	0.06
S500	4.07	21	3	1	1	5	0.06
S1000	4.12	19	1	1	2	5	0.10

Foliage

1985 P	-	-	-	2.25	21.2	<2.5	<0.5
S500	-	-	-	3.78	35.6	<2.5	<0.5
S1000	-	-	-	4.08	37.2	<2.5	<0.5

... continued

Table 5. Soil and foliar metal concentrations for Montreathmont, Ardross and Ringwood-Uddens Forest experiments ... continued

	pH	Cr	Ni	Cu	Zn	Pb	Cd
				mg/kg			
Ringwood Uddens							
Organic/litter layer							
0	4.0	-	5	7	41	57	0.33
P	4.0	-	5	8	42	57	0.71
S1 (200 m³/ha)	4.0	-	7	44	97	118	0.69
S2 (400 m³/ha)	3.9	-	8	82	125	160	0.86
SR (1000 m³/ha)	4.3	-	12	256	356	148	2.63
Soil 0-150 mm horizon							
0	3.7	-	3	2	8	14	<1
P	3.7	-	4	3	12	17	<1
S1	3.8	-	4	3	10	15	<1
S2	3.8	-	4	3	11	14	<1
SR	3.9	-	4	4	19	18	<1
Foliage							
1987 0	-	-		6.62	5.57	73.0	2.81 0.20
P	-	-		7.20	5.39	63.6	3.19 0.22
S1	-	-		5.45	4.48	70.5	3.99 0.30
S2	-	-		4.46	4.00	71.3	3.36 0.20
SR	-	-		3.42	3.62	68.5	3.72 0.24
EC Directive limit values for agricultural soil (0-250mm)		100-200 (provisional)	30-75	50-140	150-300	50-300	1-3
Draft UK Code of Practice for agricultural soil (0-75 mm)		600 (provisional)	80	130	330	300	5

Montreathmont and Ardross foliar metal results reproduced from reports of M L Berrow, The Macaulay Land Use Research Institute, Aberdeen.

Results from the pole stage stand at Ringwood Uddens (Table 5) show a similar pattern to that found at Montreathmont. Levels of copper, zinc and lead in the soil are all significantly higher in the litter layer on the sludge-treated plots, as are chromium and cadmium at the repeat treatment rate (6). However, again there is little evidence of movement down the soil profile with only zinc elevated in the 0-150 mm layer at the highest rate. All three sites have shown the pattern of sludge metals being adsorbed in the litter/ organic layer with little movement down the soil profile.

Metal levels in the foliage in 1987 at Ringwood Uddens show a similar lack of uptake into mature trees as was found at Montreathmont, with only lead slightly elevated. The foliar metal results for all the experiments are similar to published values for "normal" concentrations in foliage (13) and there is no evidence of phytotoxicity. On the contrary, the copper uptake at Ardross may be beneficial as the foliar levels in the non-sludge treatments are marginal with the trees showing copper deficiency growth deformities not present on the sludge-treated plots.

It is therefore suggested that, although forest soils generally have lower pH values, there is an adequate safety tolerance providing metal application rates and the subsequent concentrations in the soil do not exceed the values for the use of sludge in agriculture. For the majority of sludges this constraint is unlikely to be restrictive for the application regimes envisaged for forestry.

5. VEGETATION EFFECTS

In certain circumstances it is desirable for sewage sludge applications to modify the ground vegetation (e.g. early establishment), whereas in others there was concern that sludge could be detrimental to the forest environment, for example under a forest canopy.

On many heathland sites heather grows prolifically and inhibits the uptake of nitrogen by Sitka spruce. Consequently, the herbicides 2,4-D or glyphosate are frequently used to suppress the growth of heather on afforestation and restocking sites. At Ardross the two sludge application rates caused a rapid die-back of heather, which was replaced by various species of grasses and herbs on over 80-90% of the S1000 and 40-60% of the S500 plots (14). Over several years heather has gradually recolonised the lower rate plots but not the higher rate ones. This long term effect on the site has improved growth in the early establishment phase before canopy closure, possibly eliminating the need for herbicide applications.

The impoverished soils of former opencast coal sites support only a sparse ground cover with problems of soil erosion and low availability of nutrients. In the trials at Clydesdale (1986 and 1987) the sludge applications produced a dramatic improvement in the growth and density of the sward, substantially reducing the risks of runoff from further liquid sludge applications. Problems of weed competition on young trees have not been manifest but need to be evaluated further for preplanting applications.

Conversely, there were concerns that applications to pole stage stands could damage the ground vegetation under the forest canopy. At Montreathmont the high application rates, up to 716 m^3/ha, were in part selected to investigate this possibility. Ground vegetation, consisting of ferns and grasses, was smothered by the sludge to a depth of up to 70 mm in March 1982. In 7 to 21 days this dried to a crisp mat and on the low and medium rate treatment plots the original vegetation recovered by August that year. At the high rate only the ferns re-emerged and it was not until the following year that the full range of species diversity was re-established (7). Thus, at the application rates which would now be recommended, namely 200 m^3/ha per annum, no long term effects should occur.

6. HEALTH IMPLICATIONS

There are special health implications for using sludge in forestry, both for forestry personnel and for the general public who have access to afforested land for recreational purposes.

All trees are hand planted and so forestry staff have intimate contact with the soil during this operation. Thus, bacteria die-off rates were investigated at the Ardross preplanting site by placing varying mixtures of liquid undigested sludge and soil in trays on the site. Samples were taken at 1, 2, 3, 7, 14 and 21 days after the applications and analysed for faecal indicator organisms. The levels declined rapidly and approached the limit of detection by the fourteenth day (2). However, to provide a greater margin of safety it is recommended that sludge is not applied to a site in the three months prior to planting.

Conditions obviously differ under the canopy of a mature forest with greater shading and higher moisture levels. At Montreathmont thermotolerent coliform levels fell by 80% within 21 days whereas viable heterotrophs, indicative of soil and herbage bacteria adapted to the site and sludge, increased (15). It is therefore provisionally recommended that public exclusion notices are placed round sites for a period of three months after sludge is applied. Further work on the survival of bacterial pathogens, viruses and parasites is required to determine the effects of soil moisture content, temperature, sunlight, pH, organic matter and antagonistic soil microflora in order to confirm this exclusion period. In terms of public inconvenience it is similar to constraints applied during other forestry operations such as felling, thinning and herbicide treatment, and would cover a minimal area of the total forestry land.

7. WATER QUALITY

Sewage sludge applications must not cause adverse effects on the quality of natural waters so, where circumstances permitted, the quality of the site drainage from the forest experiments has been monitored. Open drains were constructed to isolate the treatment areas and intercept the surface and superficial groundwater for sampling. The results for the Ardross, Clydesdale (1986) and Montreathmont sites are summarised in Table 6.

At Ardross an interceptor drain successively receives the water from an unploughed control area, a ploughed/rock phosphate treated area and then the 24 treatment plots, a third of which received sludge applications. Thus, changes in the quality along the interceptor are indicative of changes in the drainage from a ploughed area and subsequently the treatments.

Sludge was applied in July 1983 and even with the high application rates, 364 and 728 m^3/ha, there was no significant movement of water from the site until the snow melted in February 1984. Monitoring then proceeded for 17 months but flow was only experienced during prolonged rainfall when the site was at field capacity. The concentration of ammoniacal nitrogen leaching from the site increased by 0.1 mgN/l due to ploughing and by a further 0.4 mgN/l as a result of ploughing and the various treatments. A similar increase was observed for nitrate from the treatment plots. Only 20% of the area had received sludge applications, but extrapolation to an area completely treated gives a predicted increase in ammoniacal and nitrate nitrogen of only 2 mgN/l in the runoff under weather conditions producing spate flows in watercourses. Drainage from the area ploughed and conventionally treated with rock phosphate increased the concentrations of all forms of phosphate, in particular soluble phosphate by 23 µgP/l. Sludge on the treated areas caused a similar small rise in soluble reactive and insoluble phosphorus (14).

The evidence from Ardross, and also the Teindland experiment, where there has been no surface drainage from the treated area (16), is that dry heathland sites can accept the recommended sludge application rates to give 1,000 kg organic N/ha without surface runoff, and with only low levels of nutrient leaching in the succeeding months.

Circumstances differ on the restored soils of former opencast coal sites where surface compaction, initial poor sward density and, in some circumstances, rigg and furr contouring encourage surface runoff following sludge treatments. The results for the Clydesdale 1986 trial in Table 6 show the deterioration in the site drainage quality below the treatment areas in August, two days after the applications of liquid digested sludge. The effect was more pronounced downstream of the higher treatment rate, 195 m^3/ha. The results in October, 11 weeks after the sludge was applied, showed no

Table 6. Quality of site drainage from forest experiment interceptor drains.

Experiment/ monitoring point	Sludge applied	Period	Susp. solids mg/l	BOD mg/l	Amm.N mg/l	NO_3-N mg/l	Tot.P µg/l	Sol.P µg/l	PO_4-P µg/l
Ardross									
Unploughed heathland	-	Feb. '84 Jun. '86	-	-	0.11	<0.06	31.9	17.7	3.7
Unploughed heathland plus ploughed phosphate-fertilised area	-	Feb. '84 Jun. '86	-	-	0.24	0.08	56.7	40.6	9.2
Unploughed and ploughed fertilised area plus treatment plots	Jul. '83	Feb. '84 Jun. '86	-	-	0.66	0.43	111.5	71.9	39.2
Clydesdale (1986)									
Upstream of treatment plots	-	Aug. '86 Oct. '86	6 8	1 1	<0.02 0.04	0.14 0.13	- -	- -	44 54
Downstream of S1 treatment area (94 m³/ha)	Aug. '86	Aug. '86 Oct. '86	16 8	6 1	3.31 <0.02	0.20 0.10	- -	- -	250 25
Downstream of S2 treatment area (195 m³/ha)	Aug. '86	Aug. '86 Oct. 86	47 3	27 1	33.2 <0.02	0.30 0.15	- -	- -	3140 50
Monthreathmont									
Natural control drain	-	Mar. to May '82	-	-	0.13	0.52	22	-	4.8
		Aug. '82 to Apr. '83	-	-	0.09	0.43	25	-	13.5
Interceptor below treatment plots	Mar./ Apr. '82	Mar. to May '82	-	-	11.17	0.52	799	-	234
		Aug. '82 to Apr. '83	-	-	0.13	5.22	26	-	15.5

significant difference between upstream and downstream of the treatment areas, nor any elevation in nitrate concentrations. On these sites the initial sludge application rate must be low, about 100 m³/ha, to minimise the risk of water pollution from direct runoff. Subsequently, treatment rates may be increased as the sward develops.

At Montreathmont, where sludge was applied at rates of 0, 239, 477 and 716 m^3/ha to 6 ha of pole stage Scots pine, an interceptor was constructed and the quality of its drainage compared with a natural control drain. The imperfectly drained podzol was at field capacity when the sludge was applied in March/April 1982 and direct runoff occurred from the medium and high treatment rate plots, causing high levels of BOD, COD, ammoniacal nitrogen and phosphorus in the interceptor in the subsequent three months. There was no further drainage from the site until late August, when there was then no significant difference for the next eight months for these parameters compared with the control drain. However, the nitrate levels were elevated and increased to a peak of 25 mgN/l in the interceptor in the autumn as the sludge organic nitrogen mineralised and was leached from the soil (15). Thus, on the less well drained forest soils sludge application rates must be set low to avoid direct runoff and the leaching of nutrients in the autumn but, as discussed previously, this is also necessary for avoiding anoxic conditions in the soil and damage to tree growth.

Finally, at the Clydesdale 1987 experiment site drainage quality was monitored for organic contamination plus nutrient and heavy metal leaching from a mineral soil with a pH range of 3.1 to 3.8. The site covers 3.7 ha with twenty eight 0.1 ha treatment plots. Dewatered digested sludge was applied in July 1987 to eight plots at 106 tds/ha and a further eight at 194 tds/ha. The site was then ripped at 1.2 m spacings to a depth of 0.5 m and perimeter open drains were constructed round each plot to a depth below the ripping level (3). This intensive drainage system was installed to prevent nutrient movement along the rip lines from plot to plot affecting the tree growth responses. However, it also collects the site drainage and outfalls at two points which have been monitored for 18 months. There is no cultivated untreated control for comparison but the results have been divided into two periods to compare the changes in drainage quality; see Table 7. The overall water quality is good, with no evidence of significant organic contamination except for the higher ammoniacal nitrogen levels in the first period after the sludge applications. Nitrate levels are approximately 0.4 to 0.5 mgN/l higher but other nutrients show no clear trend. Heavy metal concentrations are similarly low with only nickel and zinc having higher mean values in the first period, although it is not possible to determine whether these arise from the sludge applications or from ripping. Comparison with environmental quality standards shows the drainage water to be of good quality, with little evidence of metal leaching following the high rate sludge applications, despite the acidic conditions.

Overall, the water quality monitoring programme has shown that providing the sludge application rates are determined on the basis of the site water holding capacity and the avoidance of damage to tree root systems, then

Table 7. Site drainage quality from the Clydesdale 1987 experiment - dewatered sludge applied to a former opencast coal site prior to ripping and planting.

Monitoring point	A		B		Environmental quality standards for freshwaters (Salmonid)
Catchment area (ha)	2.8		0.9		
No. of 0.1 ha treatment plots	21		7		
No. of sludge-treated plots	5 @ 200 tds/ha 7 @ 100 tds/ha		3 @ 200 tds/ha 1 @ 100 tds/ha		
Period	Oct 87 - Jul 88	Aug 88 - Mar 89	Oct 87 - Jul 88	Aug 88 - Mar 89	
Drainage quality					
pH	4.7	5.4	4.8	5.0	6-9
Susp. solids (mg/l)	30	16	21	14	25
BOD (mg/l)	1.6	1.1	1.6	1.2	3
COD (mg/l)	26	23	17	22	-
Dis. oxygen (mg/l)	11.3	11.4	11.6	11.2	>80%
Amm.-N (mg/l)	0.67	0.22	0.46	0.09	0.78
Total oxid. N (mg/l)	0.5	0.1	0.4	0.1	-
PO_4-P (mg/l)	0.07	0.03	0.05	0.04	-
Total -P (mg/l)	0.14	0.06	0.06	0.08	-
Cadmium (µg/l)	0.31	0.32	0.37	0.24	5
Chromium (µg/l)	2.3	1.4	2.5	0.9	5-50
Copper (µg/l)	9.3	8.8	6.2	5.3	1-28
Lead (µg/l)	4.5	3.3	2.5	2.7	4-20
Nickel (µg/l)	40.9	27.8	49.3	30.0	50-200
Zinc (µg/l)	89.6	67.4	92.6	83.9	8-125

sewage sludge can be utilised without adversely affecting natural waters. A final practical point is that afforested areas tend to be on higher ground with a greater possibility of being on the catchment to a water abstraction. There must therefore be full consultation with the appropriate authorities on the protection of public, private and industrial abstractions.

8. APPLICATION TECHNIQUES

The application methods for the forest experiments and trials have been selected to give close control over the rate and evenness of the treatments.

However, they have given an insight into appropriate techniques and equipment for forestry use.

Tractor-drawn agricultural slurry tankers will only be suitable on a very limited number of sites for applying sludge prior to ploughing or after clearfelling, due to poor ground trafficability and low operating efficiency. However, conventional manure spreaders may be suitable for applying sludge cake on restored soils prior to cultivating.

Retracting reel irrigators can operate on easy terrain prior to ploughing and in pole stage stands from the avenues created for thinning. Extending the legs of the applicator carriage to increase the ground clearance allows the method to be used during the early establishment of plantations, but again only on even terrain (17).

The most flexible system for all the forestry phases is the use of manually deployed static irrigation guns fed by light weight temporary pipelines, either directly from tankers on the road system or pumped from site storage facilities.

However, there is scope for the use of more mechanised techniques using purpose-built off-road vehicles prior to planting or restocking and in thinned mature forests.

Finally, a deep ripper modified to inject simultaneously sludge down to 500 mm on a restored site proved successful at rates of 400 m^3/ha when fed by an umbilical drag hose. However, it can only operate over a limited period of the year on these difficult sites (3).

9. CONCLUSIONS

The research programme on utilising sewage sludge on forest land has determined the following:

a) Sewage sludge, both undigested and digested, can significantly benefit tree growth on podzolic, ironpan and restored soils for spruce and pine plantations. Littoral soils should also be suitable, should benefit from sludge applications and need to be evaluated;

b) Within the forest crop rotation of typically 50 years, there are operationally attractive phases for applying liquid sludge for up to 30 years which are also coincident with periods when nutrient additions would be beneficial;

c) The appropriate forest stages and application rates are:
Preplanting and after clearfelling - rates should be aimed to supply 1,000 kg organic N/ha either as cake or liquid, with the latter achieved in one or a series of applications depending on the terrain and soil drainage characteristics;

Early establishment (up to 5 or 8 year old) - sludge can be applied at 200 m^3/ha per annum but, depending on site conditions and the time of year, may need to be applied as a series of increments with appropriate drying times in between, particularly on restored soils.

Thinned pole stage stands (25-50 year old) - rates as for early establishment;

d) The application rates of potentially toxic elements should be limited to those accepted for agricultural use and concentrations in the soils, which are generally of low pH, should be similarly adhered to;

e) Sludge should not be applied in the three months prior to planting for the protection of forestry staff, and the public should be excluded from treated areas for the same period;

f) Ground vegetation can be beneficially modified by sludge for early establishment of Sitka spruce and can be enhanced on restored soils. Under a mature forest canopy, sludge applied at the recommended rates will only temporarily affect the diversity of the ground vegetation, and the use of sludges from well-screened or macerated sewage has not produced an aesthetic problem in the treated areas;

g) Sludge application regimes (rates, timing, frequency) should be constrained by the ability of the site to accept the loading without runoff. This will avoid both acute and longer term chronic adverse effects on watercourses and is consistent with the requirements for positive benefit to the trees;

h) Manually deployed rain gun systems have proved the most flexible for forestry operations but there is considerable scope for developing more mechanised specialist systems.

Forestry could therefore provide an important new outlet for the positive use of sewage sludge to the benefit of the water and forestry industries. It has the potential to provide a significant outlet in the UK and a code of practice is now being developed to encompass the findings from the research and pilot operation schemes.

ACKNOWLEDGEMENTS

The authors are grateful to the staff from Scottish Regional Councils and River Purification Boards, Regional Water Authorities, The Macaulay Land Use Research Institute and Forestry Commission Districts who have collaborated in the research, and to the Scottish Development Department for their support in part-funding the programme.

REFERENCES

(1) Forestry Commission (1988). Forestry facts and figures 1987-1988.
(2) McPhail, C. D. and C. D. Bayes (1984). Use of sewage sludge as a forest fertiliser. Ardross forest experiment - application of sludge to heathland prior to establishment of Sitka spruce. Report ER 712-M. Water Research Centre, Medmenham, UK.
(3) Bayes, C. D. and C. M. A. Taylor (1988). The use of sewage sludge in the afforestation of former opencast coal sites: Clydesdale Forest trials. Report SDD 1774-M. Water Research Centre, Medmenham, UK.
(4) Miller H. G. (1981). Forest fertilisation: some guiding concepts. Forestry, *54*, 157-167.
(5) Binns, W. O., R. D. Davis and A. G. Mugleston (1983). Preliminary results of an experiment on the use of sewage as a phosphate fertiliser in a coniferous forest. In: Processing and use of sewage sludge (P. L'Hermite and H. Ott, Eds.), pp. 318-326. D. Reidel, Dordrecht.
(6) Moffat, A. J. and J. E. Hall (in preparation). The application of sewage sludge to pole stage Corsican pine at Ringwood Forest, Dorset.
(7) McPhail, C. D. (1984). Use of sewage sludge as a forest fertiliser. Montreathmont forest experiment - application of sludge to pole-stage Scots pine. Report ER 609-M. Water Research Centre, Medmenham, UK.
(8) McIntosh, R. (1984). Fertiliser experiments in established conifer stands. Forestry Commission Record 127, HMSO.
(9) Kelly, J. M. and H. G. Miller (1985). Effect of sewage sludge on roots of Scots pine seedlings. Progress report to Scottish Development Department, Report ER 1090-M. Water Research Centre, Medmenham, UK.
(10) Kelly, J. M. and H. G. Miller (1986). Field investigation into the effects of sewage sludge on the roots of Scots pine seedlings. Department of Forestry, University of Aberdeen.
(11) Council of the European Communities (1986). Council Directive on the protection of the environment, and in particular of the soil, when sewage sludge is used in agriculture. Official Journal of the European Communtites No. L 181/6-12.
(12) Department of the Environment (1989). Code of practice for agricultural use of sewage sludge. HMSO.
(13) Keay, J. (1964). Nutrient deficiencies in conifers. Scottish Forestry, *18*, 22-29.

(14) Bayes, C. D., C. M. A. Taylor and N. M. Proctor (1987). Use of sewage sludge as a forest fertiliser. Ardross Forest experiment - tree growth and water quality monitoring. Report SDD 1471-M. Water Research Centre, Medmenham, UK.

(15) Bayes, C. D., J. M. Davis and C. M. A. Taylor (1987). Sewage sludge as a forest fertiliser: experiences to date. J. Inst. Water Pollution Control, *86*, 158-171.

(16) Bayes, C. D. and C. M. A. Taylor (1988). Use of sewage sludge as a forest fertiliser prior to replanting of clearfelled sites: Speymouth 26/87 forest experiment. Report SDD 1773-M. Water Research Centre, Medmenham, UK.

(17) Daw, A. P. and J. E. Hall (1987). Sewage sludge as a forest fertiliser: assessment of irrigation systems for the application of sludge. Report PRU 1461-M. Water Research Centre, Medmenham, UK.

Operational experiences of sludge application to forest sites in Southern Scotland

J. M. ARNOT[1], J. D. McNEILL[2] and B. F. J. WALLIS[3]

[1]Strathclyde Regional Council, Department of Sewerage, Regional Offices, Hamilton, ML3 OAJ, UK.
[2]Forestry Commission Northern Research Station, Roslin, Midlothian, EH25 9SY, UK.
[3]Borders Regional Council, Water and Drainage Services, West Grove, Melrose, TD6 9SJ, UK.

SUMMARY

This paper describes sewage sludge application experiments at two forest sites in Southern Scotland. Clydesdale is a relatively flat area on the site of a former opencast mine, whereas Glentress is a steeply sloping, upland site. The sites were chosen as the tree crops were suffering from nitrogen and phosphorus deficiency. Digested sludge was applied at Clydesdale and, as forestry is not subject to the EC Sludge Directive, raw sludge was applied at Glentress. The practical aspects of sludge application at each site are discussed and problems highlighted. Initial results are encouraging and indicate that sludge can be successfully applied to forest sites and that rapid tree growth increases can be expected.

1. INTRODUCTION

The Lanark Division of Strathclyde Regional Council in association with the Water Research Centre approached the Forestry Commission in 1986 with a view to carrying out experimental application of liquid sewage sludge to trees. The site chosen, in Clydesdale, was a former opencast mine site and the purpose of the sludge was to redress the nitrogen deficiency in the soil which was restricting tree growth. Following encouraging results from a small scale trial in 1986, two larger trial sites were identified for further applications. Subsequently the Forestry Commission and WRc have identified further sites in Scotland which would benefit from application of sewage sludge. One such site is Glentress Forest in Borders Region where a trial commenced in 1988. Both Lanark Division and Borders Region, Water and Drainage Services have long practical experience of the application of sewage sludge to agricultural land. Application to forestry, however, requires the development of different techniques and procedures, depending on the type of site, and these are discussed.

2. CLYDESDALE FOREST SITE

2.1 Site description

The operational trial area is on the former Whaupknowe opencast coal site at Wilsontown, near Forth, Lanarkshire in the Clydesdale block of Lothian and Tweed Forestry Commission forest district. Following initial trials two areas were selected for commencement of sludge application in 1987 and these are identified as Sites 3 and 5 on Figure 1. Vegetation cover on both areas was very sparse and consisted of poorly growing *Agrostis* spp., *Holcus mollis, Deschampsia caespitosa, Anthoxanthum adoratum* with occasional patches of *Juncus communis*, and *Epilobium augustifolium*. Access was very convenient with a metalled road leading into the site from the A706 north of Wilsontown.

Following the successful response to sludge at Sites 3 and 5, treatment was extended from 1988 to the other numbered sites on Figure 1.

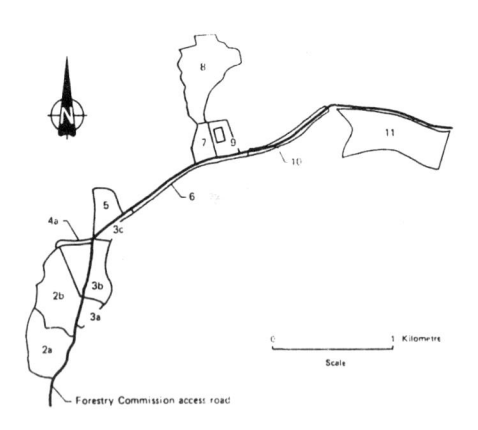

Fig. 1. Sludge spreading sites in Clydesdale Forest.

2.2 Site preparation

Following completion of coal extraction, site restoration was carried out between 1977 and 1979 using heavy earth moving machinery to replace the excavated soil. The replaced soil resulting from this operation, which is stored topsoil redistributed over mine waste material, is very stony, low in humus content and lacking in microbial activity. Where possible, slope gradients of about 1:10 were constructed during land formation in order to provide suitable drainage, but on the flatter ground, slope was achieved by forming ridges based on the old agricultural system of "riggs" and "furrs". These ridges ("riggs") were installed along 1° slope to facilitate drainage down the intervening drains ("furrs") and conformed to dimensions of 50 to 300 m in length, 14 to 30 m in width and about 1 to 1.5 m in height at ridge centres.

Due to the severe compaction caused by heavy machinery running over the site, it was necessary to tine-rip the area, including the ridges, to a depth of 0.5 to 0.7 m at 2 m spacing. This operation facilitated easier tree planting and subsequent root penetration as well as sustaining an adequate water reserve for the plants.

2.3 Crop

Planting was carried out between 1978 and 1981 using mainly pure Sitka spruce (*Picea sitchensis*), particularly on the flatter ground and the ridged areas, but other species, such as Japanese larch (*Larix kaempferi*), Scots pine (*Pinus sylvestris*) and alder, were also planted as pure crops and in mixture with the spruce, on the sloping ground. Larch and pine species are generally more tolerant of low availability of nutrients than spruce, and are often planted in mixture with Sitka spruce on nitrogen deficient upland mineral soils. The spruce derives a nutritional benefit when planted in mixture with these species and this reduces the necessity of expensive applications of inorganic nitrogen fertiliser. On this particular site, small areas of larch and spruce mixtures were planted to improve the spruce growth performance but responses were only achieved where soil conditions were more favourable than normal. In most cases there was no benefit to the spruce and the larch became the dominant species.

In 1985, yellowing foliage and declining annual height growth indicated nitrogen deficiency in the spruce. This was confirmed by foliage samples (see Table 1) and it was necessary to prescribe some form of treatment to improve health and growth.

2.4 Soil analysis

Soil samples were taken from Sites 3 and 5 prior to application of sludge to establish the background level of metals and nutrients. The samples were taken to a depth of 150 mm, air dried at 30°C and ground to pass through a 2 mm mesh sieve before analysis. The results are shown in Table 2.

Table 1. Comparison of Sitka spruce foliage analysis data from Clydesdale with average normal crops.

	Needle wt (mg)	N%	P%	K%
Clydesdale reclamation site	2.5	1.19	0.19	1.1
Average healthy crop	6.0	1.50	0.18	0.7

Table 2. Analysis of soil from Sites 3 and 5 at Clydesdale forest.

Site	3a	3b	3c	5
pH	3.8	4.3	4.2	4.4
Dried solids (%)	98.4	98.54	98.76	98.76
Mineral matter (% ds)	78.27	86.36	89.51	88.26
Volatile matter (% ds)	21.73	13.64	10.49	11.74
Nitrogen (N) (% ds)	0.34	0.21	0.13	0.17
Phosphate (P_2O_5) (% ds)	0.04	0.04	0.05	0.05
Potassium (K_2O) (% ds)	0.15	0.13	0.13	0.15
Cadmium (mg/kg ds)	0.5	0.5	0.5	0.5
Chromium (mg/kg ds)	16	21	19	21
Copper (mg/kg ds)	25	19	17	17
Iron (mg/kg ds)	22,400	20,000	23,500	20,300
Lead (mg/kg ds)	61	30	22	20
Magnesium (mg/kg ds)	945	1390	1390	1840
Mercury (mg/kg ds)	0.2	0.2	0.2	0.2
Nickel (mg/kg ds)	13	20	17	23
Zinc (mg/kg ds)	31	43	37	43
Aluminium (g/kg ds)	4,300	5,300	5,400	5,850
Calcium (mg/kg ds)	232	202	267	346

Note: Site 3 was divided into three for ease of sampling.

These results show that the soil on both areas was mainly an acid mineral soil with only background concentrations of potentially toxic elements and a low nutrient level.

The soil pH in both areas was below the EC recommended minimum for the application of sludge to agricultural land. However, the Forestry Commission were happy to allow sludge spraying to go ahead on the basis that the EC limit value for the annual addition of heavy metals to agricultural land was not exceeded.

2.5 Sewage sludge

The closest sewage works to Clydesdale Forest, capable of supplying a sufficient volume of suitable sludge for this trial, was Shotts some 16.5 km from the application site. In addition a number of loads of sludge were taken to the forest from Carbarns sewage treatment works in Wishaw, the base for the tankers used in these trials. The sludges are however similar, both being co-settled crude and surplus activated sludge anaerobically digested at 30°C for 20 days followed by consolidation in secondary digestion tanks.

Due to difficulties in producing a consistently thick sludge from the consolidation tanks at Shotts sewage treatment works the dried solids content of the sludge varied considerably during the course of the trial. In this instance lower sludge solids levels did not appear to produce any runoff at the chosen application rates.

In order to monitor the quality of the sludge being applied to each area, a sample of sludge was taken from each tanker prior to discharge. These samples were composited and the composite sample analysed. The results of the sludge analysis are shown in Table 3.

2.6 Method of sludge application

Sludge is transported to Clydesdale Forest by 13.6 m^3 rigid tankers fitted with hydraulically driven Wright Rain Manurian pumps. These pumps are capable of pumping between 30 and 50 m^3/hr against a head of 6 bar, depending on pump speed. The pump is connected by 6 m lengths of 100 mm aluminium pipe to a rain gun which can be up to 600 m from the tanker in suitable conditions.

As is normal with this method of sludge spreading, the pipes are laid out to the far end of the area being sprayed and the appropriate number of pipes are removed after two tanker loads have been discharged. This method of operation minimises the time the operators are in the sprayed area.

No major problems have been experienced using this method of spraying. The distance between the rows of trees makes laying out the lines of pipes no more difficult than in an ordinary field, although some care has had to be taken to avoid the larger rocks or the smaller trees. Occasional problems have been experienced in anchoring the rain gun in place due to the stony nature of the ground, but a firm area has always been found close to the required point to take the anchoring pins.

Table 3. Average analysis of sludge applied at Clydesdale.

	Site 5	Site 3
pH	7.0	7.0 - 7.5
Dried solids (%)	3.45	2.33
Mineral matter (% ds)	40.78	39.79
Volatile matter (% ds)	59.22	60.21
Nitrogen (Org) N (% ds)	3.46	4.03
Nitrogen (Total) N (% ds)	5.4	6.50
Phosphate P_2O_5 (% ds)	3.1	3.32
Potassium K_2O (% ds)	0.54	0.81
Volatile acids (mg/l)	260	156
Cadmium (mg/kg ds)	5	4.3
Chromium (mg/kg ds)	70	68.7
Copper (mg/kg ds)	359	432.3
Iron (mg/kg ds)	22,500	22,600
Lead (mg/kg ds)	404	388
Mercury (mg/kg ds)	8.1	6.8
Nickel (mg/kg ds)	37	38
Zinc (mg/kg ds)	92.8	93

The reaction of operators to working in the forest has so far been good as they regard it as an extension of their normal experience of working on farmland. The size and spacing of the trees allows the operators to retrieve pipes or reposition the rain gun without becoming too dirty, and waterproof leggings worn over overalls and wellingtons are normally sufficient protection.

Both sites sprayed are situated approximately 1.6 km from the main road and are reached by means of the main forest access road, which is a well constructed road capable of carrying heavy lorries without sustaining damage. All of Site 3 could be sprayed from the side of this road as could some parts of Site 5. The main part of Site 5 could, however, only be sprayed after the tankers had manoeuvred down a narrow, less firm road. This caused few problems but did increase the amount of time taken for the tankers to reach the discharge point.

2.7 Sludge application rate

Ground cover on the experimental site is sparse and there is little restriction of sludge movement across the soil during sludge application. In order to prevent runoff it was found necessary to limit application rates to 100 m^3/ha.

The area treated via the rain gun is limited by the radius of spray, which in turn depends on size of nozzle and pump pressure. The optimum size to reduce the incidence of blockages was found to be 25 mm and this provided a radius of 30 m and a dose rate of 100 m^3/ha using two 13.6 m^3 tanker loads. The circular areas are overlapped in the pattern shown in Figure 2 to provide good coverage.

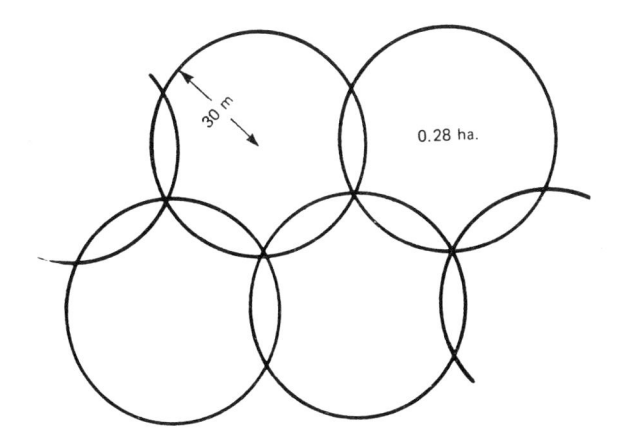

Fig. 2. Typical spray pattern at Clydesdale Forest.

In practice this theoretical application rate is difficult to achieve for a number of reasons. The wind direction and speed have a great influence on the distance sprayed by the rain gun and may carry the spray further in one direction than another. This makes it difficult to evenly cover the site and to assess actual application rates. In addition taller, spreading trees deflect spray from the rain gun, resulting in a smaller area being sprayed.

Overcompensation for these problems resulted in higher application rates than planned (see Table 4) although no problems due to runoff were experienced.

Following completion of Sites 3 and 5 a further 17 ha site (2b) was identified and application commenced in September 1988. This area was finished in a broad rigg and furr profile and the soil was well compacted. Soon after spreading commenced reports of pollution in a nearby burn were received in the evenings following applications. It appeared that the soil had become saturated following heavy rain and that sludge was being carried in surface water to the burn. Spraying was then abandoned until drier conditions prevailed.

Table 4. Sludge application data for Clydesdale.

	Site 5	Site 3	
	Application rate (kg/ha)	Application rate (kg/ha)	Annual limit value based on 10 year average (kg/ha)
Dried solids (%)	3,293	3,714	
Mineral matter (% ds)	1,600	1,478	
Volatile matter (% ds)	2,323	2,236	
Nitrogen (Org) N (% ds)	135.7	149.7	
Nitrogen (Total) N (% ds)	211.8	241.4	
Phosphate P_2O_5 (% ds)	124.4	123.3	
Potassium K_2O (% ds)	21.2	30.1	
Cadmium (mg/kg ds)	0.02	0.02	0.15
Chromium (mg/kg ds)	0.27	0.26	4.5[1]
Copper (mg/kg ds)	1.41	1.61	12
Iron (mg/kg ds)	88.27	83.94	-
Lead (mg/kg ds)	1.58	1.44	15
Mercury (mg/kg ds)	0.04	0.03	0.1
Nickel (mg/kg ds)	0.15	0.14	3
Zinc (mg/kg ds)	3.64	3.45	30
Volume of sludge applied (m³/ha)	113.7	159.4	

[1] Proposed annual limit value.

2.8 Tree response

The initial trial in 1986 showed that the foliage colour and growth of the Sitka spruce responded significantly to sewage sludge application and the same effect was noted in the areas subsequently treated. Growth increment measurements were taken in May 1989 from sample trees within the treated spruce areas to assess the crop response.

Due to applications occurring across the treated areas over an extended time span, two distinct periods were selected. These were:
a) Site 3 where sludge application was completed in December 1987;
b) Site 4 where sludge application was completed in September 1988.

Measurements were also taken from adjacent untreated areas as a control.

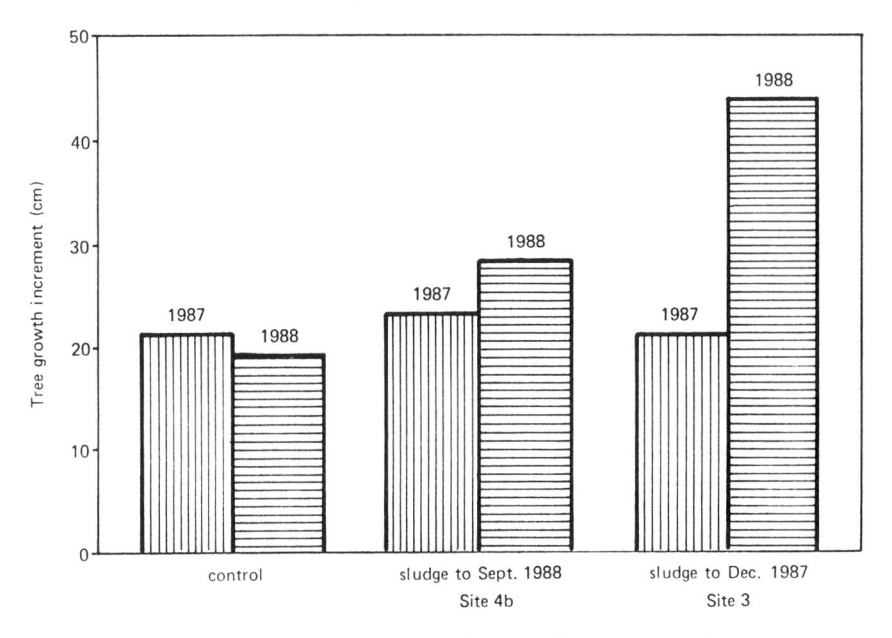

Fig. 3. Clydesdale Forest. 1987 and 1988 tree growth increments (cm) for sludge treated and untreated areas.

The area treated towards the end of the 1988 growing season revealed a slight growth improvement (24%) despite the short period of time left for growth that year. Colour was greatly improved from yellow to dark green and needle size increased slightly. Application of sludge in December 1987 produced a similar initial improvement in foliage colour and needle size, followed by an even greater response during the 1988 growing season with a height increment double that of the previous year (106%).

The growth of ground vegetation also increased with sludge application, becoming a denser and more extensive sward, particularly boosting the grass species at the expense of the less common plants. This would enable further applications at higher rates to be made with less risk of runoff. However, this may result in severe competition in areas carrying very small trees and herbicide control might prove necessary.

2.9 Costs

Costs of the operation depend on size of tanker, distance travelled, time taken for each load and number of operators required.

Larger tankers (18.2 m^3) could be used although it would be necessary to provide a separate diesel pump, thereby complicating the operation. The larger tankers would also be far less manoeuvrable on narrow forest roads.

Clydesdale Forest is 12.5 km from Shotts sewage treatment works along a main road and the spreading site is a further 4 km up the forest road. This is further than sludge would currently be taken for utilisation on farm land.

The time taken to fill tankers at Shotts sewage treatment works is about 15 minutes and this cannot be easily reduced without the installation of larger loading pumps. Discharge times depend on back pressure in the rain gun and pipework system and the frequency of blockages, which in turn depend on the nozzle size used. The numbers of tankers used has to be judged carefully to avoid queuing but the use of several tankers makes more use of the site labourer who moves pipework between loads. Present costs are estimated at £3.31/m^3 although if the operation was carried out on a more routine basis these would be reduced to £1.99/m^3.

These costs do not include any allowance for the purchase of field spraying equipment, this having been purchased over the years for field disposal operations. Typical costs for Wright Rain irrigation equipment are shown below:

Diesel driven pump	£6,137.39
6 m length of 100 mm aluminium pipe	£ 51.86
Sludge spray gun	£ 172.50
Stand for sludge gun	£ 76.15

The tankers used were purchased complete with the hydraulic driven pumps fitted. The cost of purchasing the pumps, the ancillary hydraulic equipment and fitting added approximately £8,000 to the purchase price.

2.10 General comments

The Clyde River Purification Board were kept fully informed of operations taking place in Clydesdale Forest and no objections were raised by the Board providing suitable precautions were taken.

The forest area is close to the village of Forth and is frequented by walkers, and within the forest boundary is a working coal mine. Therefore a large number of people have passed those areas sprayed by sludge but to date no complaints have been received.

2.11 Future work

There are many sites in Strathclyde Region where colliery tips and old industrial sites are being returned to forestry. The soil on these sites is generally poor and would benefit from the application of sludge. These sites may offer an outlet for large volumes of sludge in future years.

Since this work was carried out, the 5.3 ha site of the old Lanarkshire steel works close to the centre of Motherwell has been sprayed with digested sludge in an effort to promote tree growth as part of the site reclamation programme.

One possible problem is that once trees are established side branches spread to form the 'thicket' stage and access for sludge spreading is difficult. Work needs to be done to develop new equipment and techniques for this situation.

3. GLENTRESS FOREST SITE

3.1 Site description

This site is at the head of the Horsburgh Valley in Peeblesshire, within Borders Region, and forms part of Glentress Forest owned by the Forestry Commission. Access is by 5 km of unmetalled road up a steady incline to a height of 350 m above OD. The site slopes steeply above and below the road and is drained by small burns which converge in the valley to form the Hope Burn.

3.2 Crop

The crop is Sitka spruce planted in 1976 on the steep valley slopes. The site was space furrow-ploughed prior to planting but has been revegetated by the indigenous ground vegetation *Calluna vulgaris* (heather). The crop is suffering severely from nitrogen and phosphorus deficiency caused by competition from *Calluna* and low mineralisation rates in the soil. The soil type is upland brown earth and iron pan. Average tree height is 80 cm.

The Forestry Commission now consider this to be marginal land for planting pure spruce as application of nutrients in the form of urea and phosphate is expensive and can only be carried out by helicopter.

3.3 Sewage sludge

Sewage sludge for the trial is supplied from Peebles sewage works, which is approximately 7 km from the site. This works provides conventional treatment for a domestic population of 6,800 plus a small industrial contribution mainly from woollen finishing.

Average sludge production is approximately 50 - 60 m^3/week of mixed raw/humus sludge at 3.5 - 4.0% ds.

The sludge is fairly homogenous. Influent sewage is fine screened and macerated screenings are returned to the flow. A sludge stirrer is shortly to be installed in the sludge storage tank which has two weeks' capacity.

The previous arrangement for this sludge involved tankering it 14 km to another works where it was dewatered, stored and finally spread on agricultural land. The total operating cost of that arrangement was approximately £8/m^3 wet sludge.

3.4 Preparatory work

In view of the unfamiliarity in the Region with this type of sludge utilisation a number of bodies were consulted or informed prior to trials commencing.

Nature Conservancy Council. The Hope Burn is a tributary of the River Tweed and the river banks within the catchment are defined as a site of special scientific interest (SSSI). There were no objections as the location was remote and of no particular interest to the Conservancy.

District Council. The Environmental Health Department was contacted to ensure that there were no private water supplies in the area which may become polluted. This was confirmed to be the case.

Tweed River Purification Board. The Board already had information on water quality in this catchment and agreed to monitor the situation regarding any potential pollution of minor watercourses.

Local representatives. Although the site is remote, the public are allowed pedestrian access and tankers pass a popular picnic area on their way to the site. Local representatives and the members of the Water and Drainage Committee were therefore kept informed in case they received queries from members of the public. Warning signs were also positioned at the site so that people knew what was taking place.

3.5 Sludge application

The Forestry Commission identified three areas sloping from the road, two on the west side and one on the east side of the valley head. The areas were 1.2 ha, 6 ha and 5 ha respectively and were bounded by burns. Marker tapes were then fixed to identify the limits of each area. The areas extended between 60 and 180 m down the slope.

It was decided, on the basis of previous experience, to apply sludge at a rate up to 200 m³/ha, which would provide a sludge outlet for Peebles for eight months. Sludge is transported using a 5.5 m³ or 8.5 m³ vacuum tanker or a 9 m³ tanker with built-in progressive cavity pump. The round trip from Peebles works, including loading, takes approximately 40 minutes.

3.6 Direct application by tanker

This trial commenced in November 1988 and carried on until January 1989 when operations had to cease due to the adverse effect of the tanker traffic on the forestry road. This was particularly serious when the road began to thaw after a period of frost.

Sludge was applied by pumping through a side jet whilst the tanker was in motion. Initial experiments indicated a cast of 35 m from the tanker but, in practice, sludge was deposited on a 20 m wide strip from the bottom of the road banking downwards. As the tanker moved along the road, pumping was stopped when the end markers were reached.

During the trial period the strip on the east side received only a few loads as the prevailing westerly wind caused sludge to spray back towards the tanker. The other two areas received more than the design application due to the smaller area covered, although some of the sludge has moved 10 to 15 m down the slope in the furrows.

The heavy application rate and greasy nature of the sludge resulted in a heavy coating on the trees, which became unsightly due to adhesion of small pieces of paper and plastic. The period during and after spreading coincided with a period of low rainfall and it is to be hoped that the sludge coating will eventually be washed off by the combined effect of wind and rain. By June 1989 new growth had begun to appear on the treated trees.

The effect of the application on this strip will not be evident until late summer 1989 but the trial demonstrated to local Forestry Commission and Regional Council staff the principle of application of sewage sludge to forestry. This particular method of application is, however, of limited value unless there are suitable tracks available through the forest.

3.7 Application by rain gun

A rain gun together with 6 m lengths of lightweight 100 mm diameter aluminium tubing were purchased and a trial using these began on 3 May 1989. The first area chosen was on the east side of the valley and the pipework was laid out down the slope from the road. Sludge was pumped using the on-board tanker pump and resulted in a 30 m radius of spray through the rain gun, equivalent to an application area of 0.28 ha. Therefore five or six 9 m³ tanker loads per rain gun position would supply the required quantity.

The smaller vacuum tankers were also connected to the spray system and covered approximately one third of this area with a radius of 15 m. Tankers took approximately 15 minutes to discharge each load. Difficulties with this method of application soon became evident. The spray did not give a true circle due to ground slope and wind effects and an initial misjudgement of distances caused overspray to a burn during the first application. A buffer area must therefore be maintained. Once the required quantity has been sprayed the equipment has to be moved and resiting requires careful judgement to fully cover the area. This was achieved as shown in Figure 2.

During application the trees become heavily coated and moving equipment among the trees is unpleasant. The rate of sludge production enables one week's sludge to be applied in one day, allowing the equipment to be left to dry for a week. Nevertheless suitable waterproof and lightweight protective clothing must be supplied.

To avoid this problem the purchase of a rain gun which only operates through 180° in a downhill direction is being considered, although this will mean moving the equipment more often.

Training of operators is also essential so that they fully understand the requirements of the forester and the need to protect watercourses.

The Tweed River Purification Board are monitoring the catchment and up to now no deleterious effects on water quality have been noted.

In spite of the problems, this method is practical and once operatives are used to the somewhat unusual working conditions it will gradually be accepted and become a routine operation. The Forestry Commission are so far pleased with the application method although benefits will not become evident until 1990.

Indications from experimental work in Northern Scotland on a similar heathland site suggest that a similar application of sludge will eradicate the *Calluna* at Glentress, thereby reducing competition for nutrients, and provide a growth response in the Sitka spruce crop lasting five years or more. If this is the case then further applications will be extended to other areas, giving an economic alternative to several applications of inorganic fertiliser.

3.8 Future work

A further site has been identified at Elibank which is at a low elevation and situated 15 km from Peebles works. It is an easy journey and would be accessible throughout the winter, unlike the Horsburgh site. It is also near Walkerburn sewage works where Peebles sludge was transported for dewatering, and sludge from that works could also be applied.

The crop is a thinned Scots pine stand (pole stage), 40 years old, on a gently sloping grass-dominated site. This is a freely draining site (brown earth and iron pan) where the crop is fairly slow growing and would benefit from application of nitrogen. It is proposed to apply sludge via the rain gun system.

There is also considerable potential for utilsation of sludge in forestry (private and Commission) in other parts of the Region and this will be particularly important in areas where the present method of utilisation on agricultural land is limited.

4. CONCLUSIONS

Application of sewage sludge to forestry is a practical alternative to utilisation on agricultural land or other disposal methods. There appear to be considerable potential benefits to the forest industry in terms of accelerating tree growth and making savings on the use of inorganic fertilisers.

Forestry is not as yet subject to the EC Directive for sewage sludge and as such there are no restrictions on the use of untreated sludge nor on times and rates of application. It would, however, be prudent not to exceed the EC limit value for the annual addition of heavy metal to agricultural land.

More work needs to be done on gaining practical experience in order to make the operation easier, and acceptable to staff, and to ensure that there is no environmental damage. Large areas of forest are available, many within a reasonable distance of sewage works, and they have the potential as an outlet for large volumes of sludge.

ACKNOWLEDGEMENTS

The authors wish to thank their respective Authorities for permission to present this paper and their colleagues who have assisted them in these projects. The views expressed are those of the authors and do not necessarily reflect those of their Authorities.

US forestry uses of municipal sewage sludge

C. G. NICHOLS

Reclamation, Reuse and Conservation Administrator,
County Sanitation Districts of Orange County,
PO Box 8127, Fountain Valley, California, USA.

SUMMARY

In the United States, forest application of processed municipal sewage sludge has occurred for the past 25 years. Major projects have occurred in the midwest, southeast and northwest. In each case tree growth was significantly improved through the nutrients available in sludge. Metro Seattle in Washington state has recycled all of its sludge to land for the past 16 years, with a large portion being applied to forests. Extensive research was performed by the University of Washington College of Forest Resources on the fertilisation effects, the ecosystem impacts, and the operating methods to implement forest sludge application. Metro operates a forestry land application programme which includes local authority permitting and thorough contingency planning. US federal regulations over these activities are in a state of flux although more stringent regulations are expected to be adopted within two years. Metro continues to search for new, cost-effective sludge management techniques including sludge drying, application spray nozzle elevators and light-weight hose reel delivery systems.

1. INTRODUCTION

With the onset of the federal Clean Water Act in the 1970s, the United States (US) embarked on a strategy of significantly upgrading wastewater treatment throughout the nation. A result has been large increases in sludge quantities to be managed. A variety of techniques are being used including forest fertilisation with processed municipal sewage sludge. Metro Seattle and others have explored the many facets of sludge forest fertilisation and concluded that a well managed forest application system can be a safe and effective sludge management method.

2. SUMMARY OF SELECTED US FORESTRY SLUDGE APPLICATION PROJECTS

Over the past 25 years forestry sludge application projects have occurred in eight states and Puerto Rico, as listed in Table 1. Seattle Metro's experience with forestry sludge application will be described in some detail.

In northern Michigan the United States Environmental Protection Agency (EPA), Michigan Department of Natural Resources (DNR) and Michigan State University began studies of sludge in forest ecosystems in the mid-1970s (1). Their work began with small research plots, and moved into larger demonstration and research projects followed by operational scale application programmes (2). At a number of different sites, over 4,000 m^3 anaerobically digested liquid sludge were applied to four forest types including big tooth aspen (*Populus grandidentata* Michigan X), red and white oak (*Quercus rubra* L and *Q. alba* L), jack and red pine (*Pinus bankseana* Lamb. and *P. resinosa* Ait.) and red and sugar maple (*Acer rubrum* L. and *A saccharum* Marsh.).

Table 1. Locations of US forestry sludge application projects.

Pennsylvania	Puerto Rico
Georgia	California
South Carolina	Oregon
Tennessee	Washington
Michigan	

Sludge application rates on these sites were approximately 9 tds/ha. The costs of these application demonstrations were comparable to those of agricultural sludge application at $304/tds. Preliminary findings of short-term growth response were encouraging. Aspen biomass was 57 % greater than controls, basal area increased by 48 % and ground level diameter increased by 23 %. Sludge fertilised aspen mortality was higher than controls but this was brought about through increased elk browsing of the nutrionally enhanced aspen foliage.

Increased growth of oak, pine and maple was significantly higher for sludge fertilised trees compared to controls. Table 2 summarises these results.

Application techniques on these projects consisted of transporting liquid anaerobically digested sludge, ranging from 2.6% to 5.1% ds, using a 32,000 litre tank truck up to 80 km to the forest site. Sludge was transferred to an all-terrain application vehicle with an 8,300 litre tank using a vacuum-pressure system. Application onto the forest was accomplished by driving the vehicle along prepared forest trails and side-casting the sludge as much as 10 metres. Uniformity of sludge application was achieved using a three nozzle system.

Table 2. Michigan tree sludge fertiliser growth response.

	Oak	Pine	Maple
Basal area growth increase over control56% 1981-1985	56%	36%	56%
Diameter growth increase over control 1981-1985	78%	25%	48%

In the southeast US sewage sludge has been used for reclamation and reforestation. Many demonstrations have occurred in Tennessee, South Carolina and Georgia beginning in the mid-1970s (3). On demonstration scale plots, seven varieties of tree species were grown including Loblolly pine, shortleaf pine, Virginia pine, sweetgum, sycamore, green ash, and yellow poplar.

Sludge application rates on these sites ranged from 34 to 68 tds/ha. Also in these trials, the findings of short-term growth responses significantly favoured sludge fertilisation and reclamation when compared to commercial fertiliser or controls. Table 3 summarises the growth response results.

Application techniques on these projects consisted of clearing and grading the sites to be reclaimed, delivering dewatered sludge using dump trucks, and spreading and incorporating the sludge into the soil by discing 15 to 20 cm. In many cases, due to the compact nature of the soils, deep discing was practiced to a depth of 46 to 92 cm. The depth of 46 cm was found to be optimum. After these activities tree saplings were planted. As is often the case the seedlings suffered from weed and shrub competition which must be controlled for successful reforestation.

3. OVERVIEW OF METRO SEATTLE SLUDGE MANAGEMENT PROGRAMME

In the Pacific Northwest, Metro and the University of Washington began studies of sludge in forest ecosystems in 1973. Metro, a regional govern-

Table 3. Summary of tree growth response to sludge fertilisation in the southeast USA.

	Loblolly pine	Shortleaf pine	Virginia pine	Sweet gum	Syc- amore	Green ash	Yellow poplar
Height growth increase over commercial fertiliser	56%	46%	67%	489%	-	-	-
Diameter growth increase over commercial fertiliser	66%	51%	101%	453%	-	-	-
Biomass growth increase over commercial fertiliser	42%	-	-	123%	148%	278%	661%

ment in the Seattle-King County area, provides wastewater treatment, water quality management and transit services to nearly one million people. At its five wastewater treatment plants it processes an average of 7.7 m³/second and generates approximately 18,150 tds/year of anaerobically digested, dewatered primary and secondary sludge. Within the Metro service area, the agency administers an aggressive industrial pretreatment programme and works with the water supply authority to help maintain a high quality sludge low in trace metals and organics. For successful land application programmes, high quality sludge is the takeoff point.

Metro's sludge management programme has evolved through several stages over the past 16 years. During the 1970s it was a time of research and demonstration, focussing on silviculture fertilisation, soil improvement and composting. In the early 1980s Metro undertook an area-wide comprehensive sludge management policy and plan development. This led to the 1983 adoption of Metro's current sludge management plan consisting of a diversified non-food chain land application programme with 65% of Metro's sludge targeted for forestry fertilisation, 25% targeted for soil improvement and reclamation projects and 10% targeted for composting leading to horticultural marketing and distribution. Since 1983 Metro has implemented the adopted recycling policy having applied over 612,000 wet tonnes of sludge to thousands of hectares of public and private land throughout Washington state.

Implementing this programme has been partly attained using an active programme of public information, community relations and regional cooperation. Information on the wide range of technical scientific issues touched by sludge management was rewritten into brochures, newsletters and media packages. These documents form the basic tools which Metro staff use when explaining sludge application projects to interested and concerned citizens. Staff specialising in community relations organise and facilitate open houses and community meetings to help inform the local host communities of upcoming activities. Often these staff spend time with the project landowner and/or sponsor to gain a deeper appreciation for local needs and concerns so that Metro can sensitively achieve its sludge management project. Regional cooperation occurs through the 100 member Regional Sludge Management Committee. This group draws its membership from municipal sludge generators throughout the state of Washington who believe mutual cooperation ultimately leads to mutual benefits. The committee is actively involved in professional and public education and works for regional joint sludge management projects.

4. OVERVIEW OF FORESTRY SLUDGE RESEARCH

The Metro-University of Washington research effort began in 1973 at the university's Charles Lathrop Pack Demonstration Forest near Mount Rainier. Researchers focussed on answering three questions:
a) Could the beneficial ingredients in municipal sludge enhance tree growth in the same manner that sludge increased the yield of agricultural crops?
b) What were the ecosystem impacts from sludge applications?
c) While implementing such a concept, how would sludge be applied in a forest setting, what would be the site and system design and what would be the application cost?

Research results were positive in answering all three questions. Growth response of coastal Douglas fir was significantly improved by sludge fertilisation compared to either urea fertilisation or no fertilisation. Diameter, height and volume all increased significantly with little impact on wood quality. Sludge fertilised trees grew similarly to trees on high quality fertile sites. Trees from the sludge fertilised sites are suitable for the domestic lumber and pulp markets but also for the much more lucrative China, Korean and Japanese export markets. Researchers project commercially acceptable trees can be harvested in 15 to 20% less time when sludge fertilised.

Evaluation of the forest ecosystem showed that a properly designed and operated sludge fertilisation project was environmentally safe and posed a very low public health risk (4). These research results have formed the foundation of EPA's Sludge Process Design Manual (5), Washington DOE's Municipal Sludge Best Management Practices (6), and EPA's Preproposal Draft Sludge Management Technical Regulations for Land Application Systems (7). Research conclusions showed that nitrogen was usually the limiting design criterion for sludge applications, pathogens die off rapidly achieving background levels in a few months, trace organics degrade quickly and trace metals are stable and immobile remaining tightly bound to the sludge-soil particles (8). All of these observations lead to the conclusions of very low risk for public health, groundwater, surface water, soil character and wildlife.

Considerable research effort was invested to determine the optimum application rates and timing for the forest system. Two philosophies were explored:

a) Annual sludge applications that meet yearly tree nutrient requirements similar to agriculture systems;
b) An application greater than the annual nutrient requirements followed by several years without application. This approach would recognise the nutrient storage capacity of the system, especially considering a large portion of sludge nitrogen exists in the relatively immobile organic form.

Hopefully microbial degradation would slowly convert the organic nitrogen into the more useful ammonium component over several years. Results have shown that this philosophy does work. Sludge can be applied to a forest at the rate of 34 to 45 tds/ha with a five to seven year resting period between reapplications. During this time the microbial nitrogen cycle takes over and slowly releases usable nitrogen into the forest ecosystem, maintaining the fertilisation effect over the span of years. A secondary benefit of this approach allows a forestry site to return to recreational uses after a one year access restriction which ensures site stabilisation.

5. OPERATING METHODS

Development of sludge application systems in the Pacific Northwest has reflected the unique site conditions found there. In most cases sludge is applied to already established stands which limits site access as compared to annually harvested agricultural sites. Sludge must be propelled into or over the forest while maintaining uniform sludge applications. Equipment must be capable of negotiating steep and uneven terrain. Operating methods were evaluated for clearcuts, young tree plantations and mature tree plantations. Application techniques considered were set irrigation systems, travelling spray guns, manure spreader systems, dumping and spreading systems and all-terrain vehicles with a top-mounted spray cannon.

With the help of the university, Metro settled on a system where dewatered sludge cake at 18 to 25% dry solids is trucked as much as 96 km from the treatment plant to the forest site and sludge is rewatered at the site to 10 to 13% solids to allow pump spraying (9). The rewatered sludge is temporarily stored, then transferred to the all-terrain application vehicles for application to the forest. This system averages 154 wet tonnes per day applied to forest sites.

In detail the system begins operation by dewatered sludge being rewet using a portable mix-pump unit. This 70,000 litre tank was fitted with 2-7.5 kW propeller mixers and 2-18.6 kW sludge pumps. Water for rewetting the sludge was added prior to sludge being placed in the tank. Using an automatic controller, in approximately 15 minutes 18 tonnes of sludge was emptied into the tank, mixed with the water and pumped from the tank into storage. At regular intervals a 28,400 to 36,000 litre transfer tanker withdraws sludge from the storage tanks. The transfer tanker, pulled by a diesel truck tractor, moves off to application sites within a 10.5 km radius. Two transfer units operate to supply sludge to two all-terrain application vehicles. Metro owns a total of five all-terrain application vehicles with a capacity ranging from 6,800 to 9,500 litres. Two vehicles have stationary spray cannons while three units have elevating spray cannons, rising as much as 10.4 m above the ground.

Sludge applications were completed in four or five lifts to minimise sealing of the forest floor and potential sludge movement during rainfall. This approach also facilitates uniform applications in an area. A period of time was allowed between each lift which allows the applied sludge to stabilise or dry. Stabilisation periods may be two or three days during the summer or up to four weeks during the wet season.

Metro's operations include adjustments for inclement weather such as sustained or intense rain. Two problems may occur during inclement weather as the soil system becomes saturated, increasing the surface water runoff potential and impairing the soil trafficability.

Also, the programme acknowledges that occasionally an accident may occur which could result in sludge being placed where it is not desirable. Before a project begins, an emergency response plan is designed and activated. This plan anticipates immediate spill containment and cleanup.

In 1988, Metro's cost of operation for the forest fertilisation totalled $31.31 per wet tonne. These costs were split with 27% for hauling the sludge from the plants to the sites and the remaining 73% for the on-site application activity. In the application activity, costs were apportioned as shown in Table 4.

Table 4. Breakdown of forestry application costs.

Labour	40%
Equipment maintenance	23%
Trails installation	18%
Dilution water	8%
Diesel fuel	4%
Mix site preparation	4%
Site security	2%

6. PROJECT MANAGEMENT

Achieving land application project success is a challenging exercise. Effective planning, site design and permitting is crucial in achieving this success. The philosophy in managing sludge land application projects is to do whatever is needed to satisfy the local permittors. The permit system provides checks and balances which should help prevent problems, provides an educational opportunity and establishes credibility with the agency and the public.

Site selection and planning begins with knowing the project's success criteria and may include, for example, local, state and federal regulations, local land use and zoning codes and local public and political concerns. Landowner needs, site location and economics of use are among some of the other important criteria. At a planning level it might be desirable to do desktop analyses followed by windshield site surveys. Choosing a site may include detailed site checks after preliminary site screening.

A detailed site check would include obtaining quality data on site topography, water resources, property boundaries, operating areas, monitoring sites, soil fertility, soil trafficability and vegetation. This information can be used to create a site design which reflects the challenges of each new site. The data will also be useful when brainstorming on contingency plans for a site. A list of typical contingency topics is provided in Table 5.

Table 5. Site management contingency planning.

1.	Transportation
2.	Application
3.	Environmental monitoring
4.	Site management quality control
5.	Storm impact management
6.	Safety and first aid
7.	Public access

With the detailed site information in hand, the sludge application rate may be defined. Typically the controlling factor in application rate is nitrogen management. Metro uses a philosophy of balancing nitrogen applied to the nitrogen uptake of the site while eliminating nitrogen losses into the groundwater. In most cases in western Washington State, for the Douglas fir forests, this approach results in application rates in the order of 22 to 45 tds/ha.

7. REGULATIONS

Sludge management regulations in the US came into being in 1978 with the adoption of the federal law, Resource Conservation and Recovery Act (RCRA). A part of this bill was the Code of Federal Regulations (CFR) Part 257 guiding sludge management. The sludge management portion of these laws focussed only on the land application alternative, leaving sludge incineration, ocean disposal and distribution of compost or dried products unattended. For the land application portion the law sets broad management practices as the standard and limited single and lifetime applications of cadmium to agricultural sites.

On a local level, in 1981 the state of Washington established minimum functional standards for sludge management and published sludge best management practices guidelines. At the same time the state clarified that local, county-level public health departments should issue sludge management permits.

Most recently, EPA has proposed an extensive revision to sludge management regulations. These regulations are expected to be finalised in 1991 or 1992 and they will provide a comprehensive administrative and technical framework for all sludge management. These actions will result in the elimination of ocean sludge disposal during the 1990s. The administrative portion will allow states to certify their adequacy in managing sludge programmes. On the technical side EPA will set numerical standards for a variety of criteria for nutrients, microbes, trace metals and trace organics. To date there has been substantial controversy over draft numerical criteria which were developed using risk assessment models. For land application alternatives the concern seems to focus on the levels of copper in sludge.

8. NEW TECHNOLOGY

Looking to the future, Metro is not standing still in its search for cost-effective sludge management. Pilot tests of a sludge drying system will commence shortly while more effective direct forest applications techniques continue to be developed.

A sludge drying system pilot test at 0.9 to 2.7 tds/day will be performed in 1989. A privatisation contract was recently awarded to a Pacific Northwest firm to install, operate and then dismantle a pilot-scale jacketed indirect steam dryer. Sludge will be dried to 90 to 95% dry solids, pelletised and marketed. The dry sludge may be used in agriculture or on forests in a fashion similar to urea fertilisation. The contractor may supplement the Metro sludge with additional nutrients to respond to market needs.

More effective direct forest application techniques are currently being worked on. A system for raising the application spray nozzle 10.4 m into the air has been nearly perfected. This device allows sludge fertilisation over a greater range of forest heights than previously possible. Tree heights as much as 15 m can now receive sludge.

Finally, Metro is experimenting with a different sludge delivery system which replaces the large heavy all-terrain application vehicles with a smaller, lighter hose reel system. Hose reel systems are common in agricultural irrigation but face unique challenges in forest fertilisation. Continuous hose systems work best on straight, flat sites with minimal obstructions. Clearly forests do not offer such ideal site conditions, although it is believed that it will still be cost-effective to amend site designs to allow for such constraints. On the positive side, these reel systems can weigh much less than the current application vehicles thereby allowing utilisation of more site and soil types over a broader range of climatological seasons than possible at present. Also it is expected that application productivity will increase noticeably. For the current application system, over a 10 hour work day the vehicles are operated for about 8.5 hours. Of that time about one third or 2.8 hours is actual sludge fertiliser spraying while the remaining time is spent in transit to and from the transfer tanker or loading at the transfer tanker. With a hose reel system, the continuous hose provides a direct line to the transfer tanker, thus eliminating the transit time from the application cycle. With this system it is expected that the actual sludge fertiliser spraying time will be doubled to more than 5 hours per day.

REFERENCES

(1) Brockway, D. G. and P. V. Nguyen (1986). Municipal sludge application in forests of northern Michigan: a case study. In: The forest alternative for treatment and utilisation of municipal and industrial wastes (D. W. Cole, C. L. Henry and W. L. Nutter, Eds.), pp. 477-496. University Washington Press, Seattle.

(2) Hart, J. B. et al. (1988). Silvicultural use of wastewater sludge. J. of Forestry, 17-24.

(3) Berry, C. R. (1986). Reclamation of severely devastated sites with dried sewage sludge in the southeast. In: The forest alternative for treatment and utilisation of municipal and industrial wastes (D. W. Cole, C. L. Henry and W. L. Nutter, Eds.), pp. 497-507. University Washington Press, Seattle.

(4) Munger, S. (1986). Forest land application of municipal sludge: the risk assessment process. In: The forest alternative for treatment and utilisation of municipal and industrial wastes (D. W. Cole, C. L. Henry and W. L. Nutter, Eds.), pp. 117-124. University Washington Press, Seattle.

(5) USEPA (1983). Process design manual for land application of municipal sludge, EPA - 625/1-83-016.

(6) Washington State Department of Ecology (1982). Best management practices for use of municipal sewage sludge, WDOE, 82-12.

(7) USEPA (1989). Proposed draft sludge management technical reglations, land application and distribution and marketing of sewage sludge.

(8) Henry, C. L. and D. W. Cole (1986). Pack forest sludge demonstration program: history and current activities. In: The forest alternative for treatment and utilization of municipal and industrial wastes (D. W. Cole, C. L. Henry and W. L. Nutter, Eds.), pp. 461-471. University Washington Press, Seattle.

(9) Henry, C. L., C. G. Nichols and T. J. Chang (1986). Technology and costs of forest sludge applications. In: The forest alternative for treatment and utilization of municipal and industrial wastes (D. W. Cole, C. L. Henry and W. L. Nutter, Eds.), pp. 356-366, University Washington Press, Seattle.

Utilisation of dehydrated sludge from Marseille's purification station in forestry

G. LAVERGNE

Société du Métro de Marseille, 44 Ave Alexandra Dumas, 13008 Marseille, France.

SUMMARY

The area around Marseille is at risk of forest fires, and from the need to reforest affected areas the city of Marseille embarked on a project to utilise the sludge from the new sewage treatment works in reforestation. The most appropriate sludge treatment was found to be digestion followed by thermal drying. A series of lysimeter, container and field trials were carried out from 1977 to assess effects on different tree species, leachate quality and heavy metal aspects as well as the practical aspects of how to cultivate and spread sludge on the bare limestone. Agreements were drawn up with the land owners and in June 1989 approval was given for the first phase of 100 ha to be treated.

1. INTRODUCTION

The technical equipment selected for the treatment of sewage sludge is the result of a prolonged effort of study and experimentation, undertaken in order to define the most appropriate treatment process and guided by the disposal methods already selected. This decision sequence is explained below in detail, from the first studies to the implementation of the utilisation of dewatered sludge in forestry.

Every day, the Marseille treatment plant receives 3,000 cubic metres of sewage which has to be treated. Among the different methods of sludge disposal, the city of Marseille has opted for its utilisation in agriculture or forestry.

This decision originates from the sudden awareness, in 1975, of the problems raised by the destruction of Provençal forest by fire, particularly with regard to prevention, fire fighting, restoration of soils, and reforestation. The project described here deals with the two last objectives (restoration of soils and reforestation), and has fortunately gained continuous political support.

Although utilisation of dewatered sludge in agriculture is common (20% of the sludge produced in France, 40% in West Germany, 75% in the Netherlands), the use in forestry is practically unknown, with only a few recent examples in Canada and in the United States.

Nevertheless, this option appears highly advantageous; firstly by using a substance rich in organic matter and nutrients (especially phosphorus), and secondly by reforesting under optimal conditions a part of the bare limestone massifs surrounding the city of Marseille.

Several research studies have been undertaken, firstly to prove the feasibility of this method of utilisation and to select the most appropriate treatment process, and secondly to start in one of the areas concerned, the Plateau de Carpiagne, preliminary discussions prior to contract negotiation and the corresponding construction work.

The details of this research have been published widely (see reference list) and this paper deals only with the most important and significant results.

2. TRIALS IN LYSIMETERS AND CONTAINERS (1977)

Undertaken by the Société du Canal de Provence (SCP), the objective of this research was to study the behaviour of plants according to the different types and quantities of sludge, and to analyse, over several years, the quality of the leachates. The main point of these experiments was the analysis of the quantity and quality of the leachates:
a) More than half the volume of the precipitation is retained by the soil and does not leach out;
b) Contamination by faecal microorganisms is low and completely undetectable one year after spreading;
c) The greatest form of pollution is essentially by nitrates. During the first two years, their level was higher than the level permitted for drinking water (50 mg/l); subsequently, this level diminished rapidly;
d) The concentrations of heavy metals detected in the leachates were low. They were always markedly lower than the maximum concentrations permitted for drinking water.

These first conclusions appear to be very encouraging, even though the amount of dewatered sludge applied was much greater than would be used in agriculture (300 tds/ha compared to 5-10 tds/ha).

3. TRIALS ON THE PLATEAU DE LA MURE (1979)

The objectives of these trials, also conducted by the SCP, were to test the techniques for site preparation and for spreading the sludge, and also to study the behaviour *in situ* of different tree species. The main goal of these experiments was the identification of suitable sludge treatment methods for this type of use, and also to study the behaviour of soil.

It was found that there was a significant increase in the total content of organic matter (2-3 times) and nitrogen (3 times), which decreased later through mineralisation and nitrification. There was also a significant increase in phosphorus assimilable by plants (5 times), which represents a considerable improvement in these types of soil, which usually lack this element.

As far as the heavy metals are concerned, the limestone immobilises them by making them practically insoluble and thus they cannot be assimilated by plants. Moreover, the increase in their concentration is only appreciable in the soil layer (15 cm) in which the sludge has been spread and incorporated. The levels reached were not high.

From this second experiment, it appears that the dewatered sludge always produced better results than lime-treated sludge. The effect of the sludge diminished in the third year, but it helped in the rapid establishment of the vegetation, which is an important factor. The spreading of sludge had a positive effect on most soil characteristics (lowering of the pH, and increases in total carbon and nitrogen contents and in the content of assimilable phosphorus). The conifer trees tested tolerated the treatment and reacted more positively than the deciduous species tested.

These results were less encouraging than the previous ones. Firstly the lime treatment of raw sludge is known to be economical, and secondly conifers are generally more susceptible to fire than deciduous species. Nevertheless the experiments were continued as planned in the initial programme.

4. TRIALS ON THE PLATEAU DE CARPIAGNE (1981)

After the evaluation of the preceding results, the process of thermal treatment of the sludge was selected (thickening of raw sludge, digestion, thermal treatment and dewatering). It offers a number of advantages. Due to the coagulation of the suspended colloidal matter, the conditioned sludge can undergo a subsequent, more intensive dewatering such that the product obtained is drier, has less volume and is therefore cheaper to transport and spread. There is no addition of chemical reagents, since lime or ferric chloride can compromise or reduce the agronomic quality of the product. Total sterilisation is possible and its use does not require any particular hygiene precautions. During treatment the organic matter is so intensively stabilised that the final product does not release any nitrogen into the soil for several months. The important problem of the leaching of nitrates is therefore solved. The release of nitrogen is low and slow. After dehydration the sludge thus obtained has an excellent physical appearance (black, dry, granular and odourless), which makes a favourable impression on the users.

The objective of the experiments conducted on 2.5 hectares of the Plateau de Carpiagne was a full-scale test, over a long period of time, of the most suitable sludge in relation to all the significant parameters concerning its utilisation in forestry.

The site was prepared by crossed subsoil cultivation every metre at a depth of 0.8 m in order to break up the limestone blocks near the surface. The largest blocks were gathered and put into rows by a Fleco raker. A rock crusher was used for preparing the top layer of soil in order to facilitate the incorporation of the sludge (by eliminating stones) and vehicle movement (transport and spreading of sludge, plantation maintenance).

For transporting and spreading sludge, articulated vehicles were able to drive and unload directly onto the plots; the spreading was done by a loader and finished by an agricultural tractor pulling a board. Application rates were from 0 to 400 tds/ha. Ploughing in of the sludge was carried out by a disc harrow in one or several passes, depending on the thickness of the sludge layer to be incorporated.

Planting was done several months after the preparation of the soil because of the behaviour of the sludge, which temporarily immobilises the nitrogen in the soil. The distance between the planting holes (of about 100 litres, scooped out by mechanical shovel) was 2 m along the row and 3 m between the rows. Dead plants were replaced after the first six months.

The results added to those of the previous studies. The moisture retention capacity of the soil increased from 500 m^3/ha to 1,000 m^3/ha for the soil treated with 400 tds/ha. The Alep pine was not adversely affected by the large applications of sludge as far as the establishment of the trees was concerned. On the other hand, the survival rate for a species like Robinia (False Acacia) diminished from 53% to 38% in soils treated with 400 tds/ha. After four growing seasons the sludge applied had a favourable effect on the growth rate of the pine tree and Robinia, but no effect on Cypress. The Alep pine trees, which received large amounts of sludge, seemed visually to be more sturdy and vigorous than those of the control plots.

The choice of the thermal treatment process after digestion was confirmed in 1985 and the construction of the treatment plant began immediately (an investment of about 250 million francs).

5. THE QUALITY OF THE "COMPOST"

The main problem following these initial studies was that the product "compost" was not as yet in production (here "compost" means dewatered thermal treated sludge and is only a "commercial" name to help to sell the project and the product) and a trial could not produce a reasonable quantity. It was only possible to indicate the amounts of nutrients in the sludge of a similar nature produced by the treatment works of Achères (1984).

Carbon	255	g/kg ds
N total	14	g/kg ds
P_2O_5 total	79	g/kg ds
Nitrate	24	g/kg ds
K_2O	1.6	g/kg ds
MgO	13	g/kg ds

The physico-chemical treatment method used in Marseille for the sewage should increase the total phosphorus content. However, because this treatment method has the advantage of efficient heavy metal removal from wastewater, it increases their concentrations in the compost. In France the Standard U 44.041 gives maximum limit values for heavy metal concentrations in sludge when these are used in agriculture. For sludge utilisation in forestry Standard U 44.041 is not compulsory, but has been accepted in the negotiations with future users and for the public consultation.

The analytical results obtained from the compost now produced in Marseille are as follows (6 samples, June 1989):

	Mean value (mg/kg)	Standard U 44.041 limit values
Zn	1,100	3,000
Cu	510	1,000
Pb	300	800
Cr	170	1,000
Ni	30	200
Cd	14	20
Hg	4	10
Se	3	100

6. STUDIES OF THE SITE TO BE REFORESTED

6.1 Impact on ground water

The previous studies had already shown that the problems of pollution caused by leachates were linked to the concentration of nitrates. In our opinion, the choice of the process of thermal treatment definitely solves this problem at its source. Nevertheless, the works supervisor wanted to obtain as much information as possible on the hydrogeology of the limestone massif area to be reforested. These studies on the Plateau de Carpiagne were entrusted to the BRGM (Bureau de Recherches Géologiques et Minières). They had several aims:
a) To obtain more information on the hydrogeology and the hydrodynamics of the massif;
b) To establish the existing quality of ground water;
c) To define the subsequent monitoring programme to be set up.

In the case of the massif de Carpiagne, the problem is the absence of any accessible points to the ground water. It was therefore necessary to drill very deep boreholes and it was found that practically everywhere the ground water level was close to sea level. Such studies are costly and in the case of Carpiagne they unfortunately showed that it would not be possible to use a potential source of fresh water below the massif, which was a major factor when considering this site as the place to begin the first large-scale operations.

6.2 Towards large-scale reforestation

In 1987, the city of Marseille entered into the final stage of these research studies, for all fundamental and technical problems had been solved. This final stage of studies will not be described here; it consists of analyses, observations, and measures and proposals relating to the site. These were carried out by the Office National des Forêts, CEMAGREF and the Faculté des Sciences et Techniques de St Jerome.

The different aspects studied were topography, climate, land use and ownership, vegetation, pedology and landscape.

These studies have enabled public dialogue to begin before commencing with the operations. Present French law does not seem to regulate such operations, but the city of Marseille conducted an environmental impact assessment and has given this wide publicity by making the document publicly available and distributing it to local councillors and administrative offices. This public consultation was concluded on 20 December 1988 with the go-ahead from the Departmental Commission of Sites and Landscapes meeting in a group called "For the Protection of Nature".

6.3 Negotiations with the land owners

In general the city of Marseille does not own the land affected by reforestation. This land has been of no interest for agriculture and forestry for a long time. The parcels of land are generally in common ownership or owned by the State, Department or other institutions.

For each parcel of land, an agreement was drawn up between the owner and the city of Marseille. This agreement defines the nature of the operation, the methods of application, the resources available for controlling the quality of work, the responsibilities of each party and the limits to the intervention by the city of Marseille during the maintenance period.

The main terms of these agreements are the following:
a) Analysis for nutrients in the compost every 6000 tds during the operation;
b) Analysis for the heavy metals in Standard U 44.041 in the compost every 3000 tds during the operation;
c) Stopping the spreading of compost if the concentration of one heavy metal exceeds the limit fixed by this standard;
d) The owner must maintain the site as woodland for a period of five years after reforestation;
e) The quality of the soils must be monitored by analysis before treatment and after ten years. The analysis consists of the statistical treatment on values found in twenty five samples taken in a random manner over each 10 ha reforested;
f) Forestry quality control of an area of 0.5 ha for each 10 ha treated, and annual follow up during the five year maintenance period.

7. ESTABLISHING THE REFORESTATION OPERATION

Considering all the limitations, the land likely to be selected for reforestation is about 240 ha on the Plateau de Carpiagne. The planned arrangements comprise:
a) Progressive construction of tracks for access, or the improvement of existing ones;
b) The provision of temporary storage areas for the "compost" near to the land to be treated;
c) Arrangements for fire protection;
d) The preparation of the soil in the following sequence: crossed subsoil cultivation every metre at a depth of a least 0.70 m followed by a bull-

dozer equipped with a Fleco rake, which gathers the largest blocks in a row, then a cultivator of the "chisel" type followed by a heavy rock crusher, then spreading of the compost before ploughing in, for which a particularly heavy and robust plough is necessary. For satisfactory mechanical preparation of the soil, especially the rock crushing, it is necessary to do this in the dry weather period, preferably from May to September. On slopes over about 12°, to limit possible erosion or movement of the compost, it is preferable to cultivate the soil along the contour lines when the largest rocks are gathered into windrows, the optimum distance between rows being 75 m;

e) The planting of an appropriate mixture of species and the sowing of grass seed;

f) Maintenance of the reforested areas for several years.

In June 1989, Marseille city council approved the contract for the first phase of 100 ha to be treated, corresponding to five years of sludge production. The contract is composed of two parts: one for the preparation of the land with heavy equipment and one for planting and maintenance of the trees.

The tenders for contract are now invited. Total investment, including preparation of soil, transport of compost, spreading and planting of trees, is estimated at about 100,000 F/ha, all taxes included. It corresponds to 200 F/tds since the planned application rate of compost for reforestation averages 500 tds/ha.

8. CONCLUSION

The utilisation of sludge described above is, of course, more expensive than simple dumping of the dewatered sludge, since local transport is cheap and the cost of disposal in a municipal landfill is negligible. However, of the total cost of the operation, the largest part is due to reforestation, and the cost of the incorporation of the sludge compost as an "extra" is marginal.

The works supervisor, according to land availability and existing financial means, can spread the reforestation scheme over a longer time period, with the aim of gradually reforesting a vast natural landscape where certain parts could become areas for relaxation and leisure, like the Vallon de Chalabran.

Marseille is not the only city in the Department which could contribute to reforestation. It is possible that the demonstration made by Marseille will create a demand for "compost", whose incorporation, if done correctly, is not costly and greatly improves the characteristics of the soil.

REFERENCES

(1) Study of the Agence d'Urbanisme of Marseille (1979). First selection of sites suitable for sludge application.
(2) Société du Canal de Provence (1983). Experimental study of the utilisation of sludge from treatment works in forestry. Tests in lysimeters. Summary report.
(3) Société du Canal de Provence (1982). Experimental study of the utilisation of sludge from treatment works in forestry. Tests on the Plateau de la Mure.
(4) Compagnie Nationale d'Aménagement de la Région du Bas Rhône et du Languedoc (1986). Economic study of the utilisation in agriculture of thermically treated sludge.
(5) Société du Canal de Provence. Three reports (1984, 1985, 1986). Analytic follow-up study of the experimental plots of Carpiagne.
(6) CEMAGREF group Aix en Provence. Experimental reforestation of Carpiagne. Installation and observance reports.
(7) Bureau de Recherches Géologiques et Minières (1988). Hydrogeological study of the Plateau de Carpiagne.
(8) Office National des Forêts, CEMAGREF, Faculté de St Jérome, Clarac (landscape engineer DPLG) (1988). Preliminary study for reforestation of the Plateau de Carpiagne.
(9) Laboratoire de la SERAM (1988). Reports of analyses of micropollutants in the sludge of the treatment works of Marseille.
(10) Reforestation of Carpiagnes Plateau (1988). Environmental impact study of the treatment works of Marseille. Complementary document No. 3.

Long-term effects of sewage sludge application in a conifer plantation on a sandy soil

S. E. OLESEN and H. S. MARK

Hedeselskabet, Danish Land Development Service,
PO Box 110, 8800 Viborg, Denmark.

SUMMARY

In 1973-74, the Danish Land Development Service (Hedeselskabet) initiated an experiment with the surface application of 45 tds/ha of anaerobic digested sewage sludge to a sandy soil in a 75 year old spruce plantation. After Cu fertilisation in 1976, the application of sludge increased tree growth for approximately four and a half years to levels comparable to full forest fertilisation. Three and a half years after sludge application, approximately 65% of the added N remained in the sludge layer and 20% had been immobilised in the raw humus layer. Approximately 50% of the P content was removed from the sludge layer and was found almost entirely in the top soil. Heavy metals, with the exception of Zn and Pb, were found almost entirely in the sludge and raw humus layers.

After one and a half to two years, leaching of NO_3-N increased concentrations in the ground water to an average level of 10 mg NO_3/l, decreasing to initial levels after 7 years. There was no significant leaching of P to the ground water. Full-scale sludge application of 25 tds/ha on neighbouring areas showed similar trends. Before reforestation in 1988, the experimental areas were logged, leaving clearcut and shelterwood plots. Clearcutting caused an increase in NO_3-N in the ground water samples taken from the sludge-applied plots. The concentrations are still increasing. The increase of NO_3-N on shelterwood plots occurred approximately 6 months later and was less significant. With the exception of Zn, mobility of heavy metals has not been affected significantly. Investigations of the plots are to continue until 1993.

1. INTRODUCTION

Depending on the type and decomposition process, sewage sludge contains large amounts of available plant nutrients, mainly nitrogen (N) and phosphorus (P). As well as plant nutrients, sludge also contains heavy metals and a characteristic load of pathogens. There have been objections to the application of sludge to farmland because of the potential risk of accumulation of harmful heavy metals in both the soil and plants. The creation of pathogenic reservoirs and pathways of re-entry into the disease cycle also limits the use of sewage sludge on crops. The use of sewage sludge is less objectionable in forests than agriculture as only small amounts of forest vegetation become directly incorporated into the human food chain.

Investigations have shown that plant nutrients from sewage sludge, especially nitrogen and phosphorus, increase the growth of forest vegetation (1 and 2). Current work in the United Kingdom shows a general trend toward enhanced foliar nutrient levels and improved growth rates or both (3) for sludge-treated forests.

Because incorporation of sludge into forest soils is physically difficult, large scale applications onto the soil surface are more practical, although large amounts of sludge may cause increased leaching of nitrates and heavy metals to the ground water.

From 1973 to 1989, the long-term effects of sludge applications on a sandy podzolic soil in a spruce plantation (*Picea abies*), planted in 1903, have been investigated by the Danish Land Development Service. Results of the investigations until 1983 have been published in (4) and (5). The recent investigations are concentrated on leaching of nutrients and heavy metals to the ground water after reforestation in 1988.

The major objective of the investigations has been to determine the long-term effects of sludge applications on:
a) The growth of the vegetation;
b) The leaching of nutrients and heavy metals to the ground water;
c) The survival rates for pathogens and viruses.

The investigations should be used to establish guidelines for using municipal sewage sludge as an organic fertiliser in spruce plantations on sandy soils. Normally, the determining factor for tree growth is the amount of available nitrogen.

2. EXPERIMENTATION

2.1 Experimental design, sludge application

Sludge was applied to a 12,000 m² forested area consisting of two control plots, four plots receiving non-industrially loaded sludge (I) and four plots receiving industrially contaminated sludge (II). Each plot measured 1,200 m². The experimental site is Hesselvig Plantation situated near Herning in Jutland.

In 1973-74, 800 m³ anaerobic digested wet sludge per hectare were applied to eight of the plots. This corresponded to approximately 45 tds, 1,300 kg N and 690 kg P per hectare, together with smaller amounts of other plant nutrients. With the exception of Pb, the industrially contaminated sludge contained the largest amounts of heavy metals, especially Cr and Zn (Table 1). Four of the plots were supplemented with soft limestone (20 t/ha).

Since 1973, the effects of sludge application on tree growth and leaching of N, P, and heavy metals to the ground water have been studied.

Table 1. Chemical composition and calculated application rates of elements in sewage sludge applied in Hesselvig Plantation (4).

Determinand	Sludge I		Sludge II	
	Composition	Addition (kg/ha)	Composition	Addition (kg/ha)
pH	6.1	-	5.9	-
Dry matter (%)	6.4	50,800	5.3	42,400
Loss on ignition (%)	54.2	27,500	62.0	26,300
Organic C (%)	25.5	12,900	33.7	14,300
Total N (mg/kg)	23,000	1,170	33,000	1,400
Water soluble N (mg/kg)	1,800	91	1,900	80
Total P (mg/kg)	13,100	665	16,600	700
Inorganic P (mg/kg)	10,800	550	11,400	480
Total K (mg/kg)	1,500	76	2,400	102
Total Mg (mg/kg)	2,000	102	2,100	89
Total Ca (mg/kg)	16,000	810	25,200	1,070
Pb (mg/kg)	270	13.7	250	10.6
Cd (mg/kg)	6.4	0.3	12	0.5
Cr (mg/kg)	28	1.4	1,400	59
Ni (mg/kg)	34	1.7	370	15.7
Cu (mg/kg)	156	7.9	320	13.6
Zn (mg/kg)	1,320	67	9,600	407
Co (mg/kg)	23	1.2	25	1.1

In the spring of 1988, five plots with spruce trees were clearcut and replanted in ploughed furrows with Norway spruce (*Picea abies*). The other five plots were partially cleared while trees from the former stand remained as shelterwood. The degree of leaching of N and P from the plots to the ground water as an effect of clearcutting and shelterwood cutting was investigated. The mobility of other plant nutrients and heavy metals was also of interest.

During the period 1982-1986, a 200 hectare stand close to the experimental plots received a full-scale application of type II (industrially contaminated) sewage sludge. The application amounted to 25 tds/ha. To support the previous analyses of the experimental plots, the viral and bacterial survival rates in both sludge and ground water were analysed by the Public Municipal Laboratory, Herning, and the Department of Veterinary Virology and Immunology at the Royal Veterinary and Agricultural University of Copenhagen.

2.2 Soil and ground water

Sludge was applied to a sandy podzolic soil. Soil horizons and some physical and chemical properties are shown in Table 2.

Table 2. Soil horizons and some physical and chemical properties of the raw humus and the mineral soil (4).

Soil horizon	Depth (cm)	Bulk density (g/cm^3)	Org. matter (%)	Sand 0.2-2mm (%)	Total N	Total P (mg/kg)	Inorg. P	pH[(1)]
Raw humus, O	0-11	0.15	80		10,700	533	91	3.0
Leached layer, E	11-18	-	-	-	-	-	-	-
Aluminium/ iron layer, Bs	18-35	1.52	3.0	75.4	930	110	10	3.3
Subsoil, C	50	1.52	0.5	89.6	110	80	18	4.2
	200	1.52	0.1	78.6	30	35	18	4.6
	400	1.52	0	90.8	15	30	12	5.1

[(1)] pH in 0.01M CaCl$_2$.

The raw humus layer has a low rate of humification. It is nutrient deficient and has a low pH (3.0). The nutrients are present mainly as organic compounds, and are hence slowly available. The C:N:P ratio is approximately 600:31:1. This indicates that the humus should have a high capacity to immobilise N and P applied with the sludge, although mineralisation together with the immobilisation of N and P in the rather nutrient deficient raw humus is expected to be very slow. The mineral soil has a low pH and is extremely nutrient deficient.

The ground water level is at a depth of 2-3 m below the soil surface, sloping 0.09% to a nearby stream. The horizontal flow velocity is approximately 20 m per year.

2.3 Sampling and analytical methods

Tree growth: Foliar samples from the upper whorls of nine trees from each plot were analysed for the macro plant nutrients N, P, K, Ca and Mg. Contents of Cu, Ni, Cr and Zn were also measured. Analytical methods are described in (4) and (5). Tree growth was evaluated from measurements of the basal area at a height of 1.3 m.

Sludge and soil: Samples of the sludge layer were taken before sludge application and again 1.5, 3.5 and 4.5 years later. Samples of raw humus and subsoil were taken for chemical analysis before sludge application and again 3.5 years later. Dry matter content and loss of ignition were determined by heating to 100°C and 600°C, respectively. The samples were analysed for pH (0.01M $CaCl_2$), inorganic P and water soluble N according to (6). Total N and P were measured as described for analysis of foliar samples. K, Ca, Mg, Pb, Cd, Cu, Cr, Ni and Zn were analysed as described in (4) and (5). Soil water was sampled according to methods described by Linnér (7).

Ground water: Samples of the ground water were collected via PVC pipes, one on each plot, extending from the soil surface to approximately 1 m below the ground water level. The lower 50 cm of the pipes were perforated, and they were sealed at the bottom. Ground water was pumped up by means of a narrow plastic tube connected to a vacuum pump. For purposes of statistical analysis, the number of PVC pipes for sampling was increased in 1988 to five 90 mm PVC pipes on each plot. A newly constructed system made it possible to sample the upper 10-20 cm of the ground water influx (the most recent influx of ground water to the aquifer). The water samples were analysed for NO_3, NH_4, Pb, Cd, Cu, Cr, Ni and Zn according to (8).

Pathogens: Samples of the sludge and upper humus layer were examined for salmonella and helminth ova every second month for the first 17 months after sludge application (9). The upper layer of sludge and humus of the stand receiving a full-scale sludge application during the period 1982-86 was examined for decay of pathogenic bacteria and viruses. Decay of viruses was examined according to (10) and (11) and described in (12). Ground water was examined for pathogenic bacteria.

3. RESULTS AND DISCUSSION

3.1 Foliar nutrient levels and tree growth

The N and P content of the spruce needles was improved by the application of sludge. The N concentration as a percentage of dry matter increased from a minimum level for tree growth of 1.5% dry matter to an average optimum level of 2.1% dry matter by 1975. During the following five years, the N concentrations slowly decreased again to a minimum level. The trend can be seen in Figure 1. There was no noticeable increase in the concentration of K, Ca, and Mg in the foliar samples from the sludge-treated plots. On the contrary, the concentration of K was lower than for the control plots. Following K fertilisation in 1982, concentrations of K in foliar samples from control and sludge-applied plots were at the same level. The concentration of the heavy metals Zn, Ni and Cr increased due to sludge application but did not exceed those normally found.

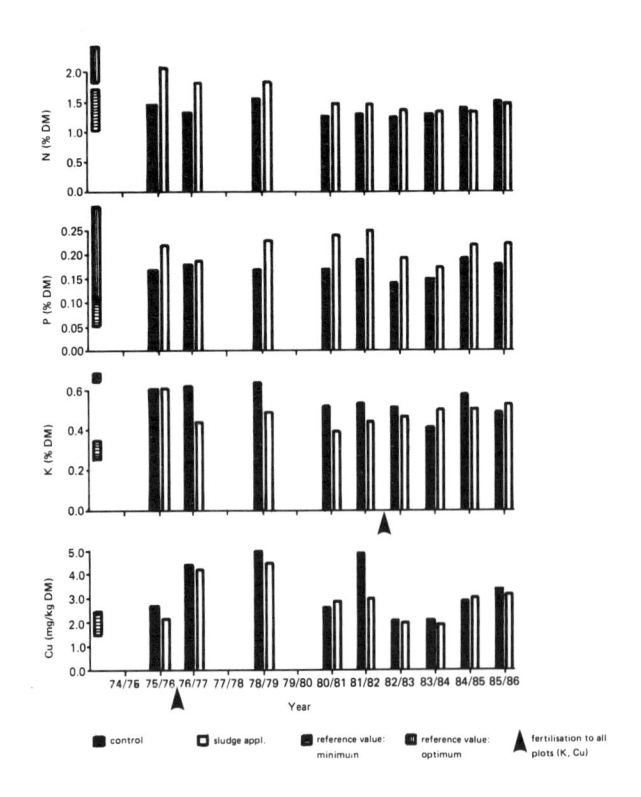

Fig. 1. Foliar analysis. Concentration of N, P, K and Cu in foliar samples from 1975-1986 (averages for sludge I and II). For reference values see (4) and (13).

In 1974-75, the application of sludge had a negative effect on growth despite increased foliar concentrations of N and P. The relative basal area increment at a height of 1.3 m (Figure 2) decreased due to a Cu deficiency caused by excess concentrations of N particularly in the industrially contaminated sludge (II). Shoot die-back occurred within these plots. Earlier trials showed that N, P, and K fertilisers had a positive effect on tree growth only when the Cu concentration exceeded 2.5 mg/kg DM (14). Copper was applied aerially to all plots in 1975.

By 1976-78, tree growth had increased. The average basal area increased by 30-70% relative to the control plots, similar to the effects of a full forest fertilisation (15). After 1978, relative basal area increments decreased to a level slightly above the control (100). The foliar analyses in 1981 showed a deficiency in K. The application of K in 1982 had no positive effect on the relative basal area increment. By 1981, the Cu concentration had decreased to a level slightly above the 1975 level. Although the N content in both control and sludge-applied plots had decreased to a minimum level, sludge still had an obvious effect on the foliar N and P content. After 1981, it is likely that a Cu or N deficiency, or both, have again been responsible for the lack of a fertilising effect.

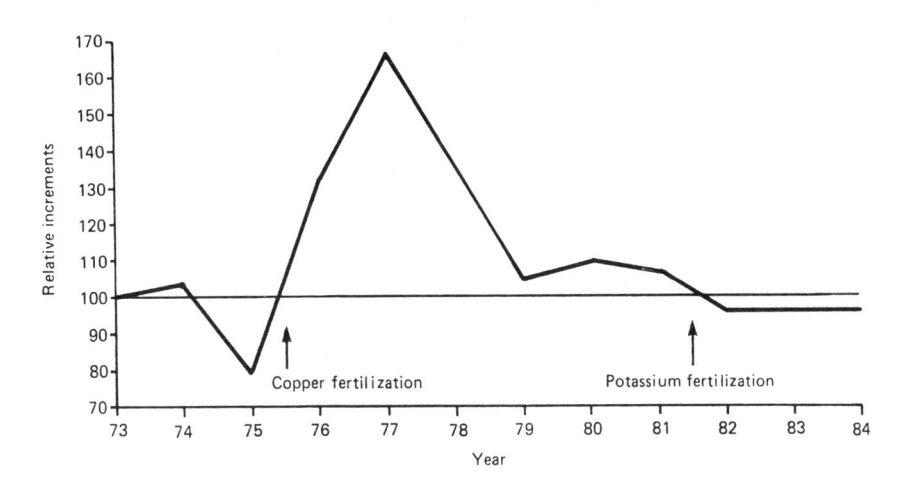

Fig. 2. The basal area increment for spruce on sludge-treated plots (average for sludges I and II) relative to the basal area increment for spruce on control plots.

3.2 Nutrient application and balance

Approximately 1,300 kg total N/ha (1,200 kg in the form of organic N) had been added by the sludge, an increase of approximately 32% of the initial content in the top soil (0-25 cm). After a 3.5 year period, 850 kg total N/ha (65%) were still present in the sludge layer, whereas 450 kg total N/ha (35%) had been removed from the sludge. Approximately 250 kg total N/ha (20%) had become fixed in the raw humus and mineral soil, 100 kg total N/ha had been leached to the ground water, and approximately 100 kg total N/ha had been removed by the vegetation. Nitrification and denitrification rates were not examined, and the magnitude of the effect of N released to the atmosphere by denitrification in relation to the amount of leaching to the ground water is unknown, but on a sandy soil denitrification above the ground water level is considered to be very small.

Approximately 690 kg total P/ha were applied with the sludge. Inorganic P amounted to 520 kg/ha. After 3.5 years, 315 kg total P/ha had been released from the sludge layer and were found almost entirely in the top soil; 140 kg total P/ha were in the raw humus layer and 160 kg total P/ha in the upper 20 cm of the mineral soil. Table 3 shows the nutrient balance for N and P over a 3.5 year period.

Table 3. N and P balance 3.5 years after application of sewage sludge (average of sludges I and II) (16).

	Total N (kg/ha)	Total P (kg/ha)
Initial application	1,300	690
Recovery after 3.5 years	450	315
- in raw humus	225	140 (50% inorganic)
- in mineral soil (13-35 cm)	25	160 (100% inorganic)
- removed by trees	100	14
- leached	100	1

Approximately 14 kg total P/ha were estimated to have been removed by the vegetation, and 1 kg total P/ha leached to the ground water. In the raw humus, approximately half of the supplemented total P had been immobilised in the organic fraction, whereas all P in the mineral soil was inorganic.

C:N:P ratios in the raw humus had decreased from 600:18:1 to 440:17:1. Microorganisms will continue to immobilise inorganic P until the C:P ratio is approximately 200:1 (17). After 3.5 years, there were still 330 kg inorganic P (50% of the total P application) in the sludge and raw humus layer, providing a satisfactory P supply for many years to come.

Similarly, the C:N ratio indicates that immobilisation of N will continue.

However, on a long-term basis, the N which has been immobilised in the raw humus layer will slowly become available. The long-term mineralisation of N and the content of inorganic P in the sludge and raw humus layer provide a long-term source of nutrients for the trees. The fertilising capacity of N in the sludge as a rate of mineralisation was estimated from the loss on ignition. Because the ash content remained fairly constant, it was assumed that loss in the bulk of the sludge was due to the decomposition of organic matter.

The mineralisation can be approximated as a first order rate (4). After 4.5 years, approximately 30% of the organic fraction (loss by ignition) of both types of sludge (I and II) had been mineralised. If the N release from sludge follows the mineralisation trend, it may be calculated that the sludge layer after 3.5 years will supply 30-60 kg N/ha per year. Nitrogen will still be fixed as organic N in the raw humus. After five years, the content of ammonium and nitrate in the sludge layer continued to show significant nitrification. Fixed N is slowly released during mineralisation. The analysis of foliar samples (see Figure 1) confirmed that a fertilising effect continued for approximately seven years.

3.3 Leaching of nitrate, ammonium and phosphate to the ground water

Soil water and ground water analyses showed that N had been leached from the sludge and upper horizons of the soil profile. Concentrations of NO_3-N and NH_4-N in the soil water at a depth of 50 cm increased rapidly six months after sludge application and reached maximum concentrations of 30 and 40 mg/l, respectively. After reaching a maximum, the downward trend was similar to that for mineralisation. Concentrations had decreased to near previous levels after a period of seven years.

Maximum concentrations of NO_3-N in the ground water were reached 1.5-2 years after sludge application, and a maximum level of approximately 10 mg NO_3-N/l (Figure 3) was maintained for 3-4 years, gradually decreasing to a level of 4-5 mg NO_3-N/l 6.5 years after sludge application (Figure 3). In the period from 1975 to 1978, the NO_3-N concentrations only occasionally exceeded 10 mg NO_3-N/l. Concentrations of NO_3-N and NH_4-N for the control plots were 0.3-3.2 mg NO_3-N/l and 0.01-0.3 mg NH_4-N/l. Maximum concentrations of 3-4 mg NH_4-N/l had been reached approximately one year after sludge application. This level decreased rapidly to initial levels of 0.02 mg NH_4-N/l after three years.

The mobility of P was limited. Following sludge application, PO_4-P concentrations in the ground water did not exceed 0.5 mg PO_4-P/l, the average being 0.06 mg PO_4-P/l. In general, most of the total P was found as inorganic P.

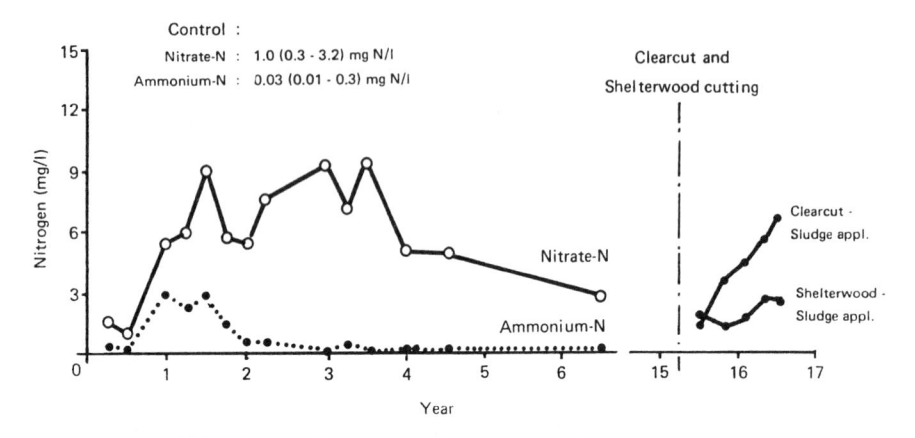

Fig. 3. Nitrate and ammonium concentrations in the ground water for the periods 1973-80 and June 1988 to July 1989 (average for sludges I and II).

Investigations of the experimental plots continued in June 1988. Ground water samples were regularly taken for analysis of N, P, and heavy metals. In October 1988, the number of samples per plot was increased for statistical analysis from one to five. The amount of data is still limited so that only mean data values are presented here with no distinction between sludge types. It should be noted that after June 1988 the samples contained only the upper 10-20 cm of ground water influx, while the earlier samples contained water from approximately 50 cm below the ground water level.

In the spring of 1988, five plots were clearcut and five were left with shelterwood. The effect of clearcutting and shelterwood on N leaching to the ground water is shown in more detail in Figure 4. Initial levels of NO_3-N in the ground water samples taken from clearcut and shelterwood plots were 1.3 and 1.6 mg NO_3-N/l, respectively. These levels are similar to initial concentrations in the ground water in 1973-74 before sludge applications. On both the clearcut and shelterwood plots, the ground water samples showed a significant increase in the NO_3-N concentration. The trend is most significant on the clearcut sludge applied plots where NO_3-N concentrations increased immediately and progressively from 1.3 mg NO_3- N/l in June 1988 to 6.5 mg NO_3-N/l in July 1989. The trend for the shelterwood plots was generally less significant. It seemed to be delayed by approximately six months relative to clearcut plots and had apparently already reached maximum concentrations (2.5 mg NO_3-N/l) by March 1989, stabilising from March to July 1989. Shelterwood control plots showed rather constant concentration levels. As expected, the N concentrations in the ground water samples taken from clearcut control plots increased, following the same trend as for clearcut sludge-treated plots, the only difference being in concentration levels. The trend ended abruptly in March 1989 due to reduced N leaching and seemed to reach levels similar to shelterwood sludge-treated plots.

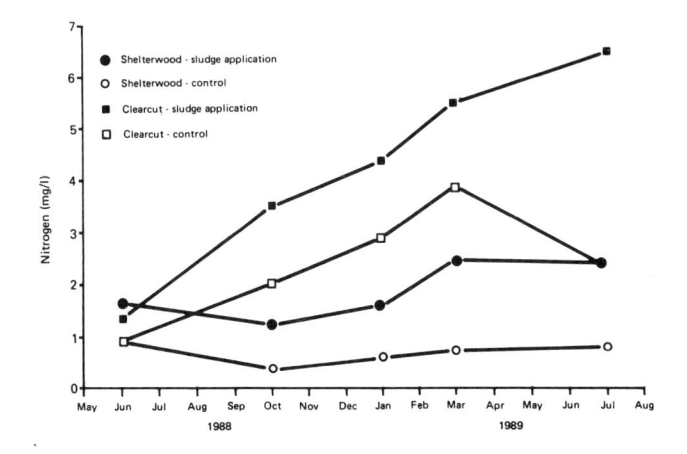

Fig. 4. NO_3-N concentrations in ground water after reforestation in spring 1988, for clearcut and shelterwood plots (average of sludges I and II).

By July 1989, NO_3-N concentrations of the ground water samples taken from clearcut sludge-treated plots continued to increase, so that it may be too early to make conclusions from the results. By July 1989, it seems that differences in N concentrations on the clearcut control plots in comparison to the sludge-applied shelterwood plots have been reduced remarkably. However, the NO_3-N concentrations in ground water for all plots are still relatively low. They are considerably lower than NO_3-N concentrations in soil solution (taken at a depth of 60 cm) one year after replanting with *Picea abies* on a similar area without sludge application and with the slash pushed together onto future trails (Loevenholm, Denmark) (18).

The greater loss of N to the ground water after clearcutting is expected to decline again to initial levels with the onset of weed growth (18, 19). To date there is no significant difference in leaching of NH_4-N from the different plots. Concentrations of 0.005-0.9 mg NH_4-N/l have been found. Initial phosphorus levels in June 1988 were 0.08 to 0.23 mg total P/l increasing to a maximum of 0.84 mg total P/l on clearcut sludge-treated plots. There has not been a significant trend as for NO_3-N, but the latest analyses in July 1989 showed higher average concentrations of total P in ground water samples taken from the clearcut sludge-treated plots than from shelterwood plots.

3.4 Heavy metals in soil and ground water

The increase in the amount of heavy metals present in the raw humus layer after the application of sludge varied from 2 (Pb) to 140 (Cr) times the amount present before application. Analyses of the concentrations of heavy metals in the sludge, raw humus and mineral soil were made before sludge application and 3.5 years after sludge application (Table 4).

After 3.5 years, Cd, Cr, Ni and Co were almost entirely recovered in the sludge and raw humus layer. However, Zn and Pb had a greater mobility, which is apparent in the concentrations at a depth of 50 cm. There was a significant increase in the Zn concentration in the ground water. One and a half years after the application of sludge, the Zn concentration was approximately 0.04 mg Zn/l for the control plots and approximately 0.5 mg Zn/l for sludge-applied plots. After 6.5 years, the concentration levels had decreased to <0.01 and 0.15 mg Zn/l, respectively.

The investigation of heavy metal concentrations in ground water for the clearcut and shelterwood plots after reforestation in 1988 showed that only Zn in larger amounts had been leached to the ground water.

The sludge application amounted to 67 and 407 kg Zn/ha for sludge I and sludge II respectively. As shown in Figure 5, the increase in Zn concentrations is most significant in ground water samples taken from plots applied with sludge II. The increase from these plots was immediate, and concentrations rose to an average maximum of 1.4 mg Zn/l in July 1989 and are still increasing. An increase in the plots applied with sludge I was delayed and less significant due to the smaller load, yet average concentrations did not exceed 0.21 mg Zn/l. Zinc concentrations in control plots did not show any significant trend.

With regard to the shelterwood and clearcut plots, leaching of Zn seems to follow a trend as for NO_3-N. Increases in Zn concentrations for shelterwood sludge-applied plots are, in general, less or at least delayed relative to clearcut plots.

Logging did not seem to have any substantial effect on the mobility of other heavy metals (Pb, Cd, Co, Cr and Ni). The concentration levels of Ni, Cr and Co on control plots were generally higher than on the same plots for the period 1973-78, but there was no significant trend toward an increase in the levels for clearcut and shelterwood plots receiving sludge. For example, the Cd concentration of approximately 1 µg Cd/l remained very constant during the investigation period for all plots.

Table 4. Content of heavy metals (Pb, Cd, Cr, Co, Ni and Zn) in sludge and soil before and 3.5 years after application of heavy metal enriched sludge (II) (4).

Metal	Horizon	Initial content (mg/kg)	(kg/ha)	After 3.5 years (mg/kg)	(kg/ha)
Pb	Sludge	250	10.6	120	4.1
	Raw humus, O	49	7.8	86	13.8
	50 cm, C	2.0		24	
Cd	Sludge	12	0.51	13.6	0.44
	Raw humus, O	0.8	0.13	1.2	0.19
	50cm, C	0.2		0.4	
Cr	Sludge	1,400	59	1,760	56
	Raw humus, O	2.7	0.43	48	7.7
	50 cm, C	2.4		2.4	
Zn	Sludge	9,600	407	3,920	125
	Raw humus, O	39	2.0	3,360	538
	50 cm, C	2.5		17	
Co	Sludge	370	15.7	150	4.8
	Raw humus, O	2.3	0.37	120	19.2
	50 cm, C	1.4		2.8	
Ni	Sludge	25	1.1	18	0.58
	Raw humus, O	1.1	1.18	6.0	0.96
	50 cm, C	1.4		4.0	

These results confirm results from other investigations. Much of the heavy metal input from sludge application is fixed by the soil, especially in the raw humus layer (20). As concluded by Bayes et al. (3), the risk of toxicity or pollution is limited for sludge application at rates appropriate to forest fertilising. However, with respect to sludge type, sludge applied too heavily can be detrimental (21). The effect on soil biology and hence decomposition and nutrient recycling requires further study.

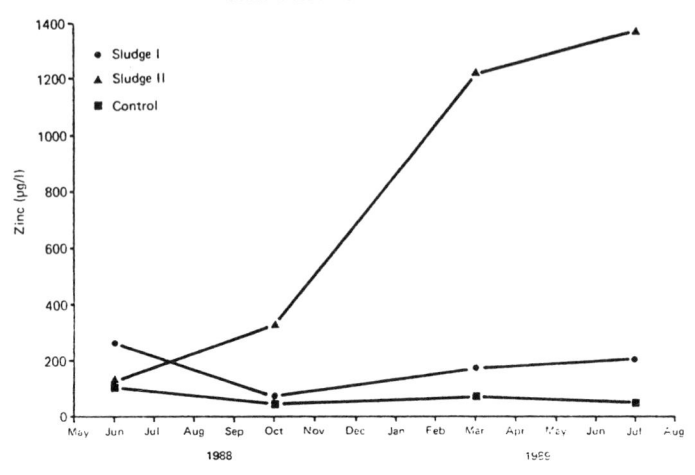

Fig. 5. Concentrations of Zn in the ground water samples taken from plots after reforestation in spring 1988 (averages of sludge II and sludge I, respectively).

3.5 Pathogens and viruses

Analysis of the industrially contaminated sludge revealed eight different types of salmonella and 900-1,700 helminth ova per 100 g wet sludge. The survival rates of these pathogens in the soil surface sludge layer are shown in Table 5.

Salmonella was present up to 2.5 months after sludge application. After 11 months, the helminth ova which had embryonated contained a living larva, but after 15 months only a few of these had survived. Transmission experiments involving the transfer of embryonated ova to piglets from a minimum disease unit have not as yet been successful.

On the neighbouring areas receiving a full-scale sludge application from 1982 to 1986, decay of pathogens and viruses have been studied. In general, survival of salmonella and *Clostridium perfringens* has been estimated to be 1-3 months. In a few tests, survival was 5-8 months, especially where the sludge had been applied in the autumn in a cold and humid microclimate. Bacteria were not found in the ground water samples where sludge had been applied. It has been concluded that sludge applications have neither immediate nor long-term effects on the pollution of the ground water by bacteria. Bayes et al. (3) have reported similar conclusions.

After sludge application, the viral content of one of the plots was investigated. From a series of eight surface sludge samples, viruses were identified in 41 of the 54 tests, amounting to 0.5-40 infectious units per g dry matter. Major units were poliovirus, coxsackievirus and adenovirus. Viral content is shown in Table 6.

Table 5. Salmonella and helminth ova survival on forest soils in months after application of sludge (9).

Months after application	Salmonella[1] (total samples/ positive)	Helminth ova per 10 g soil	Comments
1	8/8	169	Not embryonated, alive
2.5	8/2	366	
5	8/0	260	
7	8/0	128	
9	8/0	76	
11	8/0	163	Embryonated, alive
13	8/0	118	
15	8/0	89	Only a few alive
17		Transmission experiment[2]	No take

[1] *S. saint Paul, S. typhimurium, S. infantis, S. makumira, S. oranienburg, S. kiel, S. ohio, S. heidelberg.*

[2] Transmission experiment to 3 piglets from a minimum disease unit (150 helminth ova/ piglet).

Apparently when sludge had been applied in the autumn there was no significant decay in the number of viruses after approximately the first 14 weeks. After 22 weeks, viruses could not be identified in the sludge layer.

In general, tests proved negative for the ground water samples taken from the sludge-applied areas. One test identified two coxsackievirus B3 and one poliovirus.

4. CONCLUSIONS AND OUTLOOK

The application of 45 tds/ha of sewage sludge on plots of a spruce plantation on a sandy podzolic soil followed by Cu fertilisation increased tree growth by approximately 30-70%. This is similar to full forest fertilisation. After 4.5 years, approximately 30% of the organic fraction of the sludge had been mineralised. The mineralisation rate could be estimated as a first order process. Assuming that the N release from the sludge follows this mineralisation trend, the sludge layer would apply 30-60 kg N/ha per year. However, within the 3.5 year period, the C:N:P ratios in the raw humus layer had decreased from 600:18:1 to 440:17:1, indicating that the soil still

Table 6. Viral content after surface application of sewage sludge to the soil in Hesselvig Plantation[1] (12).

Weeks after sludge application	Infectious units/g DM						
	I	II	III	IV	V	VI	VII
1	0.9	1.0					
2			neg.	neg.	neg.		4.0
3						neg.	
4	40.0	neg.					
5			neg.	neg.	neg.		neg.
6						30.0	
7	neg.	neg.					
8			neg.	neg.	9.0		9.0
9						neg.	
10	0.5	neg.					
11			neg.	neg.	9.0		neg.
12						neg.	
13	neg.	neg.					
14			neg.	0.5	neg.		
21	8.6	neg.					
22			neg.	neg.	neg.	neg.	
23							neg.
27	neg.	neg.					
28			neg.	neg.	neg.	neg.	
29							neg.
31	neg.	neg.					
32			neg.	neg.	neg.	neg.	
33							neg.

[1]   Numbers (I-VII) show different plots in the analysed parcel where samplesa were investigated for viruses.

neg.   Viruses not detected.

had a capacity to immobilise inorganic N and P. Nevertheless, the analysis of foliar samples showed that after seven years, the sludge layer had a significant fertilising capacity through mineralisation of N and P. The nutrient release of N and P to the ground water as an effect of sludge application was limited. The concentrations of NO_3-N exceeded 10 mg/l on only a few occasions. PO_4-P concentrations did not exceed 0.5 mg PO_4-P/l, the average being 0.06 mg PO_4-P/l.

Heavy metals accumulated almost entirely in the raw humus layer with the exception of Zn, and only small amounts of Zn were leached to the ground water. However, Zn concentrations did not reach upper concentration limits recommended for drinking water.

After logging and reforestation of the experimental plots in the spring of 1988, there was increased leaching of NO_3-N from clearcut plots to the ground water. Concentrations increased progressively from 1.3 mg NO_3-N/l in June 1988 to 6.5 mg NO_3-N in July 1989. The increases in NO_3-N concentrations from shelterwood plots were delayed approximately six months, and a maximum concentration of 2.5 mg NO_3-N was reached in April 1989, decreasing by July 1989. Until then, both shelterwood and clearcut plots had shown significant trends and higher concentrations than the control plots, but the NO_3-N concentrations from sludge-applied shelterwood plots and clearcut control plots were almost equal by July 1989. It is expected that further investigations will show continuing trends.

With respect to clearcut and shelterwood plots, there is no significant upward trend in phosphorus concentrations in the ground water. However, in July 1989 there was a slight increase in the concentration of total P in the ground water from clearcut sludge-applied plots in comparison to shelterwood plots.

With the exception of Zn, logging and reforestation has not shown a significant effect on heavy metal concentrations in the ground water. For plots treated with the Zn contaminated sewage sludge II, Zn concentrations in the ground water were higher than for plots receiving sludge I. Average concentrations of 1.4 mg Zn/l were reached by July 1989 on plots receiving sludge II (407 kg Zn/ha), and the upward trend seems to be continuing. The effect of shelterwood in retaining Zn in the sludge and humus layer is greater compared to the clearcut plots. As for N and P, investigations of heavy metals in the ground water are to be continued.

The practice of applying sludge to forest soils does not seem to cause large hygienic risks. Depending on the time of the year, survival of pathogenic bacteria does not exceed one year, while viral survival does not exceed six months.

From the results of the investigations in 1973-81, a maximum application of 25 tds/ha to forest soils would be appropriate.

Full-scale sludge applications (25 and 1,000 tds/ha) were carried out on neighbouring spruce plantations in 1982-86, and the effects on vegetation, soil and ground water were studied. Approximately 1.5-3.5 years after sludge application, maximum concentrations of NO_3-N were found in the ground water. The average concentration was 9.3 mg NO_3-N/l (1.0-18 mg NO_3-N/l). The trend followed that of previous experiments. About 60% of the N-content was estimated as having been mineralised two years after sludge application and 80% four years after sludge application. These estimates seem higher than those found for the experimental plots. However, after one to two years, the fertilising effect showed as an increase in N and P concentrations in spruce needles, similar to levels found from investigations in 1975-81 after fertilising with Cu.

The conclusions from both the experimental plots and the full-scale application agree with conclusions drawn in the United Kingdom (22). Sewage sludge can be applied safely to nutrient-poor areas within forest plantations, especially after thinning. The benefits of sludge application are increased uptake and growth, but a balanced nutrient supply must be ensured, and sludge must be applied at rates appropriate to forest fertilising.

The sludge application had a very limited effect on the Cd concentrations in the ground water. However, after application of approximately 407 kg Zn/ha, the Zn concentrations increased to levels of 0.15 mg Zn/l, the highest concentrations 2.5 years after sludge application.

Experiences from both the experimental plots and the full-scale sludge application have been used to calculate estimates of sewage sludge application rates to be recommended in future fertilising of forest soils. By applying anaerobic digested sewage sludge to the soil surface, the mineralisation rate of N in the sludge is estimated to be 10% of the organic N per year. By incorporating sludge into the soil, 25% would be an appropriate estimate. Other research indicates that 15-20% of the organic N in anaerobic digested sludge will be mineralised in the first year (23, 24).

The following are recommended amounts of available N for plant growth when sewage sludge is used as fertiliser on forest soils:
a) To avoid unacceptable NO_3-N leaching to the ground water, the total water soluble N and the calculated mineralised N after the first year should not exceed 100 kg/ha;
b) With repeated sludge application, the mineralisation of N from earlier sludge applications should be taken into account.

The maximum recommended application of sludge can be calculated using the following equation:

$$R_a = \frac{A - R_p * C_{op} * K_{op} * (1 - K_{op})^n}{C_{oa} * K_{oa} + C_{ia}}$$

R_a = actual sludge rate (kg ds/ha)
R_p = previous sludge rate (kg ds/ha)

A = available N from sludge in the first year (kg N/ha)
C_{oa} = organic N in actual application (kg N/kg ds)
C_{ia} = water soluble N in actual application (kg N/kg ds)
C_{op} = organic N in previous application (kg N/kg ds)
K_{oa} = mineralising factor for actual application
K_{op} = mineralising factor for previous application
n = number of years since previous sludge application.

For example, anaerobic digested sewage sludge (II), as in Table 2, should approach 20 tds/ha when used as an initial N fertiliser on forest soils. The calculations should be used as an estimate, as decomposition rates in soil and raw humus layer vary. The maximum sludge application rate may also be restricted due to stricter government regulations regarding high concentrations of heavy metals.

REFERENCES

(1) Bockheim, J. G., T. C. Benzel, R.-L. Lu and D. A. Thiel (1988). Sludge increases pulpwood production. Biocycle, 29 (3), 57-59.
(2) Berry, C. R. (1987). Use of municipal sewage sludge for improvement of forest sites in the southeast. Research Paper No. SE-266, South Eastern Forest Experiment Station, USDA Forest Service, pp. 23.
(3) Bayes, C. D., J. M. Davies and C. M. A. Taylor (1987). Sewage sludge as a forest fertiliser: experiences to date. J. Inst. Water Pollution Control, 86, 158-171.
(4) Olesen, S. E., J. Lundberg and V. Larsen (1979). Application of sewage sludge to moorland spruce on sandy soil. Hedeselskabet, Danish Land Development Service. Research report No. 19, pp. 70 (English summary).
(5) Grant, R. O. and S. E. Olesen (1984). Sludge utilisation in spruce plantations on sandy soils. In: Utilisation of sewage sludge on land: rates of application and long-term effects of metals (S. Berglund, R. D. Davis and P. l'Hermite, Eds.), pp. 77-90. D. Reidel, Dordrecht.
(6) Fælles Arbejdsmetoder for Jordbundsanalyser (1972). Ministry of Agriculture, Denmark.
(7) Linner, H. (1972). Utrustning för uttgang av markvätska i fält. Grundforbättring, 25, 49-51.
(8) Dansk Standard and Standard Methods.
(9) Faulenborg, G. (1976). Undersøgelser over forekomst af salmo-

nella-bakterier i vådt, udrådnet slam fra biologiske rensningsanlæg for byspildevand og disse bakteriers forekomst og overlevelse i slammet efter deponering på parceller i et plantagedistrikt. Dansk Vet. Tidsskr., 59, 886-898.
(10) Lydholm, B. and A. L. Nielsen (1981). The use of a soluble polyelectrolyte for the isolation of virus from sludge. In: Virus and waste water treatment (M. Goddard and M. Butler, Eds.), Pergamon Press.
(11) Nielsen, A. L. and B. Lydholm (1983). The use of a cationic polyelectrolyte for the isolation of virus from raw and treated waste water. Report to the EEC, contract ENV-256 DK (G).
(12) Jørgensen, P. H. (1985). Virus i vand og slam. Department of Veterinary Virology and Immunology. The Royal Veterinary and Agricultural University of Copenhagen, pp. 144.
(13) Ingestad, T. (1962). Micro element nutrition of pine spruce and birch seedlings in nutrition solution. Medd. Skogsforskningsinstitut, 51, 118-119.
(14) Haveåen, O. (1964). Kobbermangel hos rødgran. Tidsskr. for Skogsbruk, 72, III.
(15) Holstener-Jørgensen, H. and H. Bryndum (1973). Preliminary results of experiments with nitrogen fertilisation of rather old Norway spruce on heathland localities in Jutland. Forst. Forsøgsv. Danmark, 33, 399-401.
(16) Larsen, V. (1981). Næringsstofbalance efter tilførsel af slam i nåletræsplantager. I "Slammets jordbrugsanvendelse II. Fokusering". Polyteknisk Forlag., 79-82.
(17) Fuller, W. H., D. R. Nielsen and R. W. Miller (1956). Some factors influencing the utilisation of phosphorus from crop residues. Soil Sci. Proc., 20, 218-224.
(18) Holstener-Jørgensen, H. and M. Krag (1988). Skovburg og miljø. Skoven, 20, 266-268.
(19) Bormann, F. H. and G. E. Likens (1979). Pattern and process in a forest ecosystem, p. 253. Springer-Verlag, New York.
(20) Chang, A. C., T. J. Logan and A. C. Page (1986). Trace element considerations of forest land applications of municipal sludges. In: The forest alternative for treatment and utilisation of municipal and industrial wastes (D. W. Cole, C. L. Henry and W. L. Nutter, Eds.), pp. 85-99. University of Washington Press, Seattle.
(21) Campbell, D. J. and P. H. Beckett (1988). The soil solution in a soil treated with digested sewage sludge. J. Soil Science, 39, 283-298.
(22) Moffat, A. J. and D. Bird (1989). The potential for using sewage sludge in forestry in England and Wales. Forestry, 62, 1-17.

(23) Fox, R. H. and J. H. Axley (1985). Nitrogen release from sewage sludge. Bulletin, Pennsylvania State University, Agricultural Experimental Station, USA, No. 851, 38-39.
(24) Parker, C. F. and L. E. Sommers (1983). Mineralisation of nitrogen in sewage sludge. J. Environmental Quality, *12*, 150-156.

DISCUSSION SESSION 2

QUESTION: R. J. UNWIN, MINISTRY OF AGRICULTURE.
Mr Arnot told us that the costs were approximately 20% higher for the sludge operation. Could he explain this and why two costs are quoted, one of £3.31 for current costs and £1.99 for what it could be if done routinely?

ANSWER: J. M. ARNOT.
The cost is worked out from the number of loads we could take per day to particular sites. Because of the increased distance between the Shotts Works and Clydesdale Site, and the Shotts Works and farmland, we are losing roughly one load a day, so it pushed our costs up by that amount. The current cost is using somewhat older tankers. If we, as we have now done, put newer tankers onto the route we stay on the road longer because of the lower maintenance costs and we have two tankers on at the same time, we reduce our costs quite a bit.

QUESTION: R. J. UNWIN, MINISTRY OF AGRICULTURE
What criteria were used in Mr Taylor's study for choosing 10 miles or 20 miles for deciding on what areas would be economic?

ANSWER: C. M. A. TAYLOR.
Most operations putting sludge to agricultural land are generally within a fairly short distance of the sewage works so we chose 10 and 20 miles as arbitrary figures. We can go back to the data if we wish to increase the tanker distance, as in certain locations the tanker distances are actually longer than that.

QUESTION: H. W. SCHERER, INST. OF AG. CHEM. BONN, FRG.
Did you have toxicity problems due to ammonia when you spread this sludge?

ANSWER: C. M. A. TAYLOR.
Even where we have been coating the trees very substantially in sludge we have had no damage at all to the tree needles or to the tree growth, apart from where we have overloaded the site hydraulically. This was the only time we have actually caused any detrimental effects on tree growth.

COMMENT: G. A. FLEMING, JOHNSTOWN CASTLE, EIRE.
On some of our peat lands in Ireland we run into copper deficiency, which shows up as a whirling of the tops of the trees. An interesting aspect was that when we sprayed the heather with herbicide near the plantations and the

heather died, there was a tremendous release of nitrogen and this accentuated the copper deficiency.

ANSWER: C. M. A. TAYLOR.
Where we have encountered copper deficiency normally in forest crops in Britain, it has generally been on very fertile flush sites where there is a lot of nitrogen available, or in experimental situations where we have put on very high rates of nitrogen which seems to upset the copper/nitrogen balance. The large needles effectively dilute the copper and they go into growth distortions and lose apical dominance. We have found that on the particular experiment at Ardross, where we are now putting on nitrogen treatments, we are beginning to run into these problems which we did not have before.

QUESTION: H. W. SCHERER, INST. OF AG. CHEM. BONN, FRG.
When you add sludge you add nitrogen and phosphorus, but the content of potassium is pretty low in the sludge. Do you add extra potassium?

ANSWER: C. M. A. TAYLOR.
We tend to only apply potassium on our deeper peat sites in Britain where we have deficiencies. We do not see those sites at this time as being suitable outlets for sludge. Where we are applying sludge at the moment is to sites that are principally phosphate and nitrogen deficient where there is usually plenty of potassium.

QUESTION: J MOORE, SCOTTISH DEVELOPMENT DEPT.
Due to the fact that the initial work has shown such accelerated tree growth rates do you anticipate a shorter crop rotation?

ANSWER: C. M. A. TAYLOR.
We would like to apply sewage sludge to pine plantations which are generally in longer rotations where we would actually be shortening the rotation, or if we kept the same rotation we would have much greater volume at the end of the day.

COMMENT: H. W. SCHERER, INST. OF AG. CHEM. BONN, FRG.
In Germany we do not use sludge in forestry because it is thought that the wood quality is affected.

ANSWER: C. M. A. TAYLOR.
It has often been said that by applying fertiliser to our tree crops makes them grow faster and less dense, resulting in less utilisable timber. We are not

really pushing growth on that fast since we are just basically removing nutrient limitations to growth. In general growth rates are still similar to those on better soil types and we have not really found any detrimental effects on normal fertiliser operations on wood quality. By the same analogy we do not anticipate any particular problems with sludge.

ANSWER: C. G. NICHOLS.
In the USA we are maybe a little further ahead in the analysis of structural quality of trees fertilised with sludge than in the UK. There is some decrease in quality of timber in terms of lower specific gravity but the observations have been not too significant, something in the order of 5-10% decrease in quality. The end result in terms of a timber product has been immaterial because it is still well within any grading classifications for a timber product, at least in the United States.

QUESTION: C. D. BAYES, WATER RESEARCH CENTRE.
Could Mr Nichols expand on the wildlife monitoring that has been done in the north west USA with his sludge to forestry land applications?

ANSWER: C. G. NICHOLS.
The University has done extensive research on wildlife in the sludge applied areas as compared to control areas. We have monitored the full range of wildlife including bird life, small rodents, insects, worms, snakes, large mammals including deer, and that has included deer which have been penned in quite substantial multi-hectare areas and have been surviving on the sludge-fertilised undergrowth. The net result is that we have not found any detrimental effects. In fact we found the opposite, that the wildlife growing on the sludge fertilised area is much more vigorous, for instance deer were typically having many more twins, in essence more fertile. In one sense the forest industry is not necessarily pleased with that because it may create more deer browse problems but often in a recreational sense this is an enhancement to the wildlife. We found no signs of build-up of contaminants in body tissues in the wildlife so in effect we feel very comfortable with the effects on wildlife.

QUESTION: P HYDES, SOUTHERN WATER.
What are the effects of sludge fertilisation on forest areas that are suffering from acid rain?

ANSWER: S. OLESEN.
I think the sludge will to some extent neutralise the acid soil so you may expect that it may ameliorate the effects of acid rain for at least a number

of years, but in a long run acid will be produced due to mineralisation of the organic matter.

QUESTION: H. W. SCHERER, INST. OF AG. CHEM. BONN, FRG.
You have shown some young trees which are covered with sludge in spring and these young trees have grown very well in summer. Is this growth caused in the same way as the foliar application of nutrients?

ANSWER: C. G. NICHOLS.
The experience that we have had with these foliar applications indicates that the sludge will wash off during the course of the spring and either goes through a drying and re-wetting cycle and flakes, or washes off and it leaves no long-term effect on the needles. In essence the sludge is moving quickly from being on the needles down to the soil where it is then available for uptake.

ANSWER: C. M. A. TAYLOR.
The high growth response may be a function of the ammonia in solution in the sludge which is freely available. We found in our experiments that the response to sludge is often quicker than to ammonium nitrate and urea which is also an ammonia source.

QUESTION: A. M. BRUCE, WATER RESEARCH CENTRE.
Could Mr Nichols expand a bit more on the pros and cons of drying of sludge before application? It is a technology which we have abandoned in the UK because of unfortunate experiences in the past but with modern drying technology we are considering re-examining that area as a possible means of economising on transportation in particular.

ANSWER: C. G. NICHOLS.
We are evaluating highly dried sludge at the moment from a public acceptance and economic perspective whereby, if we are drying the sludge to 90-95% solids, then we have in effect eliminated pathogens and to a great degree any public concerns. Although we feel very proud about the success we have had with liquid sludge, it does create many aesthetic problems. Creating a dry product in a pellet or powder form is possibly a trend that is going to be very much needed and realistic in order to have a continuing land application programme. The idea of using a dry sludge also is quite meaningful in relation to ready public access since there are increasing pressures for recreation and hunting and fishing. We are not that pleased yet with the economics.

SESSION 3

Landfill and Incineration

Co-disposal of sewage sludge and domestic waste in landfills: laboratory and field trials

N. C. BLAKEY

Water Research Centre, Henley Road, Medmenham,
PO Box 16, Marlow, Buckinghamshire, SL7 2HD, UK.

SUMMARY

Laboratory scale experiments (0.2 m³ reactors) and pilot scale field trials (approximately 1300 m³ landfill cells) have been set up to investigate the effects of co-disposal of domestic waste and sewage sludge on leachate quality and gas production. Results to date indicate that co-disposal increases the rate of waste stabilisation, effectively reducing the concentrations of organics and metals in the leachate produced. However, ammoniacal nitrogen and total phosphorus levels are significantly increased as a direct result of sludge addition in the co-disposal experiments. Gas generation rates, recorded in the laboratory reactors, are increased from 0.05 m³ methane/tonne volatile solids per day in the domestic waste only controls to 0.28-0.45 m³ methane/tonne volatile solids per day in the co-disposed wastes. In addition to sludge co-disposal, infiltration is shown to increase the short term total gas yield.

1. INTRODUCTION

In the UK disposal of sewage sludge to landfill (80 x 10³ tds/year) accounts for about 10% of all sludge disposed of to land (1). Legislative changes involving current sludge disposal practices are likely to increase pressure on waste disposal authorities and contractors to accept more sewage sludge at landfills. The work described in this paper was initiated to investigate the different aspects of this co-disposal technique.

Initially, laboratory experiments were carried out, using different infiltration rates, to examine the effects of sludge type and changes in mixing strategy on landfill leachate quality and gas production. Subsequently, pilot scale field trials have been set up principally to identify the most appropriate operational procedure for disposing of sewage sludge in domestic waste landfills but also to confirm the results from the laboratory trials, with regard to leachate quality and gas production. In addition to these principal objectives it is hoped that cost implications of the disposal option will be determined along with important information on the effects of co-disposal on the restoration procedures and subsequent after-use of the site. This paper, however, concentrates on the leachate quality and gas production data generated from both studies and attempts to summarise the main findings to date.

2. LABORATORY REACTOR DESIGN AND OPERATION

A series of six experiments (each duplicated) was carried out in 0.2 m³ stainless steel drums (Figure 1). The following co-disposal options were investigated along with a control containing domestic waste only:
a) Different sludge type (raw dewatered, primary/mixed dewatered and liquid digested);
b) Changes in mixing strategy (completely mixed, compared with layered sludge and domestic waste);
c) Infiltration rate (900 mm/year compared with 300 mm/year).

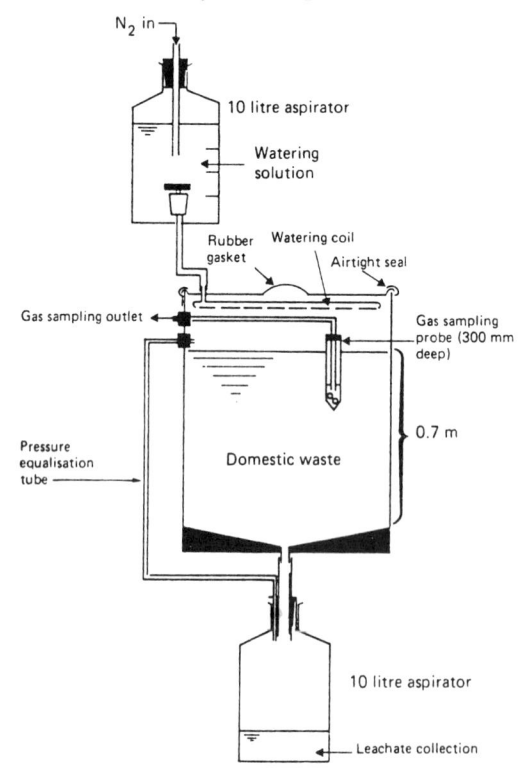

Fig. 1. Experimental unit showing irrigation, gas sampling and leachate collection systems.

Each drum was filled sequentially and in layers to ensure a uniform and comparative domestic waste mixture. Pulverised domestic waste was used since crude waste was inappropriate for the scale of experiment. A compaction rate of about 950 kg/m^3 (wet weight) was used in all the tests, simulating a density commonly achieved at operational landfills using steel wheeled compactors. The highest sludge to refuse ratio in the reactors (1:4.1 (wet weight) respectively) was chosen to represent the highest likely to be adopted in operational practice.

In the layered sludge reactors, a single depth of sludge was placed on top of 0.60 m of domestic waste and then covered with a thin layer (0.04 m) of additional domestic waste. Where sludge and refuse were mixed intimately this was done before filling the drum with each layer of waste. The chemical composition of the wastes used in the experiments is shown in Table 1.

Table 1. Analyses of waste types used in the laboratory reactors.

	A	B	C	D
Dry solids (%)	45.4	30.4	30.2	4.3
Volatile solids (%)	61	65	52	49
Total metal content (mg/kg ds)				
Fe	5,800	25,000	60,000	15,000
Mn	150	420	1,700	700
Cu	110	680	7 10	1,250
Zn	310	21,000	2,700	3,500
Pb	500	300	860	3,200
Ni	19	50	1,700	200
Cd	4	86	37	100
Cr	56	180	4,800	720

Notes: A = Pulverised domestic waste.
B = Raw dewatered sludge.
C = Primary/mixed sludge.
D = Liquid digested sludge.

Each reactor was hermetically sealed after being filled. Infiltration rates of 300 mm/year and 900 mm/year (as appropriate) were simulated by adding water twice weekly to the waste in each drum via a distribution system installed inside the reactor lid (Figure 1). Temperature in all units was maintained at 30°C to reduce the effects of significant thermal variation and to simulate temperatures commonly encountered in relatively "fresh" landfilled wastes.

Samples of leachate were analysed on a weekly basis, with bulk compositional analysis of gas every three months. Gas flow rates were recorded twice weekly, at the time of watering.

3. PILOT SCALE FIELD TRIALS

Four purpose-built test cells, each approximately 25 m long by 15 m wide and 3.5 m deep, have been constructed at a site in Huddersfield, West Yorkshire. Two of the cells are controls, containing only crude or baled domestic waste. A dewatered (centrifuge) polyelectrolyte-treated sludge was co-disposed with crude or baled domestic waste in the two remaining cells at a ratio of 1:5.5 and 1:9.3 respectively (sludge to refuse mix on a wet weight basis).

In the crude waste co-disposal cell, deliveries of residue were tipped in discrete mounds before being spread in a thin layer using a tracked machine with blade. In the baled waste co-disposal cell, bays were formed by bales of waste into which loads of sludge were tipped from a lorry. Additional bales were then worked up to the sludge disposal position, enclosing the sludge in a discrete area. Broken bales were used to infill on top of the sludge (Figure 2).

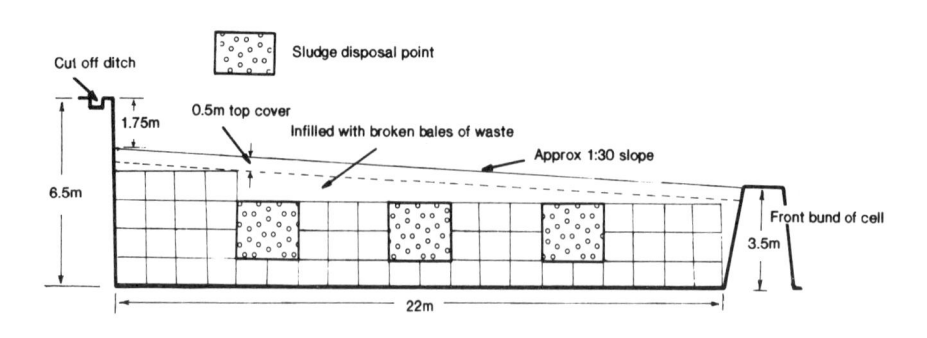

Fig. 2. Cross-section through field trial cell showing sludge disposal points for baled waste.

The chemical composition of the dewatered polyelectrolyte sludge used in these trials is presented in Table 2.

Table 2. Analysis of dewatered polyelectrolyte sludge used in the field trials.

Dry solids (%)	28.7
Volatile solids (%)	43.8
Total metal content (mg/kg ds)	
Fe	28,000
Mn	1,050
Cu	360
Zn	1,100
Pb	490
Ni	68
Cd	6
Cr	660

Immediately following filling, the experimental cells were covered with about 0.5 m of restoration material and graded to a slope of 1 in 30 before being seeded with grass. Cut-off drains were excavated between and at the back of the cells.

Samples of leachate and gas are collected from a sampling well and a series of probes respectively and have been analysed on a regular basis since the start of the experiment.

4. DISCUSSION OF RESULTS

4.1 Laboratory scale reactors

The mass of certain individual constituents leached from the laboratory co-disposal reactors (such as total organic carbon (TOC), biochemical oxygen demand (BOD), chemical oxygen demand (COD), sulphate, iron, manganese, zinc, nickel and lead) is less than that observed from the domestic waste only controls. The total mass release of other constituents such as ammoniacal nitrogen (NH_3-N) and total phosphorus was conversely found to be higher (Table 3).

To explain these observations the concentrations of COD and zinc in leachate from the codisposal reactors are used as examples to compare differences with the control tests.

Table 3. Percentage mass of material leached from the co-disposal wastes as a proportion of that leached from pulverised refuse alone (control).

	A	B1	B2	C	D
Refuse/sludge ratio (wt/wt)	4.1:1	4.1:1	4.1:1	4.8:1	9.7:1
Chemical oxygen demand (%)	-48	-64	-15	-46	-7
Ammoniacal nitrogen (%)	+10	+23	+66	+40	-8
Total phosphorus (%)	+30	+58	+203	+4	-2
Iron (%)	-57	-62	-2	-48	+5
Zinc (%)	-54	-95	-70	-71	-68
Nickel (%)	-24	-46	-4	-12	-4

Notes: Sludge added to reactors A, B2, C and D as a single layer (see Section 2).
A = Refuse and raw, dewatered sludge.
B1 = Refuse and raw, dewatered sludge (homogenous mix).
B2 = Refuse and raw, dewatered sludge (high infiltration).
C = Refuse and primary/mixed dewatered sludge.
D = Refuse and liquid, digested sludge.

Acetogenic activity within the reactors gives rise to high levels of volatile acids (C_2-C_6) within the leachate which contribute significantly to the COD concentrations shown in Figure 3. Co-disposal of sewage sludge has increased the rate of stabilisation of the waste by reducing the timescale for the onset of significant methanogenesis to about nine months (Figure 3). The corresponding increase in gas generation is shown in Figure 4. The reduction in leachate metal concentrations, exemplified by the zinc data illustrated in Figure 5, is accompanied by a corresponding reduction in sulphate concentration (Figure 6). It is concluded that a significant mechanism for metal immobilisation within the waste mass has been due to the precipitation of metal sulphides.

Conversely, the low biological uptake of phosphorus and the absence of nitrification under the prevailing anaerobic conditions within the reactors has given rise to increased leaching of ammoniacal nitrogen (NH_3-N) and total phosphorus (Table 3). This is in direct response to the additional load present in the sludge. Figure 7 shows the temporal release pattern for NH_3-N from each of the reactors. The gradual "wash-out" of NH_3-N is in marked contrast to that of the biologically mediated pattern of release for COD.

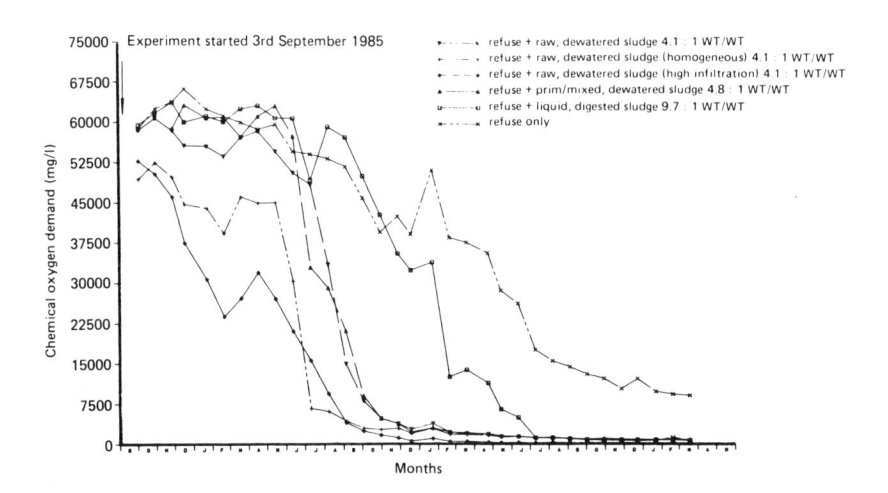

Fig. 3. Leachate chemical oxygen demand.

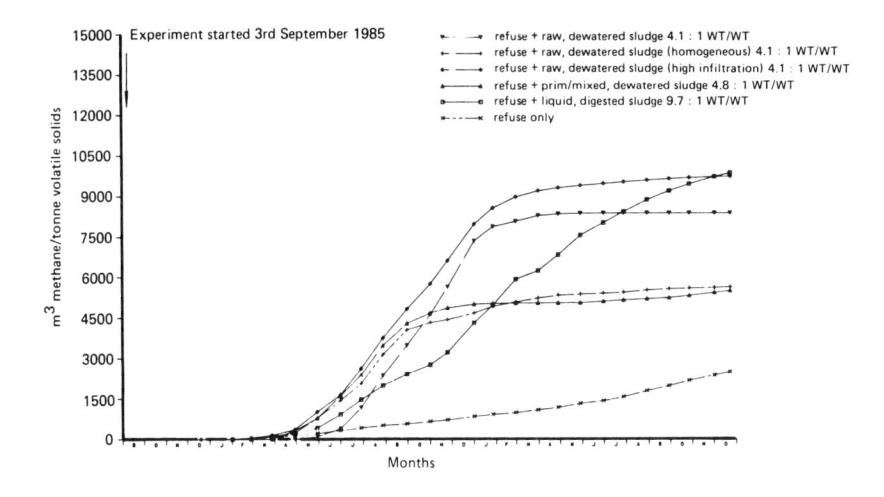

Fig. 4. Cumulative methane production rate.

The onset of methanogenesis in each reactor is accompanied by an increase in pH from about 5.7 to 7.2. The composition of gas collected from the headspace of each reactor ranged between 50 and 60% methane with the balance being carbon dioxide.

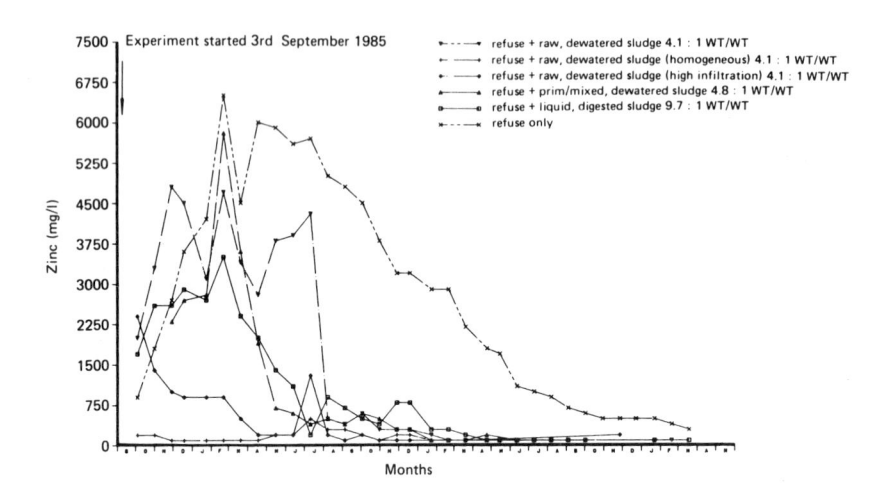

Fig. 5. Leachate zinc concentrations.

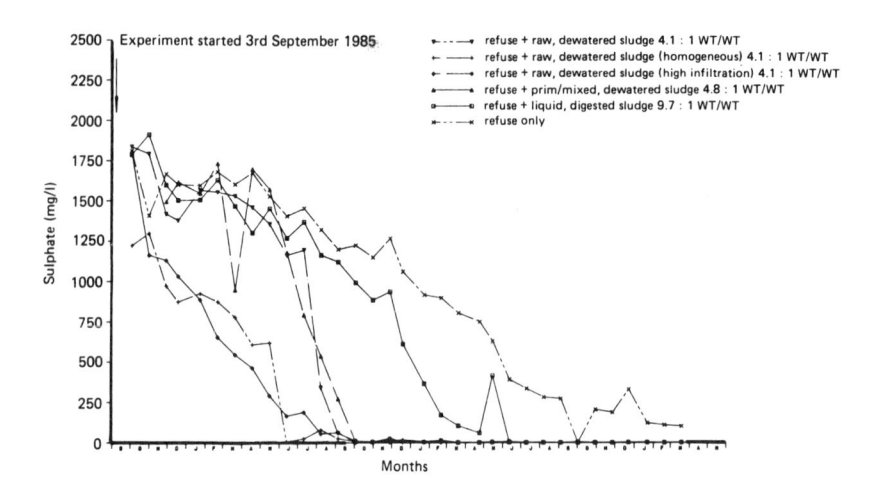

Fig. 6. Leachate sulphate concentrations.

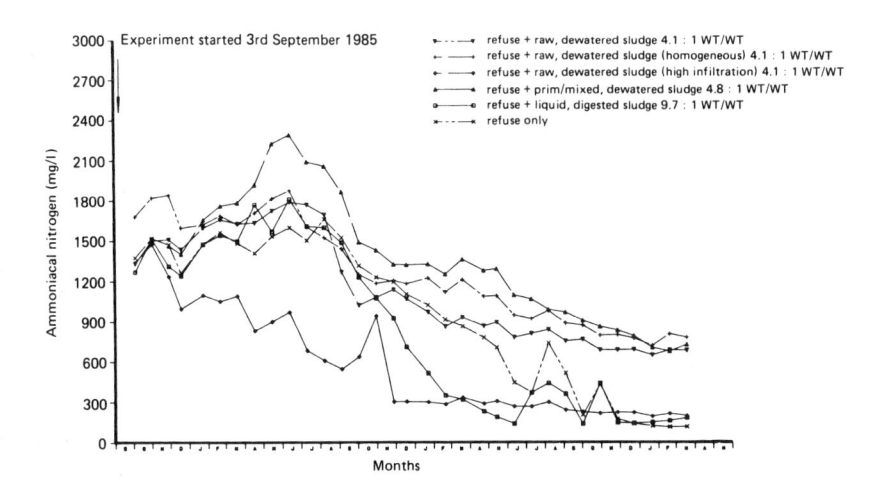

Fig. 7. Leachate ammoniacal nitrogen concentrations.

The rate of gas production is higher in the co-disposal experiments, with 0.45 and 0.28 m³/tonne volatile solids per day from the raw dewatered and liquid digested sludges respectively. This compares with a rate of 0.05 m³/tonne volatile solids per day for the domestic waste only controls. Total methane gas production in the co-disposal experiments ranges from 80-150 m³/tonne volatile solids and 36 m³/tonne volatile solids in the domestic waste only controls. In the co-disposal experiments, gas generation ceases between six to twelve months from commencement, presumably since most of the readily biodegradable material has been assimilated.

In the domestic waste only controls gas generation continues to increase twenty months after first producing gas (Figure 4).

These experimental results can be compared with those of Barlaz *et al.* (2) who found domestic waste produced 80-150 m³ methane/tonne volatile solids and Buivid *et al.* (3) who found that domestic waste produced 90 m³ methane/tonne volatile solids and 180-210 m³ methane/tonne volatile solids for co-disposed domestic waste and sewage sludge.

4.2 Pilot-scale trials

The experimental cells exhibited significant biological activity at an early stage after filling, as indicated by the gas quality analyses (Figure 8) and the chemical composition of the leachate (Table 4). Even allowing for the

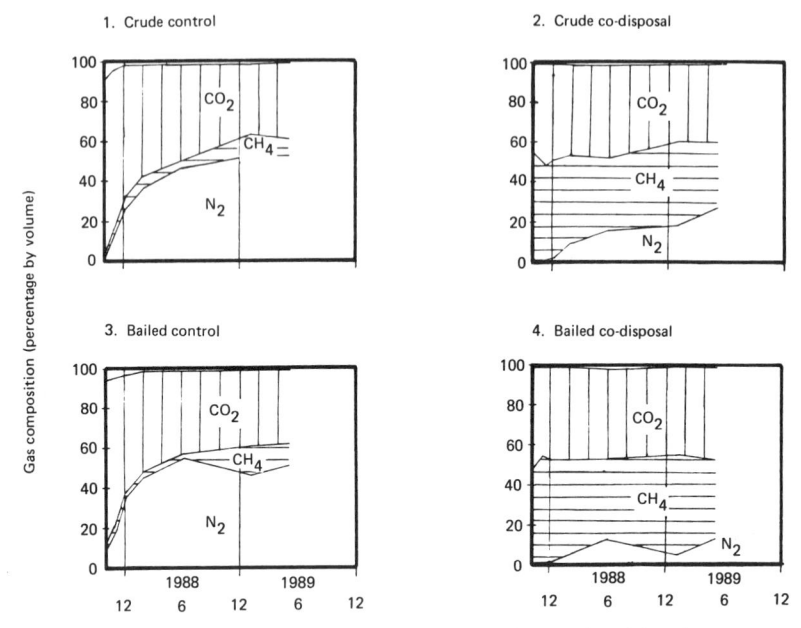

Fig. 8. Average landfill gas composition determined in six probes installed in each of the field trial cells at Huddersfield, West Yorkshire.

Table 4. A comparison of leachate in the field trial cells six months following waste emplacement (a) and twelve months from the start of leachate production (b).

Determinand (mg/l)	Crude control		Crude codisposal 5.5:1[1]		Baled control		Baled codisposal 9.3:1[1]	
	a	b	a	b	a[2]	b	a	b
COD	15,700	10,000	25,400	20,000	16,800	15,500	34,300	14,000
Amm. N	321	217	609	473	493	786	1,280	828
Total P	2.8	7.1	4.7	1.0	2.3	3.6	6.4	2.6
Sulphate	n/a	61	344	<5	20	202	142	176
Iron	464	467	610	1,410	724	351	650	165
Zinc	0.07	0.05	1.09	0.22	0.25	1.03	3.05	0.19
Nickel	0.13	0.02	0.28	<0.01	0.14	0.09	0.18	0.20

[1] Refuse/sludge ratio (wt/wt).
[2] 10 months following waste emplacement due to problems encountered sampling from the tube well at 6 months.
n/a Not analysed.

lower sludge to refuse ratio adopted (see laboratory trials ratio of 1:4, Section 2) the co-disposal cells (cells 1 and 4) exhibit the most activity, with gas analysis showing the rapid establishment of methanogenic conditions (methane up to 50% by volume). Additionally, initial leachate (6 months following waste emplacement) has comparatively higher concentration levels of waste degradation products than in the control cells, with values suggesting that biological breakdown processes have been accelerated by the addition of the sewage treatment plant sludges.

The pattern of results so far emerging compares favourably with the results obtained in the laboratory and continued monitoring, including an attempt at estimating gas yield, should confirm the longer term findings of the smaller scale laboratory experiments.

5. CONCLUSIONS

Co-disposal of domestic waste and sewage sludge significantly increases the rate of stabilisation of the co-disposed wastes, speeding up the reduction of degradable organics leached from wastes and improving the organic quality with time. In the laboratory scale reactors receiving dewatered sludge at a domestic waste to sludge ratio of 4.1:1 (wet weight) the COD was reduced by 50% in comparison to domestic waste alone. Where liquid digested sludge was used at a domestic waste to sludge ratio of 9.7:1 (wet weight) a 7% reduction in COD was achieved when compared with domestic waste alone. These results were achieved at a moderately low infiltration rate of 300 mm/year. For higher infiltration rates greater reductions are indicated.

Results from the laboratory scale reactors show that co-disposal significantly reduces the quantity and concentrations of metals leached (iron, manganese, zinc, nickel and lead), typically by about 50%. Although reductions have been identified in the field scale trials, further data are required to confirm the trends.

Co-disposal has been found to increase the leachate concentrations of ammoniacal nitrogen and phosphorus. In the laboratory trials ammoniacal nitrogen loads are increased by between 10 and 40% and total phosphorus by between 4 and 60% under moderately low infiltration conditions (300 mm/year). For higher infiltration conditions (900 mm/year) loads are increased by about 70% and 200% respectively.

During the early stages of both the laboratory reactors and the pilot scale trials, gas quality in terms of methane content is much improved as a result of sludge co-disposal. In terms of gas generation rates the laboratory co-disposal reactors yielded 0.28-0.45 m³ methane/tonne volatile solids per day compared to 0.05 m³ methane/tonne volatile solids per day from domestic waste only controls. Total gas production in the codisposal experiments ranged from 80-150 m³ methane/tonne volatile solids in comparison with 36 m³ methane/tonne volatile solids for the domestic waste only tests.

Waste stabilisation and hence gas generation has been shown to take place over a much shorter time period when sludge is co-disposed with domestic waste. Infiltration in the range 300-900 mm/hour on the other hand does not affect the rate of gas generation but improves the short term total yield.

REFERENCES

(1) Department of the Environment/National Water Council (1983). Sewage sludge survey - 1980 data. HMS0, London, UK.
(2) Barlaz, M. A., M. W. Milke and R. K. Ham (1987). Gas production parameters in sanitary landfill simulators. Waste Management and Research, 5, 27-39.
(3) Buivid, M. G., D. L. Wise, M. J. Blanchet, E. C. Remedios, B. M. Jenkins, W. F. Boyd and J. G. Pacey (1981). Fuel gas enhancement by controlled landfilling of solid waste. Resources and Conservation, 6, 3-20.

Co-disposal in MSW landfills of coal fly-ash and domestic sludge

R. COSSU, R. SERRA, P. CANNAS and G. CASU

Institute of Hydraulics, University of Cagliari,
09100 Cagliari, Sardinia, Italy.

SUMMARY

A large research programme has been set up at Cagliari University in order to study the behaviour of an ash and domestic sludge mixture in a municipal solid waste (MSW) landfill. The presence of these two materials would enable two operational problems to be solved at the same time; the moisture of the sludge reduces the dust problem present in ash disposal and the degree of drying of the sludge is increased. The present paper reports the results of batch tests which confirm that ashes are able to absorb heavy metals present in leachate from an MSW landfill and that this phenomenon is enhanced by the presence of domestic sludges.

1. INTRODUCTION

One of the more important problems in the field of solid waste disposal is that of fly-ash from coal-fuelled power plants. These residues are invariably considered as being toxic due to the significant presence of heavy metals such as Cd, Hg, Cu, Ni, Cr, Pb and As. In fact, these elements are preferentially distributed in finer-particled fractions (fly-ash) rather than coarser ones (bottom residues) due to volatilisation during combustion and subsequent condensation in the fine particles transported by flue gas (1).

Certainly, the more rigid regulations proposed by the USEPA regarding the judgement of toxicity of waste (2) negatively influence the possibility of disposal of this type of residue in sanitary landfill. Consequently, the evolution of other waste disposal systems such as incineration of MSW is jeopardised as the end-products are ashes similar to coal fly-ashes but with an even greater load of heavy metals.

The search for sites in which to develop highly-protected sanitary landfills for the disposal of combustion waste will become more and more of a problem, as demonstrated by the unwillingness of local communities to accept sanitary landfills, even for MSW alone, in their localities.

A research programme is already under way in Cagliari University to study co-disposal of coal fly-ash and MSW in sanitary landfills. To date the results have been encouraging, as they appear to demonstrate that the ash is able to speed up the process of degradation of the organic fraction of MSW without, however, releasing greater quantities of heavy metals than those present in the case of disposal of MSW only (3).

This research programme has, moreover, considered the possibility of co-disposal of fly-ash and domestic sludge in an MSW sanitary landfill. The association of these two materials in a sanitary landfill would at the same time minimise two operational problems; the dustiness of ash would be reduced, and the degree of dryness of sludges would be increased. The moisture content of the sludge is 70% under optimal conditions and causes problems for vehicles (compactors, bulldozers, lorries) working on a landfill.

Analysis of co-disposal can be carried out by means of batch and/or lysimeter tests together with tests aimed at investigating the hydraulic behaviour of waste. The aim of the former tests is to verify environmental compatibility (release of toxic substances, inhibition of biostabilisation processes) of waste disposal in a sanitary landfill whilst the latter are used to verify possible impact during operation and management of a sanitary landfill.

The present paper presents the preliminary results of an experimental test involving the co-disposal of fly-ash and domestic sludge.

2. MATERIALS AND METHODS

2.1 Batch tests

The batch tests were carried out by mixing the solid phase made up of ash alone and a mixture of ash and sludge, with a liquid phase made up of two leachates from MSW landfills. The contact between the two phases was maintained for 24 hours by means of a shaking device. During the test pH was allowed to vary. After shaking, the liquid was filtered and analysed for heavy metals (Pb, Cd, As, Cr, Zn, Mn, Cu and Ni) with an atomic absorption spectrophotometer. As the levels of heavy metals present in the solid phase and in leachate were already known and the concentration present in the combined product was verified, it was possible to determine the amount of metals released by ash, and by the mixture of ash and sludge, when leached by actual leachate from sanitary landfills.

The study of the behaviour of ash in the presence of leachate obtained from actual MSW landfills is of even greater significance for the examination of the real situation of co-disposal of ash (or ash and sludge) with MSW in landfill. If the disposal of ash in layers is considered, the liquid phase that will come into contact with the waste is the leachate derived from the over-lying layers of MSW.

The ratios of ash to sludge studied were 1:1, 2:1 and 10:1; the latter minimised the dustiness of the ash. The ash used in this study was obtained from the Sulcis Thermoelectric Plant situated in the industrial area of Portovesme (Cagliari) which has three 240 MW units fuelled by coal. These were sampled from the hoppers of electrofilters for particle-collection and the samples are representative of the quality of fly-ash produced by the plant.

The sludge was obtained from the wastewater treatment plant in Sestu, a town of about 11,000 inhabitants situated near Cagliari. Sewage treatment is by activated sludge whilst the treatment of sludge is by means of aerobic digestion. Sludge dewatering is performed on drying-beds from which the samples used in this study were obtained. The main characteristics and the metal contents of the two wastes are shown in Table 1.

The tests were carried out with a liquid/solid ratio of 20:1. The leachates used for dilution are typical of different situations. Leachate obtained from the Serdiana (Cagliari) landfill is representative of methanogenic landfill; that obtained from the Rio delle Vigne (Reggio Emilia) landfill is typical of a situation still in the acid phase.

Table 1. Characteristics of solid phase.

	Ash	Domestic sludge
Moisture (%)	0.15	88
Dry solids (%)	99.85	12
Volatile solids (%)	-	9.5
pH	12.7	-
Pb (mg/kg)	430	14.9
Cd (mg/kg)	1.15	0.206
As (mg/kg)	52	0.14
Cr (mg/kg)	10	2.45
Zn (mg/kg)	640	63.5
Mn (mg/kg)	1,100	7.53
Cu (mg/kg)	60	14.6
Ni (mg/kg)	250	2.12

The characteristics of the leachates are shown in Table 2.

Table 2. Characteristics of leachates used as eluants.

	Leachate Serdiana		Leachate Rio Vigne	
	Raw	Filtered	Raw	Filtered
pH	7.1		5.5	
COD (mg/l)	6,168		11,560	
BOD (mg/l)	3,292		6,312	
BOD/COD	0.53		0.55	
NH_4-N (mg/l)	534		300	
P total (mg/l)	0.72		0.16	
Pb (mg/l)	0.03	0.02	0.225	0.102
Cd (mg/l)	0.0005	<0.0005	0.0058	0.0028
As (mg/l)	0.88	0.75	<0.001	<0.001
Cr (mg/l)	0.078	0.078	0.34	0.217
Zn (mg/l)	0.18	0.066	3.9	2.42
Mn (mg/l)	14.5	13.13	8.2	6.62
Cu (mg/l)	<0.05	<0.05	0.31	0.09
Ni (mg/l)	0.45	0.28	0.3	0.2

2.2 Lysimeter tests

The batch tests simulate drastic conditions which are unfavourable for the release of pollutants from solid matter, as saturated conditions provide perfect contact between the liquid and solid phases. Therefore it was necessary to verify the results by means of larger scale lysimeter tests. Cylindrical lysimeters (columns) were used which were filled with two ratios of ash/sludge (2:1, 10:1) and wetted from below. The dimensions and filling of lysimeters are illustrated in Figure 1.

Up-flow condition was preferred to down-flow although the latter is more realistic for a landfill situation, as the device set up thus enables permeability of the solid phase to be determined. The biological transformations caused by the presence of sludge induce a modification in physical characteristics of the mixture and determine variations in hydraulic behaviour. Moreover, the up-flow guarantees that all material deposited in the lysimeter will be infiltrated.

The experiments were carried out using both water and actual leachate from landfill. In this way the behaviour (permeability, absorption capacity or release of pollutants) of the ash/sludge mixtures could be examined both in the presence of water alone (typical situation in an ash/sludge only landfill) and in the presence of actual MSW leachate (typical situation in co-disposal of ash/sludge in an MSW landfill). The release and reabsorption of metals (or other specific compounds such as phenols and/or polyphenols present in MSW leachate) can be compared to results obtained with the batch tests.

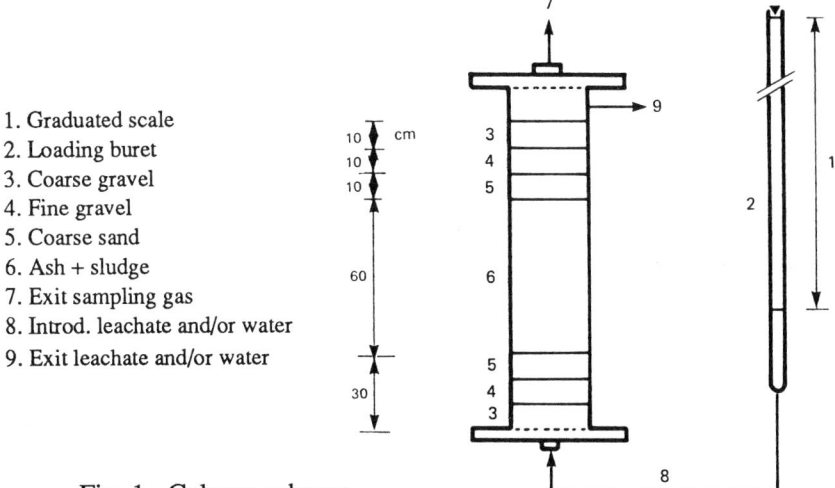

1. Graduated scale
2. Loading buret
3. Coarse gravel
4. Fine gravel
5. Coarse sand
6. Ash + sludge
7. Exit sampling gas
8. Introd. leachate and/or water
9. Exit leachate and/or water

Fig. 1. Column scheme.

It should be emphasised that the information acquired regarding permeability of the mixture over a period of time allows the evaluation of the possibility of using a layer of ash (to which sludge has been added to minimise operational problems) as a protective layer for artificial covering and as a further hydraulic barrier. Furthermore, biogas formed over a period of time in lysimeters will be sampled and the quality analysed.

The filling of lysimeters was carried out using the quantities given in Table 3:

Table 3. Quantities of ash and sludge used in the lysimeters.

Column	Quantity ash (kg)	Quantity sludge (kg)	Ratio ash/ sludge	Type of liquid	Density mixture (g/cm^3)
1	4	2	2:1	water	1.3
2	5	0.5	10:1	water	1.1
3	4	2	2:1	leachate	1.3
4	5	0.5	10:1	leachate	1.1

2.3 Anaerobic degradation test

The batch and lysimeter tests enable evaluation of the release of pollutants (heavy metals) from waste in two different situations but they cannot provide information regarding the behaviour of the mixture under anaerobic degradation. The batch tests simulate aerobic conditions whilst lysimeter tests allow evaluation of biogas formation, although this is limited due to the dimensions of columns and the difficulty encountered in extracting all gases formed.

It was considered advisable to study the degradation behaviour of sludge when mixed with varying amounts of ash. The ratios of ash/sludge studied were 10:1 and 2:1 as well as the controls of ash or sludge alone. The device used is illustrated in Figure 2.

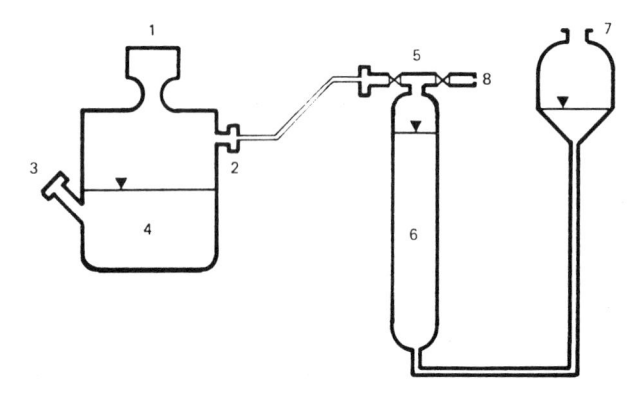

1. Stable container for solid matrix
2. Exit biogas
3. Sect for obtaining liquid sample
4. Solid waste in saturated conditions
5. Gasometer
6. Acid and salty water
7. Flask for pressure control and for emission of biogas accumulated in the gasometer
8. Opening for biogas sampling

Fig. 2. Scheme for evaluation of anaerobic biological degradation.

A container (2 litres) was prepared in which the waste was maintained in conditions of water saturation and a set burette (gasometer) was attached to measure the production of biogas by means of the displacement of an identical quantity of water. An open flask connected to this controlled pressure and allowed the accumulated biogas to be released. Samples for quality analysis could be taken from the outlet.

The container used to hold waste (digester) was prepared in such a way as to allow saturated water samples to be taken by means of aspiration into a syringe. From the quality analysis of biogas and liquid supernatant for

(volatile acids, BOD, COD, sulphide), information is obtained which is used to evaluate the trend in degradation process in various mixtures.
The four devices were filled with the amounts of ash and sludge given in Table 4.

Table 4. Quantities of ash and sludge used in the anaerobic degradation tests.

Reactor	Quantity ash (g)	Quantity sludge (g)	Ratio ash/ sludge	Quantity saturation water (litres)
1	0	150	-	1
2	300	150	2:1	1
3	410	40	10:1	1
4	450	0	-	1

3. PRELIMINARY RESULTS

The results pertaining to the batch tests are reported in full whilst the results of lysimeter and anaerobic digestion tests are still being analysed and cannot as yet be discussed. Leaching tests were performed not only on the mixture of ash and sludge but also on ash alone. The data refer to the liquid/ solid ratio 20:1 which was used to standardise the extraction test for Italian regulations (4).

Table 5 shows the results obtained in experiments using only ash and methods required by Italian law (addition of 0.5 M acetic acid at a constant pH of 5.0±0.2). Based on these values and considering the initial amounts of various metals present in the ash and leachate (filtrate), the percentage release or reabsorption by the solid matrix can be calculated on the basis of the following definitions:

$$\text{RELEASE} = \frac{\text{Quantity of metal released in liquid phase}}{\text{Quantity of metal originally present in solid phase}}$$

$$\text{REABSORPTION} = \frac{\text{Quantity of metal reabsorbed in liquid phase}}{\text{Quantity of metal originally present in solid phase}}$$

The definitions can be summarised by the following formula:

$$\%R = \frac{Q \cdot (X_2 - X_1)}{Wc \cdot Xc} \cdot 10^5 \qquad (1)$$

where:
Q = Quantity of leachate used equal to the quantity of eluant after addition of distilled water (litres).
X_1 = Concentration of metal in filtered leachate (mg/l)
X_2 = Concentration of metal in eluant (mg/l)
Xc = Concentration of metal in ash (mg/kg)
Wc = Quantity of ash (g)
10^5 = Factor necessary for dimensional respect of the formula equal to the product of 10^2 due to the percentage and 10^3 due to the changing of dimensions of Xc in (mg/g).

Table 5. Concentration of metals (mg/l) in the eluate used for experiments with different eluant phases and liquid/solid ratio 20:1.

	Water+ acetic acid	Leachate Serdiana	Leachate Rio Vigne
pH	5.2	8.6	8.4
Pb	0.03	0.018	0.016
Cd	0.011	<0.0005	0.0005
As	0.63	0.16	<0.001
Cr	-	0.083	0.045
Zn	0.4	0.082	0.21
Mn	1.05	0.1	0.31
Cu	0.1	<0.05	0.06
Ni	1.79	1.03	1.38

Table 6 shows the percentage release or reabsorption of the metals.

Table 6. Percentage release or reabsorption (-) of metals in experiments with different eluants.

	Water+ acetic acid	Leachate Serdiana	Leachate Rio Vigne
Pb	0.139	-0.0093	-0.4
Zn	1.2	0.05	-6.9
Cd	19.1	0	-3.91
As	24.2	-22.7	0
Cr	-	1	-34.4
Mn	1.9	-23.7	-11.4
Cu	3.3	0	-1
Ni	14.3	6	9.4

The results of tests using different ratios of ash/sludge are presented in Table 7 whilst the percentages of release and/or reabsorption calculated on the basis of data reported in Table 7 are presented in Table 8. In this case the results were obtained by the following formula, which takes into account metals present in the sludge:

$$\%R = \frac{Q \cdot (X_2 - X_1)}{Wc\, Xc + Wf\, Xf} \cdot 10^5 \qquad (2)$$

where:
Wf = Quantity of sludge (g).
Xf = Concentration of metals present in sludge (mg/kg).

Figures 3 to 9 show the trend of percentage release or reabsorption according to the ratio of ash/sludge used in the experiments. Trends for both methane-phase leachate (Serdiana) and acid leachate (Rio Vigne) are shown for Cr, Mn and Ni. With regard to Cd, As, Cu and Zn, the results of only one leachate are shown since the other values were either below detectable levels (Cd, Cu) or the initial quantity was extremely low (As, Zn). The trend for Pb has not been reported as the concentrations were very low.

Table 7. Concentration in mg/l of metals in eluates from experiments using ash and sludge.

Type of eluant	Leachate Serdiana			Leachate Rio Vigne		
Ratio ash/sludge	1:1	2:1	10:1	1:1	2:1	10:1
Pb	0.024	0.015	0.013	0.028	0.023	0.02
Cd	n.d.	n.d.	n.d.	0.0005	0.0005	0.0005
As	0.22	0.22	0.23	0.01	0.03	0.04
Cr	0.04	0.026	0.05	0.074	0.083	0.106
Zn	0.071	0.08	0.06	0.251	0.253	0.231
Mn	0.28	0.08	0.15	0.68	0.55	0.42
Cu	n.d.	n.d.	n.d	0.06	0.05	0.08
Ni	0.56	0.71	0.77	0.71	0.84	1.19

Table 8. Percentage release or reabsorption (-) of metals in experiments using ash and sludge. The calculations have been carried out on the basis of data in Table 7.

Type of eluant	Leachate Serdiana			Leachate Rio Vigne		
Ratio ash/sludge	1:1	2:1	10:1	1:1	2:1	10:1
Pb	0.036	-0.034	-0.036	-0.67	-0.54	-0.43
Cd	n.d.	n.d.	n.d.	-6.78	-5.5	-4.33
As	-40.6	-60.9	-20.9	0.69	1.67	1.65
Cr	-12.2	-13.9	-6.03	-46	-35.86	-23.92
Zn	0.029	0.063	-0.02	-12.3	-9.69	-7.48
Mn	-46.5	-35.4	-26.01	-21.5	-16.5	-12.4
Cu	n.d.	n.d.	n.d.	-1.61	-1.79	-0.36
Ni	4.45	5.14	4.32	7.94	7.66	8.74

n.d. = Lower than detectable value.

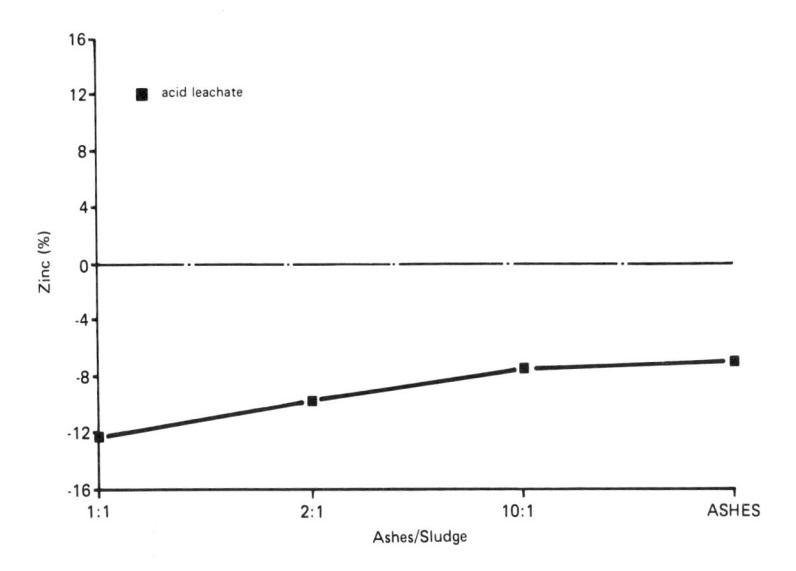

Fig. 3. Trend of Zn release in relation to ash/sludge (A/S) ratio in experiments using acid leachate (%).

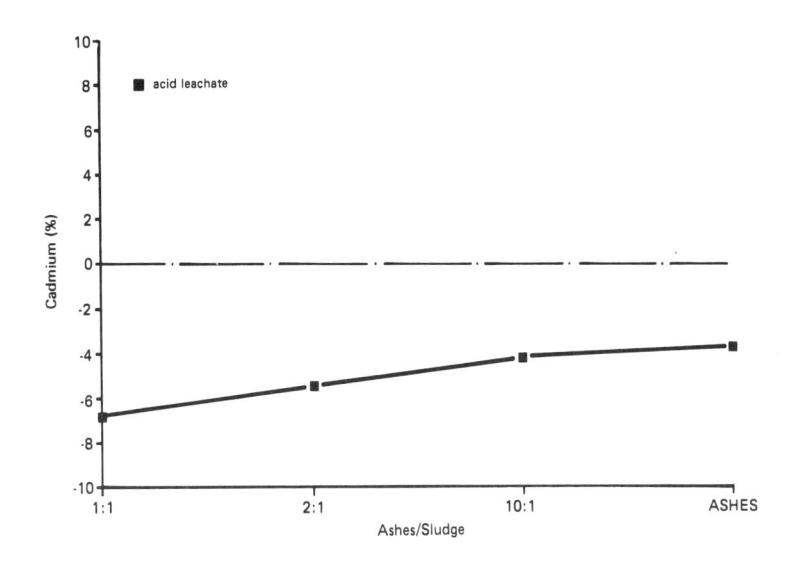

Fig. 4. Trend of Cd release in relation to A/S ratio in experiments using acid leachate (%).

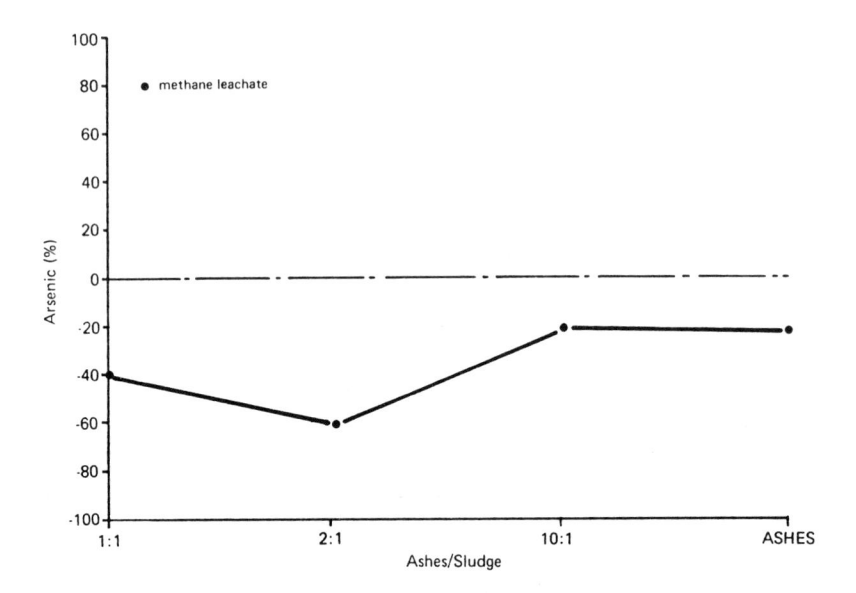

Fig. 5. Trend of As release in relation to A/S ratio in experiments using methane leachate (%).

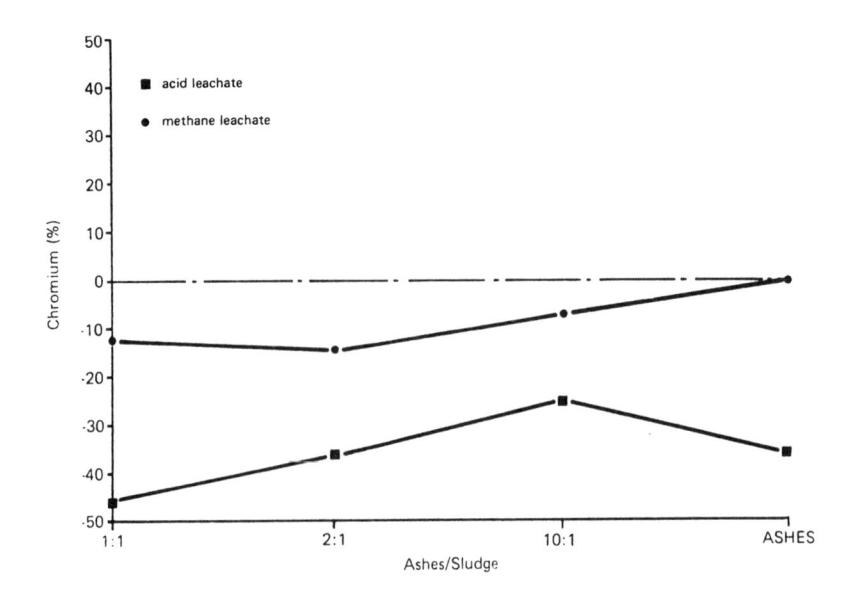

Fig. 6. Trend of Cr release in relation to A/S ratio in experiments using both leachates (%).

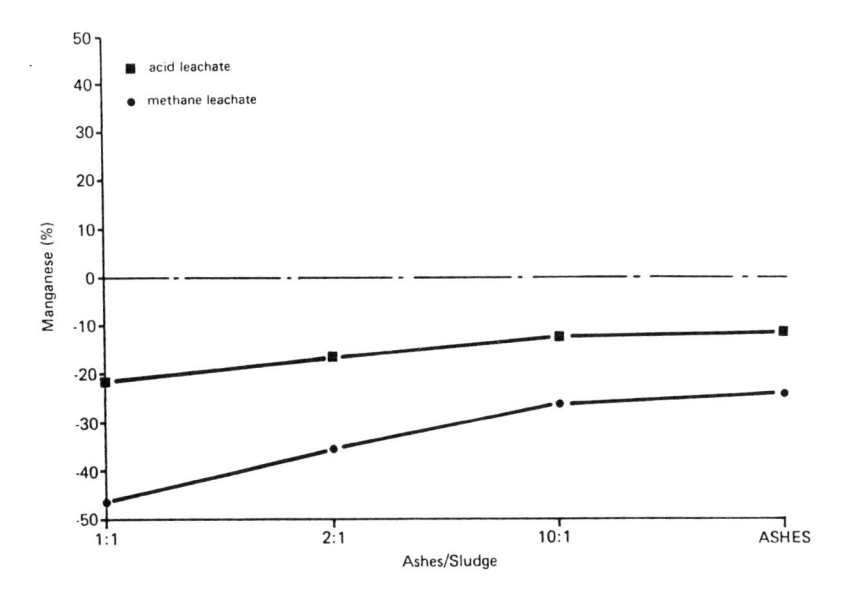

Fig. 7. Trend of Mn release in relation to A/S ratio in experiments using both leachates (%).

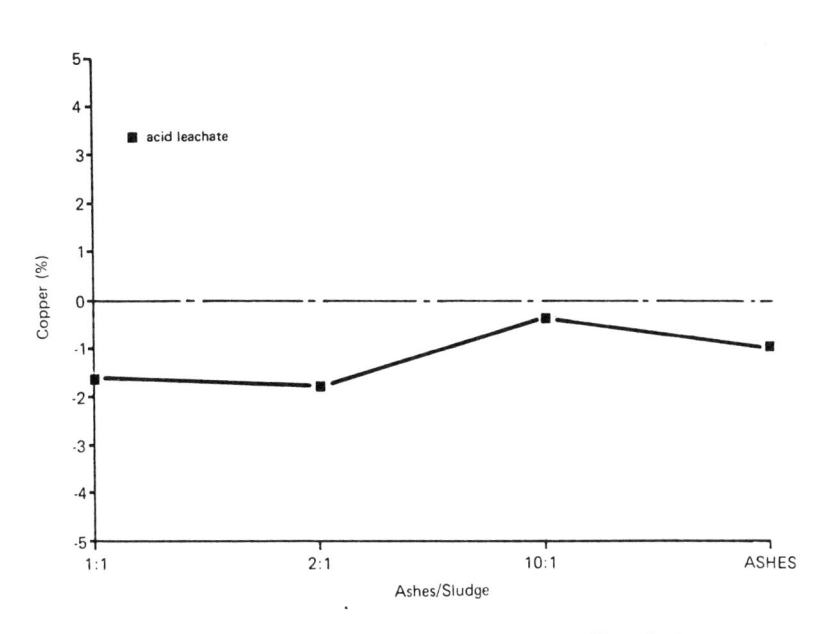

Fig. 8. Trend of Cu release in relation to A/S ratio in experiments using acid leachate (%).

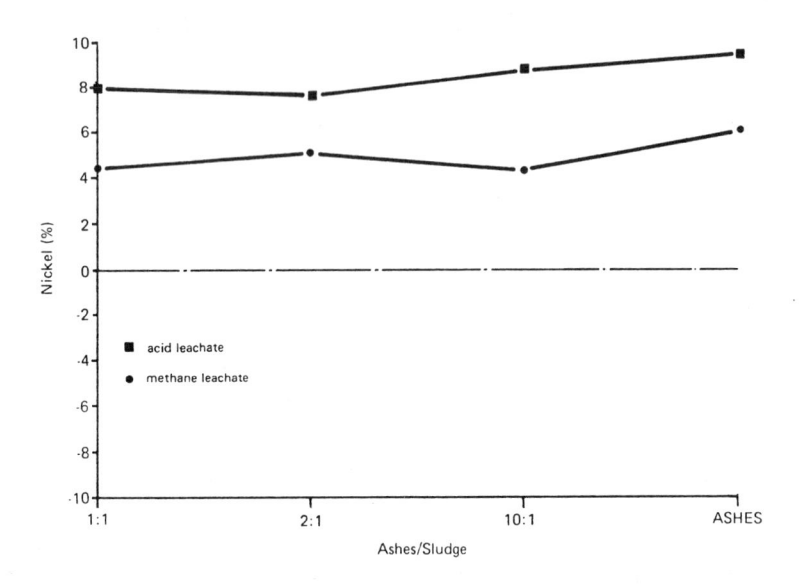

Fig. 9. Trend of Ni release in relation to A/S ratio in experiments using both leachates (%).

4. DISCUSSION

The data presented in Table 5 show that an initial pH value considerably below neutral had no influence in leachate tests on the final value, which was alkaline. Values above neutral but slightly lower than the latter (pH 7.8-8.5) were obtained in the experiment using ash and sludge.

With regard to experiments using only ash, a general increase of heavy metal concentrations in eluate resulting from tests using acid water compared to leachate are evident, even though the latter had originally certain amounts of these elements in solution. This behaviour can be linked to the method used as the leachates were allowed to vary freely their pH value whilst the test using water and acetic acid had a constant acid pH (5.0±0.2) thus creating more favourable conditions for metal solubilisation. However, the results obtained are in contrast to those of Francis *et al.* (5) who state that the eluant which better mobilises inorganic compounds is landfill leachate similar to an acetate solution with a pH of 5.0.

From Tables 5 to 8 it can be deduced that the ash and ash/sludge mixture reabsorbed a certain quantity of metals. Nickel is one of the more mobile elements, which is in accordance with results obtained by other authors (6).

The following details can be shown after further examination of the behaviour of each single element with respect to the amount of sludge present in the solid matrix.

Zinc demonstrated a changeable trend in the region of zero in the presence of methane leachate. On the other hand, acid leachate showed a more marked reabsorption in those instances where a greater quantity of sludge had been used. It is probable that the uncertain trend observed with methane leachate can be attributed to the low amount of Zn present in original leachate (Table 2).

No deductions can be drawn from the tests carried out using methane leachate in the case of Cd whilst acid leachate demonstrated a marked reabsorption in the solid matrix containing higher quantities of sludge. It should be emphasised that the results obtained are limited by low levels of Cd both in the solid matrix and in original leachate.

The behaviour of As is of interest when in contact with methane leachate which is characterised by a high concentration of this element. Reabsorption was extremely high, especially for ash/sludge ratios 1:1 and 1:2. Data pertaining to acid leachate are of little significance as the original concentration was very low.

Chromium is seen to be reabsorbed in high amounts by acid leachate and this is enhanced by the presence of domestic sludges. Manganese follows a similar trend although the higher levels of reabsorption were observed with methane leachate.

With regard to Cu the same observations can be applied as those for Cd when using methane leachate. Acid leachate showed only slight reabsorption.

Nickel was the only element which was invariably released, although this was slightly less so in the presence of sludge.

In the above discussion, the percentage of release or reabsorption and not the actual concentrations observed (Tables 5 to 7) have been considered as the latter depend on the amounts present in original leachate whilst the former do not. However, it has been shown that the effects are enhanced when the concentrations present in the original leachate are high.

5. CONCLUSIONS

A research programme has been set up at Cagliari University to study the possibility of co-disposing an ash/domestic sludge mixture in an MSW landfill.

The addition of domestic sludge with a high moisture content helps to overcome the problem of dust in a landfill for ash disposal, and the presence of ash overcomes the problem of land-filling semi-liquid sludge. It has been observed that a 10:1 ratio of ash/sludge is sufficient to avoid the dust problem when handling ashes.

The batch tests for release of elements which used actual leachate from landfills as a liquid phase showed that the majority of heavy metals are absorbed by ash and that this effect is enhanced by the addition of sludge. Therefore the contact with these wastes represents a means of immobilising metals naturally present in leachate from MSW.

Domestic sludges are thus capable of solving operational problems of disposing of coal fly-ash in an MSW landfill and of enhancing the absorption phenomena typical of this type of waste. The long-term immobilisation of metals will be verified from the results of lysimeter tests which are not yet completed.

Other ongoing tests are anaerobic digestion aimed at verifying eventual inhibition of biodegradation processes, and evaluating the influence of alkaline ashes on stabilising sludges, particularly in view of disposal of such a mixture in a mono-landfill.

ACKNOWLEDGEMENTS

The authors wish to thank the Department of Environment of Regional Government for financial assistance, the ENEL (National Agency for Energy) and ESAF (Regional Agency for Water and Waste Water) which kindly permitted respectively the sampling of ash and domestic sludge.

REFERENCES

(1) Cernuschi, S. and M. Giugliano (1986). The atmospheric mobilisation of trace elements from coal combustion. Proceedings of the 7th World Clean Air Congress, Sydney, Australia.
(2) Duranceau, P. E. (1987). USEPA's new leaching test: the toxicity characteristic leaching procedure (TCLP). Proc. Tenth Annual Madison Waste Conference, Municipal and Industrial Waste, September, 29-30.
(3) Cossu, R. and R. Serra (1989). Co-disposal of coal fly-ash and MSW in sanitary landfill. Sardinia '89 - 2nd International Landfill Symposium, Porto Conte, Italy, 9-13 October, Proc. in press.
(4) Comitato Interministeriale (1986). Modificazioni e integrazioni alle disposizioni per la prima applicazione dell'art. 4 del DPR 915/82. G.U. no. 183 del 8.8.1986 (Italian Act).

(5) Francis, C. W., M. P. Maskarinec and J. C. Goyer (1986). A laboratory extraction method to simulate co-disposal of solid wastes testing. Fourth Symposium, ASTM STP 886, Philadelphia.
(6) Young, P. J., G. Baldwin and D. C. Wilson (1984). Attenuation of heavy metals within municipal waste landfill sites. Hazardous and industrial waste symposium, ASTM STP 851, Philadelphia, 193-212.

The co-disposal of controlled waste and sewage sludge - some practical aspects

C. P. HILL

West Yorkshire Waste Management, Chantry House,
123 Kirkgate, Wakefield, West Yorkshire, WF1 1YG, UK.

SUMMARY

With future restrictions likely on the traditional use of agricultural land and the sea for the disposal of sewage sludge, there is a need to establish alternative arrangements. Co-disposal with controlled waste offers a promising alternative, but techniques employed must not prejudice the landfill disposal operation or have any adverse effects on the environment. This paper describes the techniques developed by West Yorkshire Waste Management for the landfill co-disposal of dewatered sewage sludge and controlled waste (loose and baled) and examines the accompanying environmental effects.

1. INTRODUCTION

Approximately 1.2 million tonnes of sewage sludge dry solids are disposed of every year in the United Kingdom at an annual cost of about £100m. Locally, within West Yorkshire, the level of sludge production is in the order of 77,000 tds per annum.

The disposal of such material is not always straightforward, as the more traditional methods employed are becoming increasingly more environmentally unacceptable.

Application to agricultural land is still the largest single outlet, accounting for about 40% of the total UK sludge produced. It is generally the cheapest disposal option, particularly for inland sewage treatment works, and gives benefits to the land from the soil conditioning and nutrient properties of the sludge.

However, disposal at sea which, as a disposal route, accounts for up to 30% of the national and 7% local sludge arisings in West Yorkshire, is becoming increasingly unpopular, particularly with environmental pressure groups. Indeed, EC legislation is likely to be introduced which will ban disposal at sea.

Increasing pressure is therefore being put on the water industry to develop alternative strategies for sewage sludge disposal.

Solutions which are being sought in West Yorkshire include the introduction of incineration plants and also the investigation into improved methods of co-disposal with controlled waste to landfill, a method which currently accounts for 15% of the UK sludge production.

Previous techniques of co-disposal to landfill have had limited success and have usually met with resistance from landfill site operators. However, it is now believed that with co-operation between the waste disposal operator and the sludge producer to provide an easily handleable material, and given the right conditions, co-disposal can be a satisfactory disposal option.

It is the purpose of this paper to describe the experience gained by one large municipal Waste Disposal Authority, West Yorkshire Waste Management, in the co-disposal at a landfill of sewage sludge with controlled waste.

2. AVAILABLE LANDFILL FACILITIES IN WEST YORKSHIRE

West Yorkshire Waste Management (WYWM) is the statutory Waste Disposal Authority which regulates and controls the disposal of controlled waste within West Yorkshire.

It is responsible to a voluntary Joint Committee of elected members from each of the constituent District Councils of Bradford, Calderdale, Kirklees, Leeds and Wakefield, an area which covers some 204,000 hectares with a population of just over 2 million.

In addition to its powers of regulating private waste disposal facilities, WYWM also operates its own disposal facilities. These include 16 landfill sites, 2 municipal waste incinerators, 2 high density baling plants, 7 transfer loading stations and 35 household waste sites. In total WYWM handles about 1.4 million t/year of controlled waste (i.e. household, industrial and commercial), including 5,000 t/year sewage sludge, of which over 90% is ultimately landfilled. A typical refuse analysis of household waste for West Yorkshire is given in Table 1.

The majority of household waste delivered to the landfills is loose, having had minimal compaction in the collection vehicle, but some, approximately 177,000 t/year, is in the form of high density bales. These two distinctly dissimilar forms mean that different techniques of co-disposal with sewage sludge have to be adopted at the landfill.

In general, although both forms of waste are disposed of at the same landfill, dedicated working areas have to be operated. This is because of the need for differing on-site mobile plant requirements.

Table 1. Analysis of household waste, West Yorkshire Metropolitan County Council (1982-1984).

	% by weight		
	1982	1983	1984
Vegetable/putrescible	23.3	25.5	30.8
Paper	30.2	26.7	23.6
Metals	6.4	6.6	6.0
Textiles	4.1	3.4	3.8
Glass	7.2	9.8	7.9
Plastics	6.9	6.6	8.2
Screenings <2 cm	14.5	18.0	10.1
Unclassified	7.4	3.4	9.6

Sewage sludge is currently disposed of at three landfill sites that are operated by WYWM.

3. SEWAGE SLUDGE DISPOSAL REQUIREMENTS IN WEST YORKSHIRE

There is a requirement in West Yorkshire for the local Water Authority (Yorkshire Water) to dispose of approximately 77,000 tds/year of sewage sludge. Of this total, there is an estimated need to co-dispose with controlled waste to landfill some 26,000 t/year.

The physical properties of this sludge are variable. They range from liquids with 4-8%ds to dewatered sludges with 20-30% ds.

Liquid sludges, by their very nature, are unsuitable for co-disposal. The excessive water content makes landfill working conditions extremely difficult; it also increases the likelihood of unacceptable quantities of leachate being produced.

Dewatered sludges, on the other hand, are more solid and lend themselves more favourably to a co-disposal operation, although landfill disposal is intended for solid wastes only and any liquid input will add to the operational difficulties.

The following operational procedure has been adopted for the disposal of dewatered sewage sludge which can either be in the form of pressed or centrifuged cake. A typical analysis of the latter is given in Table 2.

Table 2. Typical analysis of centrifuged sewage treatment plant residue.

Dry solids	28.7%
Volatile solids	43.8%
Chromium	660 mg/kg
Copper	360 mg/kg
Nickel	68 mg/kg
Zinc	1,100 mg/kg
Cadmium	6 mg/kg
Lead	490 mg/kg
Iron	28,000 mg/kg
Manganese	1050 mg/kg

4. OPERATIONAL CO-DISPOSAL PRACTICE

The co-disposal of sewage sludge and controlled waste has been practised in a variety of ways for some time. Methods range from the addition of controlled waste to previously formed lagoons of sewage sludge, to admixing the sludge and waste either in trenches or in thin layers between the emplaced waste. The latter method is favoured by WYWM, and described below, although the disposal of baled waste means that a modified technique is necessary.

4.1 Co-disposal of sewage sludge and baled controlled waste

The high density bales, each weighing approximately 1.2 tonnes and measuring 1 m x 1 m x 1.5 m, are delivered to the landfill in batches of 14 by articulated tractor and trailer units. They are off-loaded and stacked, two high, in a similar fashion to building blocks, by wheeled loading shovels with fork attachments.

It is not recommended that the bales are placed onto a thin layer of sludge because the loading shovels become contaminated and have subsequent difficulty with traction. It is much better to form a series of bays, using the bales to create the side and rear walls, into which the sludge can be deposited. The sludge delivery vehicle is then able to reverse into the bay and deposit its load.

Each bay, which is contiguous with the operational working face, comprises an area 3 bales wide by 2 bales deep and is able to contain approximately 20 tonnes of dewatered sludge of 28% ds. The filling sequence is depicted in Figure 1.

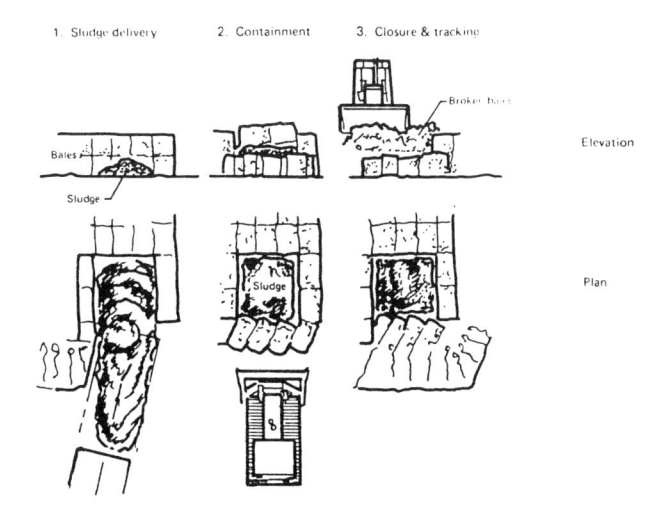

Fig. 1. Co-disposal of sludge with baled waste.

Upon filling, the open end of the bay is closed off by a further 3 bales. By manoeuvring these bales towards the rear of the bay, using the on-site mobile plant, the sludge is squeezed until its upper surface is level with the top of the adjacent lower layer of bales. The cell is then covered by pushing whole and broken bales onto the surface of the sludge and tracked in using a crawler bulldozer.

The bales, which form the cell, restrain the movement of the sludge, particularly when the top covering of waste is applied, and enable the upper surface to be consolidated, giving relatively stable ground conditions.

Using this technique, it is possible to encapsulate 20 wet tonnes of sewage sludge with around 80 tonnes of baled controlled waste, with minimal contamination of on-site mobile plant. This gives a controlled waste to sewage sludge ratio in the order of 4:1.

For operational reasons, however, it is advisable that the cells be dispersed as widely as possible across the landfill area, to avoid compounding the possibility of creating localised 'soft spots'. As such, it is recommended that the optimum waste to sludge ratio should be no less than 10:1.

4.2 Co-disposal of sewage sludge and loose controlled waste

A more conventional approach is adopted in this case. Under normal circumstances, the disposal of loose controlled waste is undertaken in accordance with the guidelines given in Waste Management Paper No. 26. Briefly, this involves the waste from the collection vehicle being deposited at the crest of the landfill tipping face. It is then subsequently pushed over the inclined face and consolidated by means of successive passes of a mobile steel-wheeled compactor. Using this method it is possible to achieve an emplaced density of up to 1 t/m^3.

A slight modification to this technique is required, however, when sewage sludge is added. Experience has shown that the optimum method of incorporating this material is to apply thin layers between successive lifts of controlled waste.

The sludge delivery vehicle deposits its load at the toe of the working face. Using the blade of the mobile compactor, the sludge is spread evenly in a thin layer approximately 0.25 m thick over the underlying waste. Loose refuse is then pushed down the inclined face onto this layer, always ensuring that there is sufficient thickness to prevent the sludge forcing its way to the surface and contaminating the compactor. The minimum thickness of loose waste required is usually 1 m, resulting in lower achievable emplaced densities.

This sequence of operation, which is shown in Figure 2, is repeated on the adjacent landfill area until the entire sludge load has been covered. Although the mobile compactor generates very high ground pressure point loading across the landfill surface, sludge should not come into contact with its wheels, except during initial spreading, provided that there is sufficient thickness of overlying refuse cover.

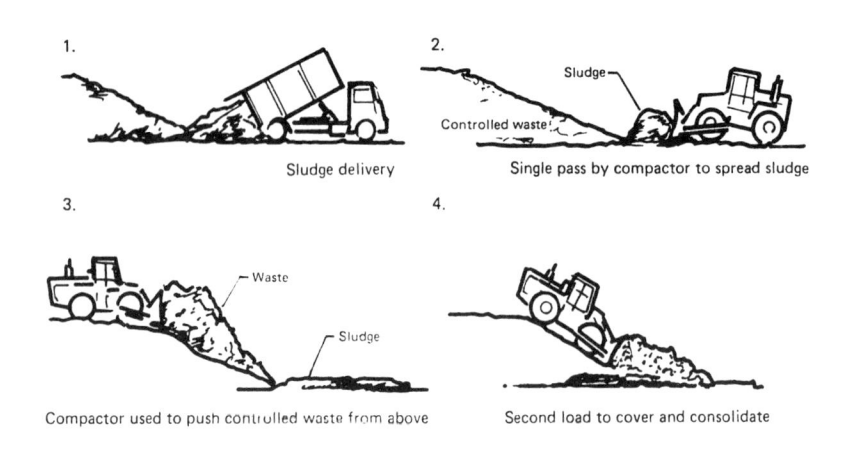

Fig. 2. Co-disposal of sludge with loose waste.

Using this technique, it is possible to co-dispose of crude controlled waste and sewage sludge in the ratio of 3:1. Compared with the co-disposal with baled controlled waste, there is not the same restriction on the location of the sludge disposal area, provided that the thickness of the sludge layer is acceptable and covered by sufficient controlled waste. However, the effect of seasonal weather conditions does have an influence in determining the acceptable quantities of sludge additions.

5. EFFECTS OF CO-DISPOSAL

The co-disposal of controlled waste and sewage sludge is not without its side effects. Whether these are beneficial or detrimental, a balanced view must be taken in assessing the overall suitability of the process as a disposal option. Parameters in need of consideration include settlement, gas and leachate production, odour and contamination of the landfill mobile plant.

The latter two can be readily assessed at the landfill by means of comments received from the site operators and local residents, and also by visual inspection of the on-site mobile plant. However, the effect on settlement, and the production of gas and leachate is more difficult to establish, particularly on a large operational landfill where the distribution of the deposited waste types may not have been uniform. This makes accurate comparisons with conventional controlled waste landfilling extremely difficult.

Nevertheless, in an attempt to measure these effects, a series of four test cells has been constructed at a landfill and filled with various mixes of controlled waste and sewage sludge, and a specific monitoring programme initiated. This work, which is part of a joint research programme involving collaboration between WYWM, the Water Research Centre, Yorkshire Water and Wakefield Metropolitan District Council is now in its second year of a three year monitoring phase.

Each cell measures approximately 25 x 15 x 4 m and contains 600 tonnes respectively of the following mixes: baled waste, baled waste and sewage sludge, loose controlled waste, and loose controlled waste and sewage sludge. In those cells where sewage sludge is present, the waste to sludge ratio is approximately 10:1.

In all cases, filling was undertaken using the method described previously although this was subject to the constraints of the cell dimensions. Each cell was completed by the application of suitable capping material and sown with grass. Provision was made for monitoring settlement, and gas and leachate production.

5.1 Settlement

By comparison with controlled waste only landfill sites, the co-disposal of controlled waste and sewage sludge tends to give rise to a greater incidence of settlement.
Furthermore this settlement appears to be more pronounced where high density bales are involved. Here the creation of low spots corresponding to the location of the sewage sludge 'pockets' is evident. Obviously, this is undesirable, particularly when the site is reaching final contour level. Consequently, it is recommended that deposits of sludge are not made in the final 5 m of a balefill site.

5.2 Landfill gas

Methane is produced as a direct result of the anaerobic decomposition of organic waste. Both controlled waste and sewage sludge are organic in character, and consequently will generate methane under the right conditions. However, it appears that the addition of sewage sludge to controlled waste has a catalytic effect on the decomposition process, accelerating the onset of methanogenesis. As a consequence, methane concentrations of around 50% by volume are evident within 2 months of the waste and sludge emplacement. By comparison, in sites taking controlled waste only, this state of decomposition can take over 12 months to reach.

Unfortunately it has not yet been possible to measure the rate of methane production in order to compare the effects of the different operational methods.

5.3 Leachate

The disposal of dewatered sludge to landfill, by its very nature, introduces liquid into the waste and affects the overall site water balance. Resultant leachate volumes are not only dependent upon the quantities of sludge disposed but, more importantly, its ratio to controlled waste. This is common to both loose and baled waste. Therefore, to minimise the likelihood of leachate production, it is essential to keep the ratio of waste to sludge above 10:1.

The quality of leachate is also influenced by the addition of sewage sludge. In general there appears to be a reduction in the concentration of certain metals (i.e. Fe and Mn), but this is accompanied by a corresponding increase in ammoniacal nitrogen and total phosphorus levels.

5.4 Odour

Co-disposal operations are normally accompanied by a release of odour from the sludge. Unfortunately, this odour seems to be specific to the treatment process at the originating sewage works and beyond the control of the landfill operator. However, its effect can be reduced by covering the sludge with waste immediately after its delivery to site. Failing this, it is essential that the sludge is stockpiled for as short a time as possible.

5.5 Mobile plant contamination

Due to the intimate mixing of the sewage sludge and controlled waste, which is necessary in the co-disposal operation, it is impossible to give assurances that the mobile plant will not become contaminated. The main causes of concern are the wheels, tracks and front blade of the plant. However, by careful site husbandry, it is possible to reduce this contamination to a minimum. It is also important that proper arrangements are made for machines to be quickly and easily washed down and cleaned, and the wash water disposed of properly.

6. CONCLUSIONS

From the experience gained by West Yorkshire Waste Management, the following conclusions can be drawn:
a) The limited trials undertaken so far appear to show that the co-disposal of sewage sludge and controlled waste (loose and baled) is a promising disposal option, but there is a need to evaluate the long term effects of the operation;
b) Close co-operation is needed between the sludge producer and landfill operator to ensure that the sludge is delivered in a state which is compatible with the disposal operation;
c) The location of the sludge deposits, particularly where baled waste is concerned, needs careful consideration to avoid undesirable low spots occurring in the final contours;
d) Methane production appears to be accelerated with co-disposal. This is of particular value where the opportunity exists for the commercial exploitation of the gas;
e) Leachate quality and quantity appear to be affected by the addition of sewage sludge to the controlled waste. The reduction in concentration of certain metals (Fe and Mn) is offset by a corresponding increase in the level of ammoniacal nitrogen and phosphorus.

ACKNOWLEDGEMENT

The author wishes to thank the Chief Waste Management Officer of West Yorkshire, K. J. Bratley, for permission to publish this paper. The views expressed are those of the author and not necessarily those of West Yorkshire Waste Management.

REFERENCES

(1) West Yorkshire Waste Management Joint Committee (1989). West Yorkshire Draft Waste Disposal Plan.
(2) Leslie, D. (1986). The co-disposal of sewage sludge and solid waste at Clifton Marsh landfill site. IWM, N. West Centre Meeting, Blackpool.
(3) Lowe, P. (1987). Incineration of sewage sludge - back to the future. Seminar on treatment and disposal of industrial waste - Institute of Environmental Management, Leeds.
(4) Department of the Environment/National Water Council (1981). Report of the Sub-Committee on the Disposal of Sewage Sludge to Land, DoE.
(5) Craft, D. G. and P. J. Maris (1988). Co-disposal of sewage treatment plant residues and controlled wastes at Brick and Tile Works, Huddersfield. Report PRS 1903-M/1, Water Research Centre, Medmenham, UK.

Environmental aspects of landfilling sewage sludge

D. BEKER[1] and J. J. van den BERG[2]

[1]RIVM, PO Box 1, 3720BA Bilthoven, The Netherlands.
[2]Grontmij, De Holle Bilt 22, De Bilt, PO Box 203,
3730 AE de Bilt, The Netherlands.

SUMMARY

The dumping of sewage sludge has to be assessed with regard to environmental impact due to the relatively high content of heavy metals, the large volume for disposal, and the absence of alternative outlets in the short term. The total quantity of sludge produced in the Netherlands is about 400,000 tds/year of which 40% is landfilled, mostly together with other wastes. It was found that the quantity of sludge being landfilled does not exceed 10% of the total amount of waste. An effect of sludge on leachate quality could not be found with regard to heavy metals and some other parameters like COD, Kj.N and Cl. The same results have been found for groundwater. Due to the relatively low contribution of polluting components from sludge to the total quantity of polluting components in the landfilled waste, a large effect was not to be expected. Supplying large quantities of sludge (25%) to other wastes did have an influence on the quality of leachate, such as lower COD and VFA concentration and higher pH value. As a result of the higher pH value, the concentrations of heavy metals were reduced. Within 6 years this difference will probably disappear, and the values of COD, VFA and pH will be in the range normally observed following stabilisation.

1. INTRODUCTION

During the last few decades there has been an increase in sewage sludge production to approximately 400,000 tds/year in the Netherlands. The expectations for the near future are that industrial activities will increase and, in the Netherlands, a small growth of population is to be expected. Therefore, the amount of sewage sludge produced will reach 500,000 - 600,000 tds by the year 2000. This excludes the expected increase in the production of sewage sludge caused by phosphate removal from sewage. At the moment in the Netherlands about 30% of the sewage sludge produced is used in agriculture. However, in the near future this outlet may come to an end, first of all because of heavy metals and other contaminants present, and, secondly, because there is competition with other organic materials such as manure and the organics in household refuse that can be used as fertiliser or for soil improvement. The demonstrated presence of organic pollutants such as dioxins and furans also does not stimulate the agricultural use of sewage sludge. If the use of sewage sludge in agriculture is not possible, there are at least in the Netherlands some alternative outlets, namely landfilling, incineration, and wet oxidation (Vertech system).

About one third of the total amount of 400,000 tds is disposed of in landfill sites, mostly together with other wastes (municipal solid waste, industrial waste etc.). Due to the large volumes of sewage sludge, there is a tendency to develop treatment techniques to reduce this volume. However, in the Netherlands at this time only incineration is a serious alternative route that has proved to be a realistic one. The environmental aspects of incineration have stimulated the development of alternative destruction techniques such as pyrolysis and wet oxidation. These developments are still in progress, but at the moment it seems that they are not a realistic solution for the sewage sludge problem in the following decade. This requires in the near future sufficient capacity for landfilling of sewage sludge and incinerator ash.

1.1 Legislation and administration

Landfill disposal has been practised in the Netherlands for centuries. Sludges have always been and still are disposed of at landfill sites. During recent years landfilling practice has been changed to meet certain environmental requirements. The general introduction of the well-engineered and well-managed sanitary landfill started around 1973 when the World Health Organisation introduced the 'Code of Practice for the Disposal of Solid Wastes on Land' (WHO Copenhagen, 1973). Based on this code, but adapted to the specific circumstances in the Netherlands, a Dutch code of practice was issued, including recommendations on site investigation, site selection, planning, construction, management, techniques and controlled tipping (1). Important legislative frameworks regarding the improvement of operating landfill sites are the Water Pollution Act (1969), Hazardous Waste Act (1976), Act on Waste Materials (1977) and the Soil Protection Act (1988).

For sewage sludge disposal the legal framework is extended by the Anti-Nuisance Act (1953), Air Pollution Act (1970) and the Fertiliser Act (1986).

The Water Pollution Act gives rules on the conditions under which dumping of sludge in surface water is prohibited. The Soil Protection Act gives rules for the use of sewage sludge, and the Anti-Nuisance Act and the Air Pollution Act give rules on acceptable emissions.

Under these Acts technical Guidelines and Provisions (permits required) were issued. However, over the years it became clear that legislation was too segregated to tackle the environmental problems efficiently. Therefore it is intended to integrate the Act on Waste Materials and the Hazardous Waste Act.

The first Guideline (under the Act on Waste Materials) with respect to landfilling was issued in 1980. With regard to the major topic, the protection of soil and groundwater, it was ruled that the bottom of the body of the landfill should be 0.5 m above the groundwater level. To avoid pollution of the surface water, an adequate drainage system (ring ditches, etc.) had to be constructed.

Since 1982, landfill sites must be granted a permit from local and provincial authorities. To determine the conditions for this permit, the authorities employ the Guideline "Controlled Landfilling', from the Ministry of Housing, Physical Planning and the Environment under the Act on Waste Materials. In 1984, on account of the Water Pollution Act, a provision was issued that determined that a permit was required for the discharge of leachate of landfill sites into surface water.

The second revised Guideline was published in 1985. This required that for new sites and enlargement of existing sites, a layer impermeable to water has to be constructed between the waste and the original soil surface.

1.2 Quantities and composition

In the Netherlands, the total quantity of waste amounts to about 115 t/year. Table 1 shows the sources of wastes and disposal methods (2). According to Table 1, about 13.5 million t/year are landfilled directly and about another 0.5 million t/year are landfilled as slag and fly-ash from incinerated waste (assuming 25% residue by weight, of which 80% is landfilled). In most Dutch landfill sites there are different types of waste present, such as from households, building and demolition, sewage treatment plants, industry, markets and sweeping. During the landfilling operation, most types of landfilled wastes are mixed with other types.

Whilst about 40% of the 400,000 tds/year of municipal and industrial sewage sludges produced in the Netherlands is landfilled, other important disposal outlets are agriculture and incineration. Due to the presence of heavy metals the use of sewage sludge in agriculture has been restricted in recent years. Table 2 shows the concentrations of heavy metals in sewage sludge (2). On account of the Fertiliser Act (1986) legislation is provided to restrict further the use of sewage sludge in agriculture. Table 3 gives the

Table 1. Estimated annual amounts (t/year) and disposal methods per waste type (source) in the Netherlands (1986).

Waste	t/year	Recycling	Incineration	Landfilling	Misc.[1]
Household waste	4,700,000	15%	34%	51%	
Bulk household waste	600,000	8%		59%	
Road sweeping/market/park waste, flotsam,			33%		
dredged slurry, pit mud	1,400,000	14%	7%	79%	
Office/shop/service waste	1,800,000	17%	22%	61%	
Building and demolition waste	7,750,000	41%	1%	58%	
Sewage sludge (wet)	4,710,000	59%	1%	10%	30%
Sewage sludge (ds)	398,000	51%	2%	43%	4%
Industrial waste[2]	15,100,000	78%	1%	21%	
Hospital waste	115,000		52%	48%	
Car tyres	65,000	23%	54%	23%	
Shredder waste (mainly car wrecks)	510,000	78%			
Phosphoric acid gypsum	2,000,000			22%	100%
Fly-ash from coal-fired power plants	530,000	94%		6%	
Hazardous waste and waste oil	1,100,000		21%	9%	70%[3]
Dredged spoil	60,000,000				100%
Surplus of manure	14,000,000	100%			
Contaminated soil	500,000	55%		45%	

[1] Mainly discharged into surface water.
[2] Excluding sewage sludge and hazardous waste.
[3] Including recycling.

maximum tolerable concentrations of heavy metals (3). Besides heavy metals, agricultural use of sludge can be restricted further due to limitations in the use of phosphate. Sewage sludges have a large range of different dry matter contents but for agricultural purposes sludge is not dewatered; other outlets require a certain degree of dewatering.

The degree of contamination of the sludge depends on the types of industry, and the basic load from household, piping, gutter, traffic, air deposition, etc. Table 2 shows that the concentration of heavy metals has decreased from 1981 to 1986.

The limited re-use of sewage sludge has resulted in a need to consider landfilling as a possible option. In the following sections landfilling of sludge will be described in detail.

Table 2. Concentrations of heavy metals in sewage sludge from municipal sewage treatment plants in the Netherlands (mg/kg ds).

Year	Cu	Cr	Zn	Pb	Cd	Ni	Hg	As
1981	494	200	1,750	450	8.3	76	3.3	6.8
1982	499	156	1,663	410	9.3	71	3.2	6.9
1983	484	133	1,562	377	7.7	59	3.0	6.1
1984	492	129	1,530	360	7.1	52	2.3	6.7
1985	461	142	1,451	348	5.1	55	2.4	6.8
1986	471	100	1,398	316	6.0	50	2.2	6.8

Table 3. Maximum tolerable concentrations of heavy metals in sewage sludge for agricultural purposes in the Netherlands (mg/kg ds).

Period	Cu	Cr	Zn	Pb	Cd	Ni	Hg	As
Up to 31/12/1990	900	500	2,000	500	5.0	100	5.0	25
From 1/1/1991 to 31/12/1994	425	350	1,400	300	3.5	70	3.5	25

2. LANDFILLING

2.1 Introduction

At the moment, some 236 landfill sites are active facilities (4). Of the 113 sites with a documented surface area, about 50% are smaller than 10 hectares. In Table 4, additional documented characteristics of the landfill sites are given which indicate that there are still a lot of small landfill sites.

In addition, there are many old (abandoned) sites. The exact number of old locations is unknown, but estimates vary from 3,200 to 4,500 (5).

It must be noted that many landfill sites have been partly constructed according to the directives mentioned above. Generally, the conditions have been improved after the introduction of the technical guidelines (1980 and 1985).

Table 4. Landfill sites according to capacity.

Capacity t/year	Number of sites	Total t/year
0 - 100,000	55	1,854,200
100,000 - 200,000	11	1,619,000
200,000 - 300,000	7	1,882,000
300,000 - 400,000	2	679,000
400,000 - 500,000	2	1,000,000
500,000 - 1,000,000	1	1,000,000
Total	78	8,034,200

2.2 Processes in the landfill body

In practice the landfilling of sewage sludge has physical and chemical/environmental aspects.

The physical properties of sewage sludge are with regard to the treatability of sludge at the landfill site, the adhesion of sewage sludge on tyres of the transport vehicles and the machines used to increase the density of the landfill body, and the influence of sewage sludge on the soil mechanical behaviour of the landfill body

A major problem of landfilling of sewage sludge is the volume. The available space for landfilling is primarily for the dumping of municipal solid waste, industrial waste, etc. From the point of view of physical planning a political and technical resistance exists against the utilisation of the landfill site for sewage sludge disposal. However, desk studies carried out by the government indicate that, even if the capacity of sludge incineration should increase by a factor of 3 - 4, the space requirements for landfill sites will increase for the next fifteen years.

The chemical aspects are related to the environmental aspects and involve the composition and decomposition of the wastes dumped, the leachability and the emissions to the environmental compartments, air, water and soil. In this paper the environmental aspects are described.

The most important water flow from a landfill site is leachate. It results from infiltration of surplus precipitation into the landfill body. During infiltration a number of processes can take place, such as filtration, adsorption, dilution, diffusion, precipitation and microbiological decay. The way these processes occur in a landfill body are very complex and poorly understood. Due to the knowledge of water treatment the microbiological processes are somewhat better described. Microbiological processes occur especially in landfills with household refuse. In the pores of newly deposited waste there is still oxygen, which will be used quickly by aerobic micro-organisms. Then the anaerobic phase starts and is characterised by a succession of processes described in Figure 1 (6).

Biopolymers
(carbohydrates, proteins, fats)

Hydrolysis (I)

Monomers
(sugars, amino acids, fatty acids)

Acidification (II)

Volatile fatty acids
Ethanol

Acetogenesis (II)

Acetate, CO_2, H_2

Methanogenesis (III)

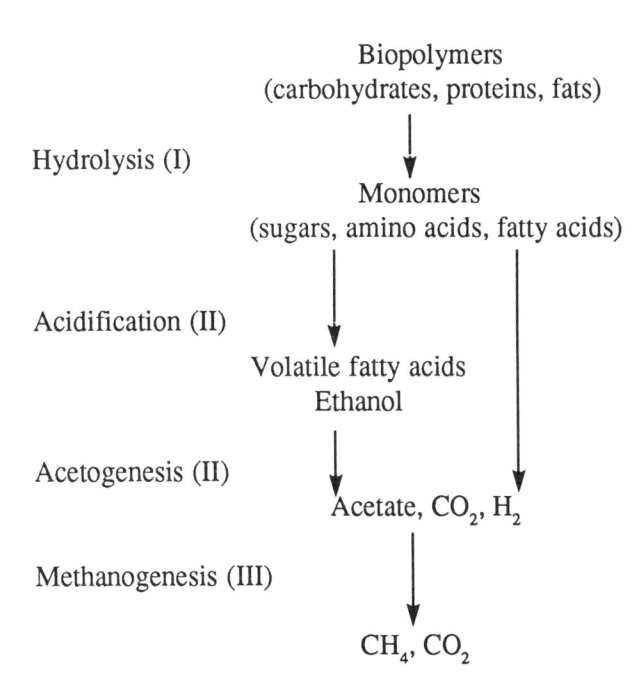

CH_4, CO_2

Fig. 1. Scheme of the conversions occurring in anaerobic degradation of organic material to methane and carbon dioxide.

Table 5. Recommended target values for discharges of treated leachate.

		Surface water		Municipal sewage system	
		A	B	C	D
COD	mg/l	100	1,000	Dependent on specific local circum-	
BOD_5^{20}	mg/l	20	20	stances such as designed capacity	
Kj.N	mg/l	20	20	and real load sewage treatment plant.	
pH		6.5 - 9		6.5 - 9	6.5 - 9
Cd	µg/l	2.5	5	5	50
Hg	µg/l	0.5	2.5	2.5	5
As	µg/l	50	50	50	50
Zn	µg/l				
Cr	µg/l				
Ni	µg/l	400	1,000	2,000	3,000
Pb	µg/l				
Cu	µg/l				
Btex[1]	µg/l	5	100	500	500
Other organic micropollutants		Research		Research	
Parameters like: PO_4^{3-}, O_2, pH Cl^-, SO_4^{2-} NO_2^-, NO_3^- smell, colour		Research, dependent on specific local circumstances; supplementary requirements may be necessary			

A Discharging treated methanogenic leachates into relatively vulnerable surface water.
B Discharging treated methanogenic leachate into relatively less vulnerable surface water.
C Discharging methanogenic leachate.
D Discharging treated acidogenic leachate and xylene.
[1] Benzene, toluene, ethylbenzene and xylene.

The acidification rate for soluble substrates (glucose, saccharose) is much higher than the methane production rate. In the leachate of landfills, for instance, the COD can consist of 90% volatile fatty acids (7).

2.3 Environmental protection

The location of landfill sites is selected in part on the basis of evaluation of soil physics and hydrogeology; sometimes risk analyses have also been performed. The waste is deposited at least 50 cm above groundwater level (average highest level occurring once per 10 years). Drainage systems and lining systems usually cover a part of the sites (i.e. extensions at existing sites). An investigation in 1987 of 30 landfill sites showed that synthetic bottom liners were installed at 12 sites; one site applied a special clay of low permeability. At seven sites natural barriers of clay or loam are present underground (8).

At most landfill sites there is a ring ditch. Sometimes a liner is installed; sometimes the water in the ditch is aerated. Various leachate treatment systems are used, such as:
a) Discharging into a municipal sewage system;
b) Recirculation;
c) Biological treatment;
d) Aerobically, with effluent discharge into surface water;
e) Anaerobically, followed by reverse osmosis, with effluent discharge into surface water and concentrate to the landfill site;
f) Physico-chemical, with effluent discharge into the municipal sewage system.

By virtue of the Water Pollution Act, landfill sites are compelled to treat their leachate to meet certain requirements before it may be discharged into surface water. Table 5 shows the recommended standards to which the water boards usually adhere (9). At the moment, there are no standards for emissions to air. Leachate and groundwater near the site are usually monitored.

3. ENVIRONMENTAL ASPECTS OF LANDFILLING SEWAGE SLUDGE

The following are important when assessing environmental aspects:
a) Emissions of polluting components to water, air and soil;
b) Scarcity of land;
c) Disturbance of landscape.
Furthermore it is important to know the effects of sludge on these aspects.

Normally sludge is landfilled together with other categories of waste. Relatively wet sludge will fill up the pores in these other wastes. The effect of sludge on item (b) and item (c) will be of minor importance. Therefore environmental aspects of landfilling sludge can be restricted to emissions of polluting components to water (surfacewater, groundwater), air and soil.

3.1 A survey of existing landfill sites

Data were collected from the provincial authorities in the Netherlands for these institutions give waste disposal permits and control landfill sites. At the moment there are about 250 landfills in use. Out of these 250 landfills, 30 were chosen as being representative of the Dutch situation (8). Old abandoned sites were rejected due to lack of data.

3.1.1 The amount of sludge landfilled In the past most sites (18 out of 30) landfilled various sludges together with other wastes. In relation to the total amount the quantity of household waste varied from 30 to 60%. The percentage of sludge disposed is in the order of 1 to 10% (wet weight). There are more sites where sludge is disposed but no data are available of the quantity involved.

The relative percentages of sludge landfilled together with other wastes at present and in the past have not changed very much and they are mostly below 10%. This is probably due to the consistency of sludge. Data on dry matter content of the different sludges are unknown.

The total quantity of waste landfilled at present in these 30 landfill sites is about 5.68 million t/year. The quantity of sludge produced in the Netherlands is about 400,000 tds of which about 15% (as sludge with 20% ds) could possibly be landfilled on the 30 sites investigated.

3.1.2 Environmental aspects When environmental aspects of landfilling sludge are considered it is necessary to know the effect of sludge on the leachate quality and quantity in relation to the other categories of waste disposed on the same site. The main components of sludge are water, salts, ash, residual organic matter and the environmentally important heavy metals and organic micropollutants. A comparison of concentrations (7, 10) and quantities of some components in municipal sewage sludge and household waste are given in Table 6.

As shown in Table 6, the addition of 5% sludge does not influence the concentrations of polluting components in household waste very much. Bearing in mind that almost every landfill site contains a lot of household waste, it is clear that it will be difficult to find any effect due to the metal content of sewage sludge on the quality of leachate.

The reason for these relatively low amounts of sludge landfilled together with other wastes, is probably due to the consistency of the sewage sludge. Too much sludge results in, for example, the subsidence of vehicles, dirty roads and sliding of wastes.

Table 6. Quantities of some components in 1 m³ of landfilled household waste and sewage sludge, and the influence of sewage sludge on these concentrations.

	Household waste	Sewage sludge	Household waste +5% sludge	Change (%)
Water (kg)	380	800 - 950	401 - 409	+6.6
Ash (kg)	230	20 - 80	220 - 223	-3.7
Organic matter (kg)	390	30 - 120	372 - 377	-4.0
As (g)	3.7	0.4 - 1.4	3.5 - 3.6	-4.0
Cu (g)	90	25 - 100	87 - 91	-1.1
Zn (g)	533	88 - 350	511 - 524	-2.9
Cd (g)	5.6	0.5 - 1.8	5.3 - 5.4	-4.5
Pb (g)	245	23 - 90	234 - 237	-3.9

3.1.3 Leachate Leachate is the waterflow emerging from the bottom of a landfill; it has not been mixed with other waterflows like run-off, groundwater or precipitation. Most qualitative data on water concern groundwater; after 1985 a (synthetic) liner was prescribed, resulting in data on leachate. There are only a few data on landfill gas so it is difficult to say anything about the effect of sludge on landfill gas. Data on leachate of landfill sites with sludges are scarce. Although much information is present in this survey it is not possible to find enough data on landfills with either the collection of leachate or the disposal of household waste with and without sludge to find an effect of sludge on the quality of leachate. Every landfill site which contains sludge also contains household waste. The composition of household waste is geographically fairly constant. This is not the case with industrial waste. So a comparison of landfill sites with about the same quantity of industrial waste (and with and without sludge) will not be useful.

In Figure 2 some results are shown of leachate quality from three landfill sites with different percentages of sludge. It shows that Kj. N (and also NH_4^+, Cl^- and SO_4^{2-}, not shown), follows the COD concentration although on a lower level. The COD concentration is about 10,000 - 15,000 mg/l, and

Kj. N and Cl⁻ about 800-1,500 mg/l. Only the concentrations of heavy metals show substantial differences; landfill 4 and landfill 12 have about the same concentrations, while landfill 13 shows much lower concentrations. The reason for this is the composition of the waste, i.e. the quantity of industrial waste like shredder waste, bottom ash and fly-ash. Here the effect of increasing quantity of sludge plays a minor role.

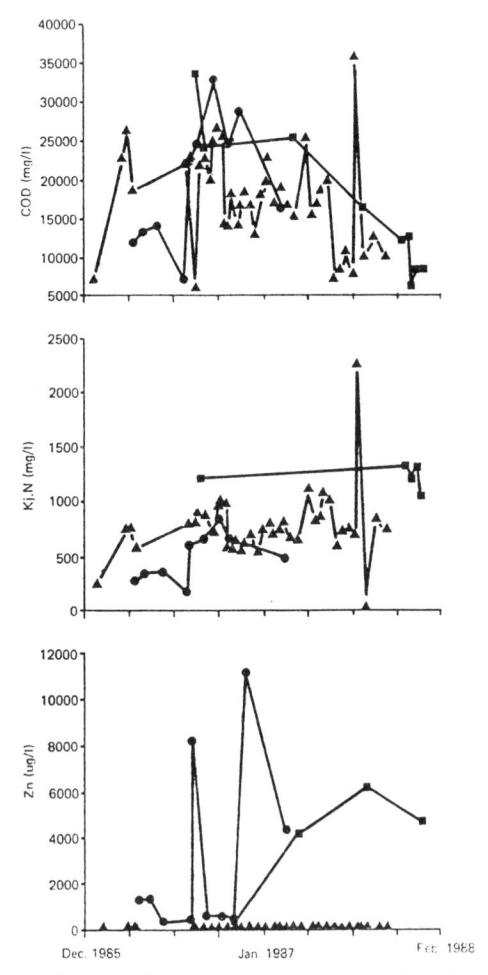

Fig. 2. COD, Kj.N and Zn concentrations in leachate of three different landfills; percentages of sludge in 4, ● 12 ■ and 13 ▲ landfills respectively are 2.50, 2.68 and 3.75.

3.1.4 Groundwater What has been said of the effect of sludge on the quality of leachate holds to a greater extent for groundwater. Since leachate migrates into the soil and groundwater, concentrations in the leachate change through dilution, adsorption, biological decay (pH-change) etc. To

Environmental aspects of landfilling sewage sludge

what degree these concentrations change depends on the geohydrological situation (depth and permeability of the different soil layers, location of different water boundaries, etc.). In Table 7 some data are shown for concentrations of different components in groundwater (downstream).

As can be seen from Table 7 there are great differences in the qualities of groundwater. The influence of quantities of sludge on the quality of groundwater was not shown. For instance, when comparing low pH (site no. 8) and high pH (site no. 6), a low pH does not automatically lead to higher heavy metal concentrations in the groundwater.

Table 7. Average concentrations[1] in groundwater downstream of landfill sites with sludge.

n^1	1	2	3	4	5	6	7	8
COD	8699	77	-	-	68	157	35	-
pH	5.0	6.9	-	-	7.1	7.6	7.3	5.1
Cl	-	1759	-	3028	197	663	46	66
Kj.N	170	16	-	-	-	20	2	-
SO_4	-	-	-	-	34	79	39	76
As	-	15	5.6	12	-	352	0.5	3.7
Cd	-	1.0	2.4	1.7	-	0.1	-	0.2
Cr	-	13	10	6.7	-	9	-	0.3
Cu	-	588	33	16	-	5.7	-	-
Hg	-	0	0.1	13	-	0	-	-
Ni	-	10	97	9	-	12	0	0
Pb	-	23	158	36	-	2.3	-	1
Zn	-	676	315	26	-	200	15	18
EoCl	-	0.4	1.4	1.3	0.4	-	-	-
Oil	-	-	-	23	0	-	-	-
n^2	19	2	2	4	4	72-80	2	3
n^3	'84-88	'85	'83	'84-86	'86	'84-87	'86	'86
n^4	'76	'80	'73	'77	'60	'70	'67	'67

[1] COD, Cl, Kj. N, SO_4 and oil in mg/l; heavy metals and EoCl in µg/l.

n^1 = number of landfill; n^2 = number of samples;
n^3 = period of sampling (year); n^4 = start of landfilling (year);
\- = no data available.

3.2 Experiment: effect of supplying sludge

In several, mostly laboratory-scale experiments, it was shown that combined disposal of sewage sludge with municipal solid waste leads to a shorter acidification period of the "landfill-reactor" (11). In the following paragraphs the results will be given of an experiment that was performed to establish the influence of sludge disposal on the behaviour of the landfill body.

In this experiment two compartments (each 50 x 50 x 4 m) were constructed with municipal solid waste. To the waste of one compartment chamber filter pressed sewage sludge was supplied in layers. The sludge had a dry solids content of 38%; the calcium content was 7.5%. The ratio of sludge/waste was 1:3.4. To measure the leachate quality, leachate samples were taken from a drainpipe connected to the drainage system of each compartment. Gas samples were taken from gas drains, two in the middle of each compartment at different heights.

3.2.1 COD, volatile fatty acids, pH and gas In Figure 3 the effects on, volatile fatty acids, pH and CH_4 are presented; the compartments were filled at Day 0. It can be seen that supplying sludge has a marked positive influence on the quality of the leachate. The production of methane indicates that both compartments are in the methanogenic phase, although with a somewhat lower concentration in the compartments without sludge. The cause of this phenomenon is not quite clear, but may be due to the heterogenity of the waste producing methane in local niches.

The reason that the compartment with sludge shows lower COD-concentrations (and VFA) in the leachate has probably to do with the stabilising effect of sludge on the volatile fatty acid production. The same effect has been shown when adding compost to household waste (11). Adding less reactive (microbiological) organic material (like aerobic/anaerobic treated organic material in sewage) to household waste decreases the production rate of volatile fatty acids (12), allowing methane producing bacteria to grow, due to a higher pH resulting in a balance between fatty acid production and consumption.

3.2.2 Metals In Figure 4 some data are given of the heavy metal concentrations in leachate. Although the data are not always consistent, it is remarkable that on average the heavy metal concentrations in the leachate of the compartment without sludge are higher than in the leachate of the compartment with sludge. This is in contrast with the higher quantity of heavy metals in the compartment with sludge. The reason for this is probably due to a lower pH in the compartment without sludge.

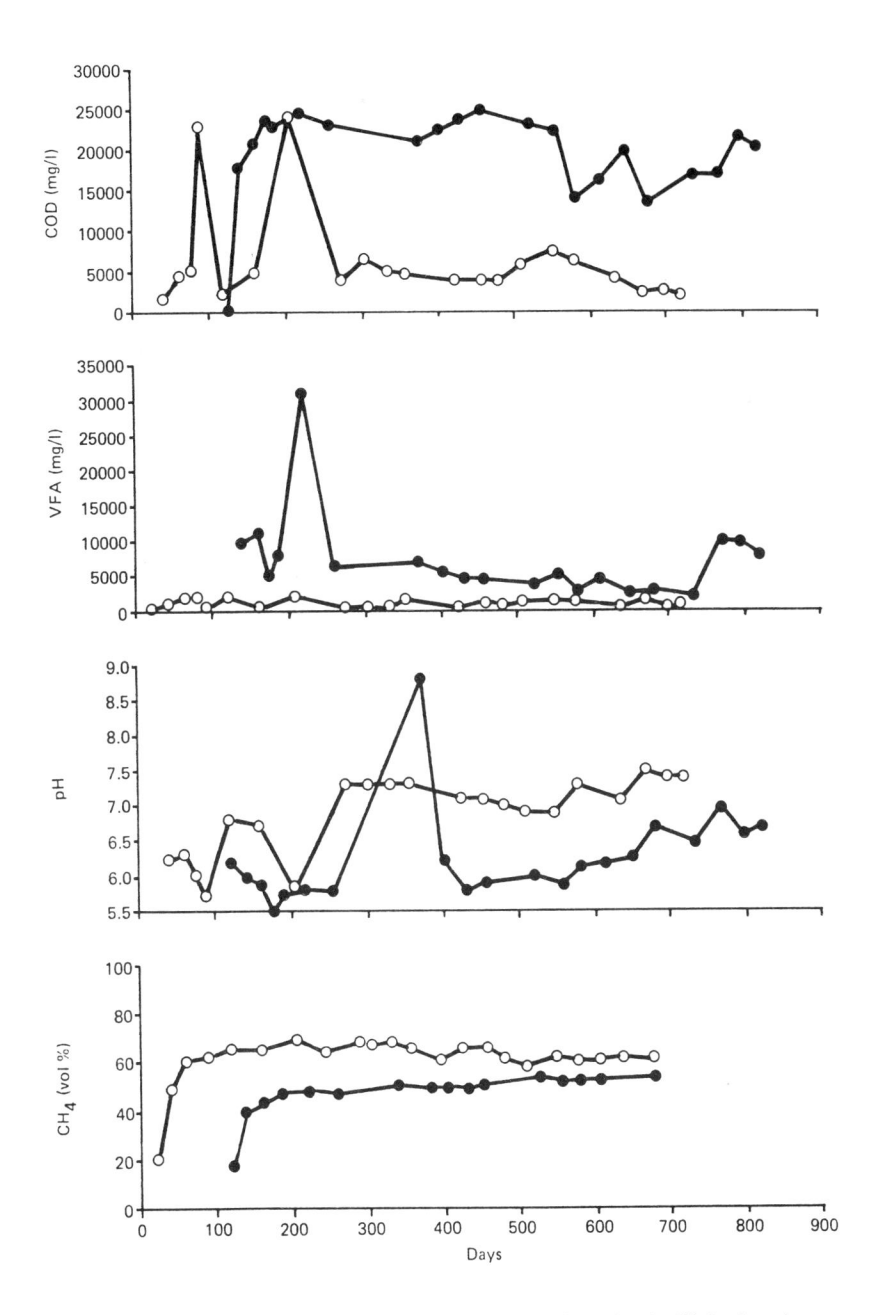

Fig. 3. COD (mg/l), VFA (mg/l), pH and CH_4 (vol. %) in leachate; ○ = compartment with sludge; ● = compartment without sludge.

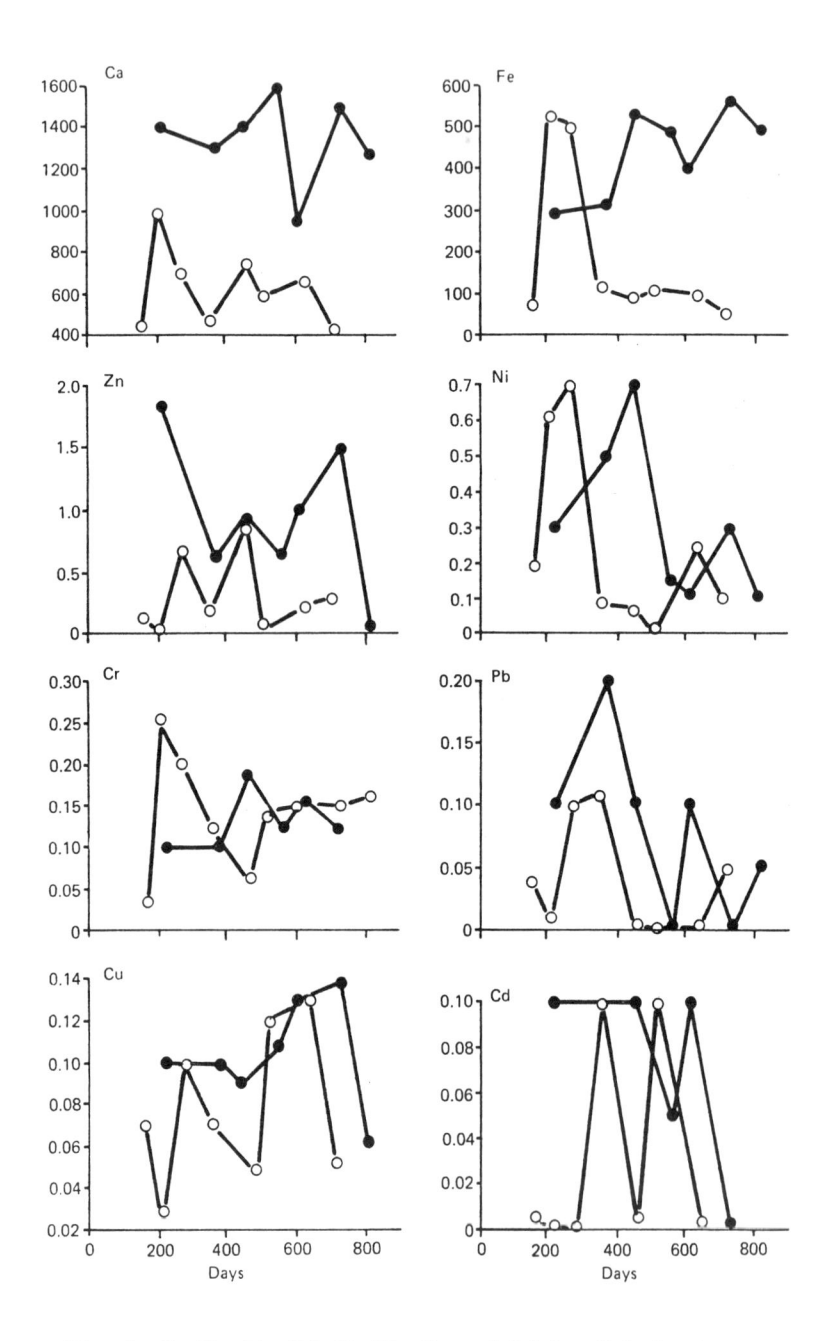

Fig. 4. Ca, Fe, Zn, Ni, Cr, Pb, Cu and Cd (mg/l) in leachate;
○ = compartment with sludge; ● = compartment without sludge.

Looking in detail at the metal concentrations, the following can be deduced. The Ca contents fluctuate with the COD. While in Compartment 1 (without sludge) the Ca concentration is probably much lower (Ca in sludge is 200,000 ppm), the Ca concentration in the leachate throughout the time is much higher. Higher pH values, due to biological processes, cause increasing CO_3^{2-} concentrations resulting in precipitation of $CaCO_3$ in the compartment with sludge, resulting in low Ca concentrations in the leachate.

Only in the beginning (< 300 days) Fe in the leachate from the compartment with sludge is higher. Dissociation reactions of Fe implies that oxidation-reduction reactions should be considered (13). Not only the pH but also the redox-potential is then important. Instead of the redox potential, reducing conditions can also be characterized by the partial oxygen pressure. At the same pH value, the Fe^{2+} concentration will be higher at a lower oxygen concentration. At the time of high Fe concentrations in the leachate of the sludge compartment, oxygen concentrations were almost zero, while later on oxygen concentrations were fluctuating (see Figure 5). The same holds for the other compartment. The mobility of the heavy metals will be restricted by precipitation as carbonates and sulphides and adsorption to hydroxides and organic material. Mobility can be increased by decrease of pH value, and the presence of organic acids (VFA), chloride, sulphate and chelating agents (14).

The presence of lower concentrations of heavy metals in the leachate of the sludge-containing compartment can be explained by assuming that the pH is the regulating factor for solubilisation of heavy metals. A high pH value causes heavy metals to be adsorbed as cations to organic material (humic components in the sludge or to VFA).

4. CONCLUSIONS

In the Netherlands a significant part (40 %) of the sewage sludge production has always been and still is landfilled. Based on a survey of landfill sites the quantity of sludge being landfilled does not exceed 10% of the total amount of waste that is landfilled. Landfill sites with sludge always contain household waste too. The quantity of household waste varies from 30 to 60% in relation to the total amount. Environmental effects of landfilling sludge should be determined, particularly the quality and quantity of polluting components in the leachate and/or groundwater. Due to the small quantity of sludge and the relatively high concentrations of heavy metals in household waste, the contribution of heavy metals from sludge to the concentration in leachate is not very high.

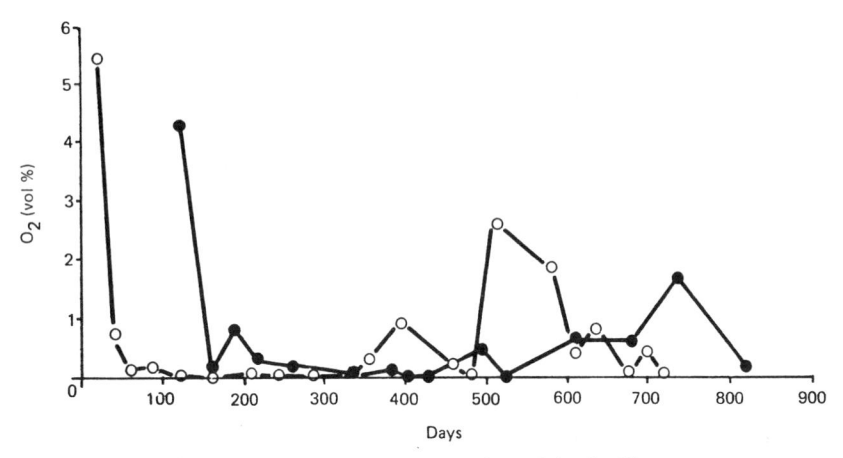

Figure 5. Content of O_2 in gas produced (vol. %);
○ = compartment with sludge;
● = compartment without sludge.

Some industrial wastes show much higher heavy metal concentrations than sewage sludge (fly-ash, bottom ash, shredder waste, grit blast) and household waste. The effect of these wastes is much more pronounced. In this survey an effect of sludge could not be demonstrated on classical parameters like COD, pH, Cl and Kj.N in leachate. The situation for groundwater is very similar to leachate.

When sewage sludge is supplied in large quantities to solid waste, the dry matter content must be high enough to prevent subsidence of vehicles. In a sludge (38% ds) to solid waste ratio of 1:3.4 the influence of sludge was very pronounced in the first 2-3 years. It appears that in this case sludge had a stabilising effect on the processes in the landfill body, with lower COD and VFA concentrations and higher pH in the leachate. As a result of the higher pH value, the concentrations of heavy metals in the leachate were decreased in spite of larger quantities of heavy metals in the landfill body.

In the long term (within 6 years), this difference will probably diminish, and the landfill body without sludge will be stabilised as a result of the degradation of organic matter in the landfill body (12). The acidification rate of the residual organic matter in the landfill body will decrease with time, so the pH will rise. There are not enough data available on the quality of leachate over long periods, but from groundwater analyses of landfill sites an effect of pH on the heavy metal content cannot be demonstrated, even in the long term.

Thus, low percentages of sludge disposed of together with other wastes had no effect on the environmental aspects of a landfill site, but higher percentages of sludge in combined disposal appear to have a positive outcome on environmental aspects, as shown in this experiment. The survey described here will be studied further; it will try to determine more precisely the effects of sludge on leachate, soil and groundwater.

REFERENCES

(1) Stichting Verwijdering Afvalstoffen (1974). Aanbevelingen voor het ontwerp, de inrichting en uitvoering van afvalstortterreinen. Amersfoort.
(2) Nagelhout, D., K. Wieringa and J. M. Joosten (1989). Afval 2000. Rapportnr. 738605002, RIVM, Bilthoven.
(3) Nederlandse Staatscourant, nr 209, 27 Oktober 1988. Staatsdrukkerij. 's-Gravenhage.
(4) Tauw Infra Consult B.V. (1987). Evaluatieonderzoek Afvalstoffenwetvergunningen van stortplaatsen, tussenrapport. Deventer.
(5) Kerkhoven Management B.V./Adviesgroep Data Process (1988). Project beheersstructuur nazorg stort- en saneringslokaties, tussenrapport.
(6) Breure, A. M. and J. G. van Andel (1987). Microbiological impact on anaerobic digestion. In: Bioenvironmental systems (D. L. Wise, Ed.), Vol II, pp. 95-113. CRC Press, Florida, U.S.A.
(7) Beker, D. (1988). Recirculatie van perkolatiewater. Rapportnr. 738471002, RIVM, Bilthoven.
(8) RIVM/LAE (1989). Inventarisatie stortplaatsen, in press. Bilthoven.
(9) CUWVO, werkgroep VI (1987). Zuivering van perkolatiewater van stortplaatsen voor voornamelijk huishoudelijke afvalstoffen. 's-Gravenhage.
(10) Projekt Priaf. Prioriteitsstelling Afvalstoffen (1988). Ministerie VROM, Leidschendam; RIVM, Bilthoven, R.P.C. B.V. Delft.
(11) Stegmann, R. (1982). Ergebnisse von Abbauuntersuchungen in Labormasstab. Gas- und Wasserhaushalt von Mülldeponien, Internationale Fachtagung, 29.9 -1.10.1982 Braunschweig. Veröffentlichungen des Instituts für Stadtbauwesen, Technische Universität Braunschweig, Heft 33.
(12) Beker, D. (1987). Control of acid phase degradation. International symposium: "Process, technology and environmental impact on sanitary landfill", Sardinia, Cagliari, Italy, 19-23 October, 1987.
(13) Bolt G. H. and M. G. M. Bruggenwert (1976). Soil chemistry. A. Basic elements. Elsevier Scientific Publishing Co, Amsterdam.
(14) Beker, D. (1977). Mobiliteit van zware metalen in de grond onder een vuilnisstort. ICW nota 959, Wageningen.

Physical aspects of landfilling of sewage sludge

J. J. van den BERG[1], P. GEUZENS[2] and R. OTTE-WITTE[3]

[1]Grontmij, De Holle Bilt 22, De Bilt, PO Box 203,
3730 AE de Bilt, The Netherlands.
[2]Nuclear Research Institute, Boeretanj 200, 2400 Mol, Belgium.
[3]University of Bochum, Lehrstuhl F. Wasserwirtschaft 52,
Postfach 102418, D-4630 Bochum,
Federal Republic of Germany.

SUMMARY

The physical properties of sewage sludge are important for landfilling. The problems intrinsic to dumping sewage sludge can be described in terms of soil mechanical stability and the bearing capacity of the dump-site body. At this moment there are no general and standardised methods of determining the properties of sewage sludge with regard to landfilling. In this paper an overview will be given of the specific requirements in a number of countries, of the physical properties for sewage sludge with respect to landfilling. Subsequently, a summary will be given of the investigations on the physical properties of sewage sludge. The aim of these investigations is to develop a practical standard to determine the suitability of sewage sludge as landfill.

1. INTRODUCTION

The physical problems of dumping sewage sludge in landfill sites involve the physical properties of sewage sludge, the adhesive properties of sludge on transport facilities, and the effects of sludge on the soil mechanical properties of the dumping site body. In this paper an overview will be given of the state of the art of dumping in West Germany, Belgium and the Netherlands. Subsequently, a summary will be given of the investigations and developments concerning the description of, and solutions to the physical problems of dumping sewage sludge.

2. STATE OF THE ART

The problems intrinsic to dumping sewage sludge are strongly related to the method of dumping. In some countries, it is possible to use abandoned mines and quarries to dump the sludge produced during the treatment of drinking water or of waste water. In general, the soil in such locations can be considered to be more or less impermeable to liquids, and the influence of the percolation water on the soil is virtually nil. Such dumping sites are generally structured as a mono-dump, that is to say they are used only for sludge disposal. Besides being dumped in mono-dumps, sludge is also disposed of at sites where domestic waste material is dumped as well. In a number of countries, the dumped sludge must comply with certain requirements with respect to its physical properties. These conditions are incorporated in directives, guidelines, regulations and in legislation.

In Belgium all sludge that is dumped at a controlled dump-site has to meet the requirements of being "congealed". The requirements concerning the structure of the dump-site with respect to the protection of the soil and the groundwater depend on the waste to be dumped on it. A classification of dumps exists. Sludge from communal waste water treatment plants is mostly disposed on a Class II dump-site together with municipal refuse. The sludge from drinking water is mostly dumped in mono-dumps or dump-sites for industrial waste (Class I) depending on the content of heavy metals. The term "congealed" is not legally defined, which is a problem for proper dump-site exploitation as well as for the controlling authorities. A research programme recently started focusses on finding a more detailed definition and foundation for the term "congealed". It includes both sewage sludge and sludge from drinking water treatment.

In France there is no strict legislation concerning landfilling of sewage sludge. However, the following guidelines are general accepted. For the sludges to be dumped on a landfill dump-site, a maximum water content of 70% is allowed. Furthermore, the content of free water has to be less than 30% (the content of free water means the quantity of liquid released by the sludge under a uniform pressure of 1 bar). Exceptions can be made for sludges with a higher content of water when the mass ratio between sludge and municipal refuse on the dump-site is limited to 10 or 20%, depending on the water content of the sludge.

In Switzerland, the sewage sludge is dumped both in mono-dumps and together with domestic waste. The bottom of the dumping site is covered with a few metres of clay on which the dewatered sludge is dumped. If the sludge is dumped together with domestic waste, the volume of the sludge may not exceed 15%. In Switzerland, the dumping of sewage sludge will no longer be allowed in the near future. Plans are being made to incinerate all sewage sludge in the furnaces of the cement industry. They are also investigating the long-term behaviour of dumped sewage sludge. The decomposition of organic matter in the sewage sludge ranges over several years, the major part taking place in the first eight to ten years after dumping. The decomposition processes influence the properties of the sewage sludge, such as its structure. At the moment, research focusses on gaining insight into the texture changes of the dumped material by means of electron microscopy tests.

The Federal Republic of Germany (FRG) does not have a general guideline concerning the requirements that sewage sludge to be dumped has to meet. Several states, however, do make so-called recommendations. According to these recommendations, the water content of the sludge to be dumped may not exceed 65%. In view of the soil mechanical stability of a dumping body, attempts must be made to dewater the dumped sludge together with domestic waste as much as possible. Furthermore, the sewage sludge to be dumped must be biologically stabilised in order to control odour pollution. Research in the FRG is aimed at defining the parameters for the dumping suitability of the sewage sludge in relation to the dumping method. Several research institutes, as well as COST 681 (1, 2), are participating in phased investigations. Within the framework of this investigation, a literature study has been conducted into the characterisation of sewage sludge with respect to its suitability for dumping on the basis of soil mechanics. This study has resulted in the formulation of a hypothetical dumping suitability criterion. According to this criterion, the shear strength of the sewage sludge to be dumped must be at least 15-20 kN/m^2.

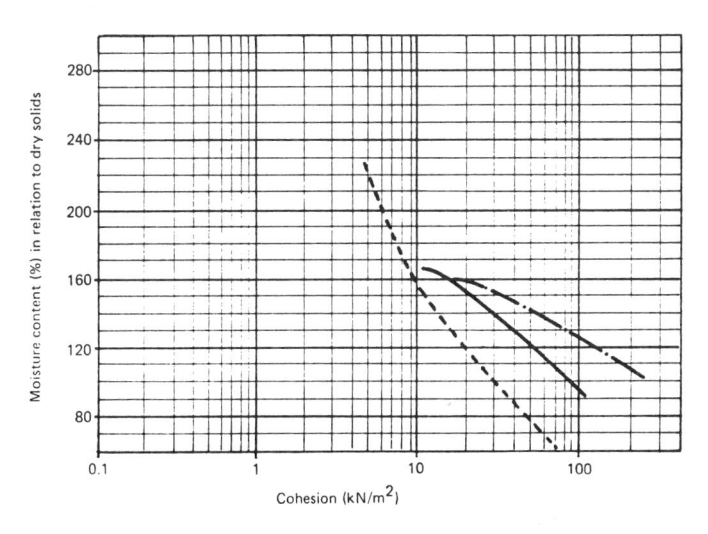

Fig. 1. Cohesion of three sludge types as a function of water content.

Subsequently, the measuring methods were selected so that shear strength could be determined, and an extensive measuring programme was carried out to test the validity of the hypothesis. The investigation that is being conducted at the moment focusses on the dumping suitability criteria with respect to the dumping method. From 1990 onwards, the dumping of organic materials (organic dry matter >10%) will no longer be allowed in some states of the FRG, because of a shortage of dump-sites.

In the Netherlands, sewage sludge to be dumped is subject to a guideline under the Act on Waste Materials. In view of odour control, the sewage sludge to be dumped must be biologically stabilised. Moreover, the sludge must be "congealed", which means that it must have a dry solids content of 35%. (Note: Whereas in Belgium the concept of "congealed" is not defined, in the Netherlands the term is related to the dry solids content of the sludge). As in the other countries mentioned, dumping liquid sewage sludge is prohibited. Research in the Netherlands focusses on obtaining a general characterisation of the sewage sludge and formulating general dumping suitability criteria for the sludge. Soil science, rheology and soil mechanics were used as approaches in this research, which, in several respects, meshes with that in the FRG and Belgium, in the sense that each country eventually opted for the soil mechanical approach. Possible parameters besides shear strength that are being considered are the bearing capacity of the dumped sewage sludge and the adhesion or adhesive force.

When sewage sludge is dumped, a large quantity of water is brought in. This water affects the soil mechanics and hydrological conditions of the landfill. By dumping sludge with a dry solids content of 5% together with domestic waste, the landfill weight is more than doubled. The applicable guideline, therefore, imposes a higher limit on the dry solids content of the sewage sludge to be dumped, amounting to not less than 35%. Thus, the sludge will often have to be dewatered mechanically after conditioning with lime and ferric chloride or with polymers.

Practical experience has shown that if the dry solids content is 35%, combined processing with domestic waste is feasible and not very critical with respect to the sludge/waste ratio. This dry solids content can easily be achieved for filter press sludge conditioned with lime and ferric chloride and for thermally conditioned sewage sludge. For sewage sludge conditioned with polymers and dewatered with a sieve belt press or a centrifuge, the weight ratio of domestic waste and sludge is approximately 7:1. If the ratios are smaller, the sludge can no longer be processed properly together with the domestic waste, and the capacity to bear vehicles will decrease. This causes compacting equipment to skid and bog down, while the sludge will stick to the wheels, resulting in a continuous extension of the dumping front.

Besides the bearing capacity of the landfill and the dumping body, consideration should also be given to the soil mechanical stability of the landfill. Gay et al. (3) conducted theoretical research into the influence of sewage sludge on the mechanical properties of a landfill. Triaxial tests were used to establish a relationship between the dry solids content of various types of sewage sludge and the shear strength. The study showed that the cohesive force of sludge with a dry matter content of 35% is approximately 10 kN/m^2 (see Figure 2).

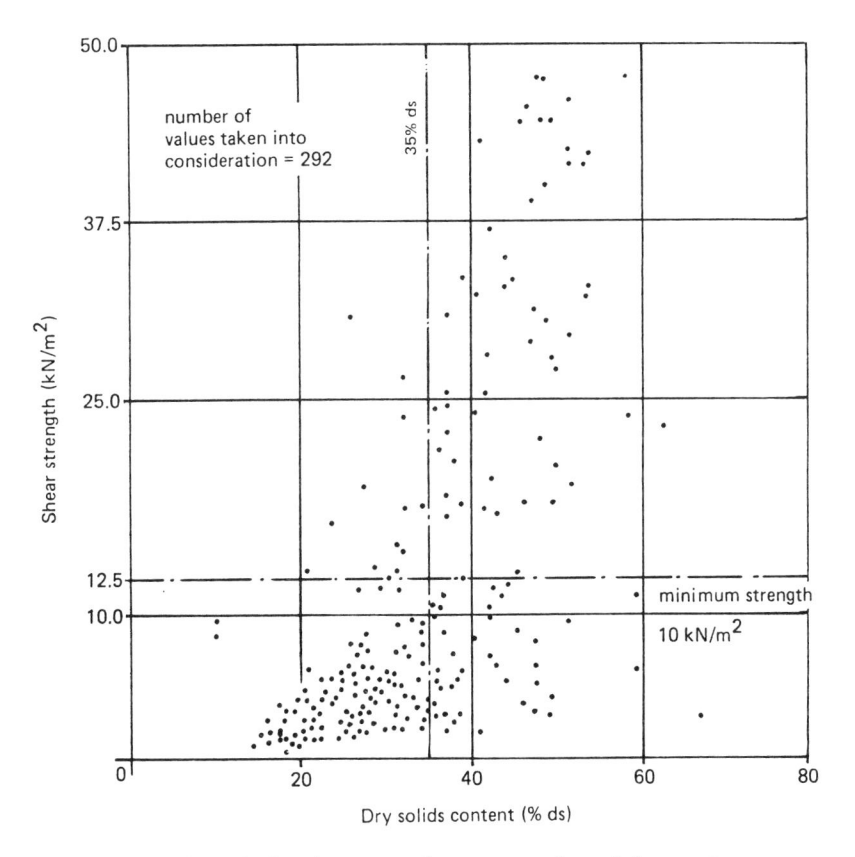

Fig. 2. Correlation between shear strength and dry matter.

Apart from cohesive force, shear strength is also influenced by friction. If friction is ignored, the cohesive strength could be adopted as the minimum shear strength. Stability computations in which the cohesive strength was taken as the minimum shear strength have shown that the stability of a landfill decreases substantially if the sludge is present in continuous horizontal layers (see Figure 3).

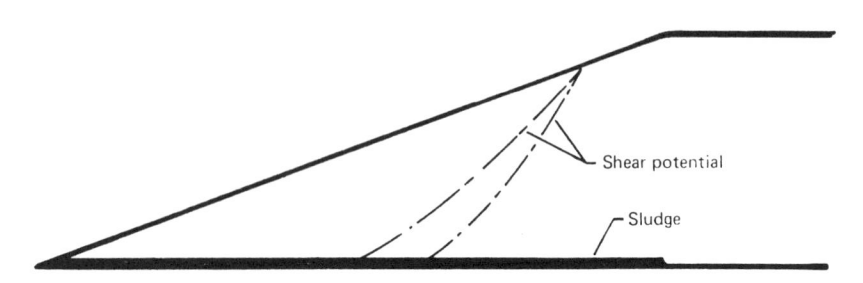

Fig. 3. Effect of a continuous sludge layer on landfill stability.

Placing sludge packages perpendicular to the potential shear line improves stability. Furthermore, applying a covering layer may also increase the slope stability. Theoretical and practical research has been carried out with respect to the soil mechanics parameters of sewage sludge. The research was mainly aimed at characterising the physical parameters of sewage sludge with respect to its soil mechanical stability, bearing capacity and adhesive force. Different theories, such as soil mechanics and rheology, were used as approaches to the problem.

It can be said that, in general, theoretical research has resulted in the insight that the soil mechanical approach has the largest sphere of application. The starting point here is that the sewage sludge to be dumped, although of heterogeneous composition, can be considered as a fine-grained soil type with a high organic matter content. The soil mechanical properties of such a material should be comparable to those of disturbed clay and silt. These types of soil are called cohesive soils, which, as opposed to non-cohesive soils like sand, also have a certain resistance under low pressure and dry conditions. Another characteristic is that their properties of resistance and deformation depend on the way they are subjected to loads. Under a sudden load, the pore water pressures will generally increase and the soil will behave frictionless and maintain a stable volume. The undrained shear strength in that situation depends on the water content, a water-saturated soil having the lowest resistance. Under slow loading, the material will show friction behaviour and there may be relatively substantial changes in volume.

3. CONSISTENCY

The consistency of cohesive soils is the degree of adhesion between (soil) grains that form the skeleton structure and the resistance against the forces that try to deform the grain structure or make it collapse. In the literature, the consistency of sludge is described in various manners. Colin (4) distinguishes a liquid and a plastic condition as well as a solid condition in which shrinkage may or may not occur. According to Graig (5) there is a distinction between a liquid, a plastic, a semi-solid and a solid condition, depending on the water content. Gay (3) defines dumpable sludge as a "cohesive soil" in which water content and grain spacing are the primary factors.

The consistency of the soil is related to the water content. If, for instance, the water content of a clay-slurry gradually decreases because of a slow dehydration, the material will go from a liquid phase by way of a plastic phase to a solid phase. The boundaries between the various phases depend on properties like loam content, mineralogical composition, and the organic matter content. The transition from one phase to another is not an abrupt one, but covers a certain range of water contents. Each method of determining consistency boundaries is, in a sense, subjective. Being based on a certain method can, however, result in a standardisation of the consistency of a certain type of soil or sludge.

A method often used in actual practice to determine the consistency of soil types is the Atterberg method. Without explaining this method further, it can be said that it uses four criteria: the liquid boundary, the plastic boundary, the plasticity index and the consistency index. Gay (3) has applied the Atterberg method to a limited number of samples of dewatered sewage sludge. On the basis of the above-mentioned criteria, the consistency of this sludge was found to be "mushy", and the sludge in general was labelled "congealed". This can be considered as illustrative for the limitations of the soil mechanical approach with respect to sewage sludge.

4. COMPRESSIBILITY

The soil mechanical behaviour of a landfill not only depends on the nature of the materials dumped, but also on their compressibility. Dumping sewage sludge generates a normal force, which is the result of the weight of the material and that of the waste on it. Under the influence of this normal stress, the accumulated sludge will show consolidation or setting. Consolidation is a (limited) reduction in volume of a water-saturated soil at the expense of (part of) the pore water. In this process, Gay (3) distinguishes between short, medium and long-term consolidation. Consolidation behaviour of materials can be determined by the degree of volume reduction as a result of the external pressure exerted on the materials.

5. SHEAR STRENGTH

The stability of a soil body is determined by the shear strength, which can be defined as the maximum value of the shear stress, which can be considered as the resultant of reaction forces that are generated when a load is placed on a soil. The reaction forces can be divided into cohesion forces and the effective properties of strength of the skeleton structure. Cohesion can be roughly divided into real and apparent cohesion which is the result of the capillary pressure of water in the pores between the different horizontal layers. The relationship between shear strength and the different reaction strengths are given in Coulomb's law.

6. BEARING CAPACITY/SUITABILITY FOR DRIVING

The bearing capacity of the dump-site must be considered in relation to the suitability for driving on the surface of the site. The bearing capacity can be characterised on the basis of the theory used in foundation technology. This can be used as a theoretical basis to find the value for the undrained shear strength during which "soil fracture" occurs.

Another approach is to determine the pressure under which a certain degree of deformation occurs. The so-called "Hildesheimer Prufstempel" has proved to be a suitable method to get an indication of the bearing capacity (6).

7. MEASURING METHODS

Measuring soil mechanical parameters such as the shear strength is done under controlled conditions in a laboratory or in the field (*in situ*). In order to interpret the results they must, in general, be related to the soil mechanical classification of the material, that is the grain structure, organic matter content, water content and Atterberg boundaries. The following can be measured:
a) Shear strength (undrained);
b) Compressibility constant;
c) Bearing capacity.

Germany, Belgium and the Netherlands are more or less consistent in the manner in which the undrained shear strength and the bearing capacity of sludge can be measured. Of all methods used to measure undrained shear strength, the laboratory motor vane appears to be the most suitable one for application to sewage sludge. The laboratory motor vane for measuring

Physical aspects of landfilling of sewage sludge

vane shear strength, which is shown in Figure 4, has an electric motor and can also be operated by hand. The vane is turned either manually with a handle or by means of the motor. The vane is connected through a torsion spring to a display which shows the angle of deflection on a 360° scale by means of a trailing pointer. The torsional moment can be calculated from the angle of deflection shown and the spring constant. The torsional moment increases as the vane turns and the pointer remains stationary at the reading of the maximum moment when the specimen fails. The value of the vane shear strength is calculated from the torsional moment at failure of the specimen, the geometrical data on the vane and the calibration table specific to the apparatus.

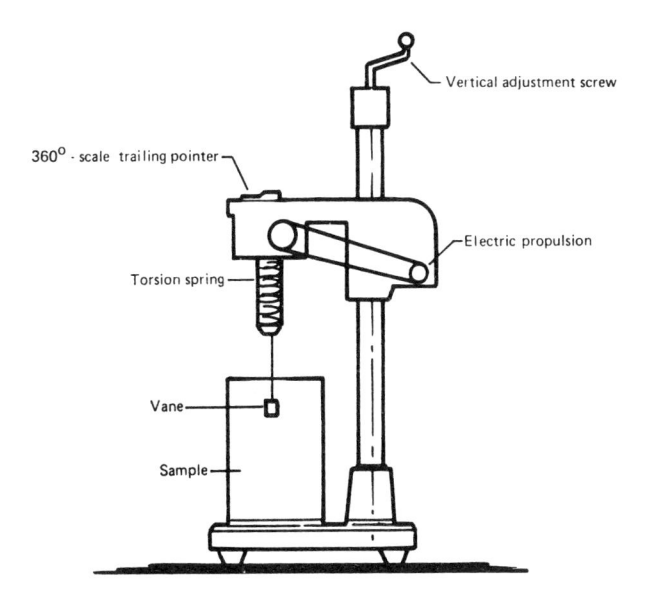

Fig. 4. Laboratory motor vane for measuring vane shear strength.

There are other methods to measure the shear strength such as the falling-cone. However, the calibration of this method with regard to sewage sludge is not known. Tests have proved that the reproducibility of measurements is poor and there is a tendency for this method not to be considered as a practical one for general use with sewage sludge.

The bearing capacity of sludge could be determined by means of a load stamp with which the impression of a stamp could be measured according to a variable load applied. The depth of an impression as a function of the load applied must be (an indication of) the bearing capacity of the sludge.

It should be noted that the calibration of the measurement devices normally takes place by measuring the undrained shear strength or the bearing capacity of a number of defined and classified types of soil. This means that the interpretation of the values measured are conditional until sufficient measurements of different types of sludge have been carried out and a frame of reference has been formed. It is of primary importance that the pre-treatment of the sample should take place in a standardised manner.

8. RELATIONSHIPS BETWEEN SOIL MECHANICAL PARAMETERS

The available literature has shown that, so far, there has been little routine research into the possible relationships between shear strength, cohesion, dry solids content, compressibility and bearing capacity.

Research in Germany and the Netherlands has shown that it is generally impossible to formulate calibration standards or reference values for the mutual comparison of the results of separate measurements for sludge. The soil mechanical classification, and therefore the properties of the sludge are determined by a large number of factors, such as the waste water composition, the load of the waste water treatment plant, the company approach towards stabilisation and sludge thickening, sludge conditioning, dewatering and, possibly, after-treatment. In a German investigation (7), more than 1,000 measurements were conducted of sludges from 70 waste water treatment plants. The relationship between the dry solids content and the shear strength was one of the elements investigated. This research showed that the shear strength often increases if the dry solids content increases, but it turned out to be impossible to prove a uniform relationship. In approximately 65% of the samples, the shear strength was <10 kN/m^2 when the dry solids content was 35%. Wolf and Neuschafer (8) conducted research into the relationships between the shear strength, the bearing capacity and the suspended solids content. Especially when the sludge had been conditioned with lime, resulting in an increase in dry solids content, they found an increase in bearing power and shearing strength. The research was conducted for sewage sludge to be dumped. It is not known whether the improvement which resulted from conditioning with these additives is permanent.

9. INFLUENCING THE DUMPING SUITABILITY PARAMETERS

Although the only requirement made of sewage sludge to be dumped involves the dry solids content, it can be said that the material to be dumped must have a maximum shear strength and bearing capacity. Further treatment of the sludge after dewatering could greatly improve the dumping suitability of the sludge. One must, however, distinguish between the dumping suitability of sludge and the behaviour of the dumped sewage sludge in the long term.

Loll (2) conducted research into the effects of lime, fly-ash and residual substances of incineration installations when adding these substances to dewatered sewage sludge. Similar research was carried out in the Netherlands, albeit on a somewhat smaller scale. By adding an inert substance like sand to the sludge, a skeleton structure is formed which can cause an increase in friction. The volume ratio between sand and sludge in that situation depends on the organic substance content, the type and quantity of the conditioning agent and the manner in which they are mixed. However, the sand/sludge ratio must be relatively large (4 - 6:1 by weight, equivalent to a volumetric ratio of 2:1) for the shear strength to increase significantly. This shows that the contribution of the internal friction to the soil mechanical stability by an increase in dry solids content is relatively small. For fly-ash, too, a relatively large ratio is needed if a significant increase in shear strength is to result, despite the large increase in dry solids content. It is to be expected, however, that the shear strength of sludge to which large amounts of sand or fly-ash have been added, will also be reasonably stable in the long term. Not only was research carried out into the application of sand and ash as a "filler", but the effect of reactive additives, such as lime, was also studied. Provided they have good water-binding properties, the addition of small quantities (10 - 15% by weight) will result in a large increase in shear strength. This is also true for calciferous bed ashes from fluid bed incinerators. It is not known yet to what extent the shear strength is a permanent result of the conditioning with reactive additives. The addition of lime will cause a considerable increase in the pH which will stop the microbial decomposition processes of the sludge.

Furthermore, there will be a considerable amount of water-binding in the initial stage. After dumping, the binding of water will decrease after some time has elapsed and the pH goes down. Subsequently, the decomposition of organic matter will start up again. It is to be expected that these processes will bring about a decrease in shear strength. The effect of the increase of the shear strength of sludge by the addition of lime or calciferous material could therefore be only short-term.

Besides adding conditioners to improve the suitability for dumping in the long or short term, two other techniques are used in the Netherlands; drying of sludge and composting of sludge (9, 10). Te Marvelde (9) describes the increase in the shear strength of sludge, measured *in situ* with a hand-operated vane, by drying the sludge under the influence of natural processes like evaporation. As drying takes place over a short period of time, hardly any decomposition of the organic fraction can occur, and it has a short-lived effect. Composting of sludge not only results in the removal of moisture, it also brings about decomposition of the organic matter. For the benefit of the composting process, additives like wood chips are added. These are partly removed by screening, but a certain proportion of these wood chips will remain behind in the composted material. Measurements of composted sewage sludge indicate that the undrained shear strength will increase to 35 to 45 kN/m^2 after composting. Eighteen months after it was dumped, composted sewage sludge had an average undrained shear strength of more than 35 kN/m^2. Insofar as there is any decrease in shear strength, this will be a gradual process. This indicates that the soil mechanical stability will also be retained in the long term.

10. CRITERIA FOR THE SUITABILITY FOR DUMPING

Besides the above-mentioned criteria of being "congealed" and of suitable dry solids content, a number of other requirements are also made of sewage sludge to be dumped. Even if it is assumed that the dumping suitability is largely determined by the soil mechanical properties of the sludge, such as the shear strength and bearing capacity, it proves impossible to formulate criteria that can be applied as standards for the dumping suitability of sludge in general. There are too many factors, which can even vary depending on treatment plant and season, that influence the factors of solidity of the sludge to be dumped. With respect to the stability of landfills, it is generally assumed on the basis of research findings that the undrained shear strength, which is determined by means of a (motor) vane, offers the best option. In view of the multiplicity of factors that influence the production of solid sewage sludge, this test must be seen as a mere indication, at least for the time being. A high value of undrained shear strength does not necessarily mean that the sludge is suitable for dumping; a low value does not necessarily mean that the sludge is not suitable for dumping.

It is necessary to create a frame of reference by means of measurements for each type of sludge, that is differentiated according to treatment plant.

The lack of a frame of reference has been an important reason why the hypothesis formulated in Germany (2, 7), which states that dumpable sewage sludge must have a dry solids content of 35% and an undrained shear strength of 10 kN/m^2, was abandoned. Large-scale research conducted in Germany has shown that only 65% of the observations have met with these criteria, while sludge types with a low dry solids content proved to be highly suitable for dumping. Secondly, the shear strength of sewage sludge is not constant, but will decrease under the influence of decomposition processes. The degree and speed at which this decrease will take place depends not only on the nature and composition of the sludge itself, but also on the dumping method.

If it is assumed that the soil mechanical stability of dumped sewage sludge must have a minimal shear strength of 10 kN/m^2, the sewage sludge to be dumped must have a shear strength of 25 to 30 kN/m^2. This implies that the sludge will always have to be conditioned.

Although it may be possible to deduce a theoretical connection between the undrained shear strength and the bearing capacity of the sewage sludge, research has not been able to prove this unequivocally for sludge types that were dewatered after conditioning. There are, however, indications of such a connection for sludge to be dumped to which lime has been added.

By adding a reactive and water-absorbing additive like lime, there will be a temporary improvement in the dumping suitability of the sludge. The quantities needed to achieve this are relatively small. To obtain an improvement of the dumping suitability in the long-term, one could consider inert substances like sand or ash from incinerators. A drawback here is that relatively large amounts will be needed; the volume to be dumped will be a factor 2 to 3 times larger than the volume of the sludge.

A processing method that could accommodate that drawback is the composting of sewage sludge. The composting process will result in a reduction of the sludge volume, although part of the volume will be the residues of the additives of the composting process.

11. FURTHER RESEARCH

It is clear that it is not possible to formulate general criteria for the dumping suitability of sewage sludge. With the exception of mono-dumps, the influence of dumped sewage sludge on the stability and bearing capacity of the dump-site will depend on the nature and composition of the remainder of the sludge to be dumped, the way in which the sludge is processed in the dump-site and the ratio between sludge and other waste materials. On the basis of this knowledge, requirements for the sewage sludge to be dumped can be formulated.

At present, the research efforts are aimed primarily at the development of dumping suitability requirements with respect to long-term behaviour in the dump-site. This research will involve the consolidating mechanisms of lime and sludge and the effect of time on these mechanisms. The Forschungs- und Material prufanstalt in Stuttgart is presently conducting a more goal-oriented study into the building-in of sludge into dump-sites and the manner in which dump-sites can be compressed. In that study, the emphasis will have to be on obtaining the greatest possible solidity of the sludge by dewatering, thus limiting the need for conditioning.

REFERENCES

(1) Otte-Witte, R. (1989). Influences on the mechanical properties of sewage sludge for disposal to landfill. In: Sewage sludge treatment and use. New developments, technological aspects and environmental effects, (A. H. Dirkzwager and P. l'Hermite (Eds.), pp. 307-324. Elsevier Applied Science, London.

(2) Loll, U. (1986). Characterisation of the physical nature of sewage sludge with particular regard to its suitability as landfill. In: Processing and use of organic sludge and liquid agricultural wastes (P. l'Hermite, Ed.), pp. 168-177. D. Reidel, Dordrecht.

(3) Gay, G. Ch. W., K. F. Henke, G. Rettenberger and O. Tabasaran (1981). Standsicherheit von Deponien fur Hausmull und Klarschlamm. Stuttgarter Berichte zur Abfallwirtschaft, Band, 14.

(4) Colin, P. (1978). Caracterisation de l'etat physique des boues. Revue documentaire et bibliographique. RH-82-31, Institut de Recherches Hydroliques, Nancy.

(5) Graig, R. F. (1978). Soil mechanics. Van Nostrand Reinhold Company.

(6) Gerschler, L. J. (1984). Verfestigung von Sonderabfallen. Handbuch Mull und Abfall, nr. 8134, E. Smidt Verlag, Berlin.

(7) Otte-Witte, R., U. Moller, W. Kassner, U. Loll and Ch. G. Gay (1984). New definition of sewage sludge depositing capabilities. Recycling International, Berlin.

(8) Wolf, P., and U. Neuschafer (1987). Langzeitverhalten eines zwischengelagerten Klarschlamms - Empfehlungen fur einheitlliche Untersuchungsmethoden. Abwassertechnik, *1*.

(9) Te Marvelde, J. H. B. (1989). Sludge treatment and tipping site Hartelmond. In: Sewage sludge treatment and use. New developments, technological aspects and environmental effects (A. H. Dirkzwager and P. l'Hermite, Eds.), pp. 361-363. Elsevier Applied Science, London.

(10) van den Berg, J. J. (1986) Composting of sewage sludge containing polyelectrolytes. In: Processing and use of organic sludge and liquid agricultural wastes (P. l'Hermite, Ed.), pp. 518-522. D. Reidel, Dordrecht..

The use of sewage sludge as a fuel for its own disposal

P. LOWE[1] and J. BOUTWOOD[2]

[1]Yorkshire Water, 38 Southgate, Wakefield,
West Yorkshire, WF1 1TR, UK.
[2]Yorkshire Water, Spenfield, 182 Otley Road, West Park, Leeds,
West Yorkshire, LS16 5PR, UK.

SUMMARY

Sewage sludge has a calorific value which allows it to be burned without the need for support fuel. This requires consideration to be given to the heat balance in designing a sewage sludge incineration plant taking into account the constraints placed on its operation to meet stricter air emission standards. Modern plants can be designed, built and operated to meet such constraints and to compete economically with other disposal routes.

1. INTRODUCTION

The choices of the treatment and disposal route for sewage sludge depend upon the local circumstances, the nature and composition of the sludge and the quantity of sludge involved. While incineration has been practised in the United Kingdom since 1968 when a multi-hearth plant was installed at Sheffield, it is not until recent years that it has been regarded as an economic alternative to land, landfill and sea disposal. The economic balance of the process is however biased towards large works where significant quantities of industrial effluents are discharged to the sewerage system and where the risk of contamination of the sludge is assessed to be significant. Pressures are being brought to bear on the other routes such as the implementation of the EC Directive for the disposal of sludge to agricultural land, together with the national and local codes of practice, stricter controls of the dumping of waste in landfill sites and the general concern about the use of the sea for marine dispersal. As these restrictions begin to take effect the costs of the associated disposal routes will increase and, providing the air emission standards can be met, incineration is likely to be a viable alternative for the next decade.

2. SEWAGE SLUDGE AS A FUEL SOURCE

Raw sewage sludge from both primary and secondary treatment stages has an organic content of about 70% in the dry solids, the other 30% being composed mainly of silica and alumina which does not have a fuel value and remains as an ash after burning. The fuel content, measured as calorific value (CV), for sewage sludge is typically about 25 MJ/kg dry ash-free solids compared to 45 MJ/kg for light fuel oil.

Of course, there will be some variability in the ash and CVs of sewage sludge at different works, particularly where a significant proportion of the sewage arises from industry. The processing of the sludge itself may have a detrimental effect on the CV. For instance, digestion of sludge will reduce the proportion of organic material and the addition of lime will not only increase the inorganic content but also reduce the overall CV because of the chemical energy used during incineration (1). However, analysis of raw sludges from several large works in Yorkshire Water (YW) has shown gross CVs to be normally in the range 23-27 MJ/kg.

Elemental analysis of the organic material in sewage sludge from two works in Yorkshire Water is given in Table 1. During incineration the carbon burns to form carbon dioxide and the hydrogen burns to produce water vapour. It should be noted that, to take account of the latent heat needed to vaporise this water formed during combustion, about 1.5 MJ/kg has to be deducted from the gross CV to give the effective net CV of the sludge. A part of the sulphur content will also be converted into sulphur dioxide in the flue gases, but part of it will stay in the ash as sulphates.

Table 1. Elemental analysis of organic content of sewage sludge.

Element	Composition by weight (%)	
	Plant 1	Plant 2
Carbon	62	60
Hydrogen	8	8
Nitrogen	4	5
Sulphur	1	1
Oxygen	25	26
Total	100	100

Autogeneous combustion, when incineration of sewage sludge is sustained without support fuel, is attained by achieving a balance between the heat required to evaporate the remaining water content of the sludge cake and to raise the flue gases to required exit temperature, and the fuel content of the organic material in the sludge. This balance can be maintained successfully by recovering heat from the flue gases to preheat combustion air to about 600°C and by dewatering the sludge cake feed to about 30%. The figures for a Yorkshire Water incinerator are given in Table 2 and illustrate this heat balance.

Table 2. Incinerator heat balance.

Heat input	MW	Heat output	MW
Preheated air	3.0	Heat losses	0.5
Sludge fuel content	10.5	Water evaporation	4.0
Fan power input	0.1	Hot flue gases	9.1
Total	13.6	Total	13.6

3. ENVIRONMENTAL PROTECTION

While design of the incineration plant to achieve autogeneous combustion is the key to ensuring its economic viability, its acceptability as a sewage sludge disposal route requires a similar design emphasis to be placed on ensuring that the environment is protected. European Community Directives on air pollution control (2) and planning control of industrial development (3) have led to new UK regulations. Since March 1989 sludge incinerators have been brought under the stricter control of the HM Inspector of Pollution (4) and since July 1988 an Environmental Statement may be required as part of planning applications for major industrial developments (5).

A comprehensive study of the sludge disposal requirements for the West Yorkshire area of Yorkshire Water (6) concluded that in order to secure the disposal routes from some 61 sewage treatment works it would be advisable to construct up to 4 sewage sludge incineration plants. In anticipation of the trend towards improved air emission standards, a decision was made to design such plants to meet the air pollution control standards adopted in West Germany (TA-Luft) (7). At the time these emission standards were more restrictive than UK regulations demanded.

Such standards inevitably place some constraints on the design of incineration plants. A key requirement to ensure effective destruction of organic material in the sludge feed is that the temperature of the exit gases must be raised to at least 800°C with residual oxygen being at least 6% by volume after incineration. The formation of carbon monoxide is the accepted indicator of incomplete combustion with the required TA-Luft limit set at 100 mg/m³.

It should be noted that such emission limits are related to standard conditions that discount the different temperatures of emission, different moisture contents and different levels of excess air; in the case of TA-Luft these are gas in its dry state at 0°C, pressure 1.013 bar, and adjusted to 11% oxygen content.

The other emission limits cover metals content as well as the acid gases of sulphur oxides, chlorides and fluorides. These limits demand flue gas cleaning equipment that requires scrubbing with alkali as well as highly efficient dedusting.

Process control and monitoring must also be extended to ensure that these strict standards are maintained. Stack emissions at one recently commissioned sewage sludge incinerator have been sampled and analysed by Warren Spring Laboratory of the Department of Trade and Industry. The results are given in Table 3 and show that the TA-Luft limits can be successfully achieved.

Table 3. Comparison of actual emissions with TA-Luft.

Monitored conditions	TA-Luft	Test 1	Test 2
Incinerator exit gases			
Temperature (°C)	min 800	891	889
Oxygen content (%)	min 6.0	7.2 - 7.7	6.8 - 7.7
Stack emissions(mg/Nm³) [1]			
Carbon monoxide	max 100	less than 1	less than 1
Dust	max 30	11.3	6.1
Sulphur oxides (as SO_2) [2]	max 100	66	21
Chlorides (as HCl) [2]	max 50	14	5

[1] Dry gases at NTP and 11% O_2
[2] Scrubber outlet pH set at 6.8 for test 1 and 7.0 for test 2.

The appearance of a visible plume from an incinerator stack can cause public concern and its elimination may be a planning requirement. Heat recovered from the incinerator flue gases can be used to prevent this occurring. Utilisation of the plant 24 hours per day is essential to its economic viability, and the presence of high-powered air fans can cause a nuisance if the noise is not attenuated; housing fans and compressors in an acoustic enclosure, and fitting silencers on air ducts can ensure noise is not a problem. There is no doubt that elimination of the environmental impact of plume and noise is necessary to make this type of plant acceptable.

During the design phase consideration must be given to the final disposal route for the ash. Although the content of metals in sewage sludge is very unlikely to make the ash hazardous enough to be defined as a Special Waste under UK regulations (8), it is still a Controlled Waste for which a suitably licensed site must be found. Even if disposal is left to a contractor, the Waste Management Authority will wish to know the composition of the ash, and in particular will be concerned to see that any dry ash is effectively damped down to avoid blown dust.

Anyone planning to build an incinerator will have to respond to concern about the possibility of trace amounts of the toxic organic substances polychloro-dibenzene-p-dioxins (PCDDs) and polychloro-dibenzo-furans (PCDFs) being emitted. A recently published report by the Department of the Environment summing up the current knowledge on dioxins in the environment (9) does provide some reassurance on incineration of sewage sludge with the following comments:

"... in the USA the levels of PCDDs and PCDFs found in the stack gases were low compared to two municipal solid waste incinerators tested using the same sampling method ... In West Germany concentrations of PCDDs and PCDFs in stack gases were below detection limits. The bottom ash samples appeared to contain virtually no chlorinated organic compounds (at a ppb detection limit). Similarly, testing of a fluidised-bed sewage sludge incinerator in Belgium indicated that only OCDF was present in all samples and all four hepta CDF isomers were found in only one sample. No detectable amounts of PCDDs and PCDFs were found in the fly-ash. It is thus not anticipated that sewage sludge incineration is a significant source of PCDDs and PCDFs in the UK."

4. PROCESS IMPLICATIONS AND PLANT SELECTION

The requirements for high exit temperatures to meet air pollution control requirements, and no support fuel to ensure economic viability, effectively rule out consideration of the Multi-Hearth design commonly chosen in the 1960s and 1970s for sewage sludge incineration. The choice of a Fluid Bed (FB) incinerator, with waste heat recovery to preheat the combustion air, is now regarded the norm for sewage sludge incineration (10). The FB incinerator does, however, require a high-powered forced draught fan to fluidise the sand bed, but its simpler refractory lining can reduce maintenance requirements.

There are a number of plant manufacturers who market FB incinerators, each with their own peculiarities in design and operation. An assessment of each system is worthwhile before the final choice is made.

Because it is so crucial that the sludge be dewatered to around 30% dry solids, to provide confidence that this can be attained trials using mobile filter presses and centrifuges from several suppliers should be carried out. If the results are short of the required 30% target an additional drying stage, using waste heat recovered from the flue gas, is necessary upstream of the incinerator.

Waste heat recovery is essential to preheat the combustion air, to provide heat to the stack for eliminating the steam plume and, if needed, to heat the sludge drier. If the latter is not required, then surplus heat can be used for office heating, or possibly power generation where the size of plant makes this economic. An example for a YW plant is given in Table 4.

Table 4. Waste heat recovery (MW).

Heat losses	0.2
Preheating combustion air	3.0
Heating stack gases to avoid plume	1.1
Outlet flue gases at 295°C	7.0
Surplus heat for office heating	1.8
Inlet flue gases at 890°C	13.1

It should be noted that all ash is carried over from a Fluid Bed incinerator through the waste heat exchanger, so the ash fusion temperature of about 1000°C will set the upper temperature limit if fouling in the heat exchanger is to be avoided.

The strict limits on dust emissions may be achieved using a multi-field electrostatic precipitator or a high efficiency venturi scrubber. The limits on acid gas emissions mean that where an electrostatic precipitator is used a scrubber must also be included. The operational advantages of collecting and storing the fly-ash in a dry state, with addition of the minimum amount of water to damp down for transport and disposal, may make it attractive to choose an electrostatic precipitator to take out about 95% of the dust load. This can then be followed by a wet scrubber using sewage works final effluent to remove the remaining ash, neutralise the acid gas using caustic soda and condense out much of the water vapour. The very different efficiency/particle size distributions of the precipitator and wet scrubber may also mean that the combined efficiencies of these units is better than the sum of them individually and results obtained using this combination at a Yorkshire site are demonstrated in Table 3. To achieve high efficiency in a scrubber the pressure drop has to be maintained at about 50 mbar and this requires a relatively powerful induced-draught fan.

Manning is one of the major elements in operating costs, thus two key requirements that should be specified are a high level of automation and potential maintenance problems being designed out where possible. As a consequence, a Process Logic Controller will be needed to control the sequence of operations during start up and shut down, with built-in safety trips to shut down the plant if pre-set conditions are overstepped.

Other major elements of operating costs are indicated in Table 5, which has figures based on the performance test for a recently commissioned sewage sludge dewatering and incinerator plant; estimates based on recent studies (11, 12) have been made for maintenance, as the true costs will only become apparent well into the life of the plant.

Table 5. Incinerator operating costs (£/tds).

Labour	11.4
Rates (estimated)	2.1
Power	8.7
Chemicals	11.0
Ash disposal	2.8
Contract maintenance (estimated)	6.5
Total	42.5

The main costs apart from labour are for polyelectrolyte chemicals used in sludge dewatering, and electrical power for the incinerator plant. It is worth seeking process guarantees for these items as well as plant capacity, and the Institute of Chemical Engineers "Red Book" model conditions of contract (13) provide for such performance guarantees to be backed up by contractual tests.

5. CONCLUSION

By designing to ensure autogeneous combustion together with low manning, and by specifying highly efficient flue gas cleaning equipment, modern sewage sludge incinerators can be cheaper to run and far cleaner than in the past. Evidence is now emerging to substantiate the view that incinerators compete with sea disposal and landfill as the Best Practical Environmental Option for contaminated sludge, and for very large works where the sludge is suitable for agricultural use, incineration may even be an alternative to land spreading.

ACKNOWLEDGEMENTS

The Authors would like to thank A. I. Ward, Director of Water Services, Yorkshire Water for permission to publish this paper.

REFERENCES

(1) Paine, R. and C. Thompson (1984). Polyelectrolyte conditioning of Sheffield sewage sludge, Wat. Sci. Tech., *16*.
(2) European Commission (1984). Directive 84/360, Official Journal of the European Commission, L188 (28.6.84).
(3) European Commission (1985). Directive 85/337, Official Journal of the European Commission, L17 (5.7.85).
(4) Health and Safety (Emissions into the Atmosphere) (Amendment) Regulations (1989). SI 1989: 319.
(5) Town and Country Planning (Assessment of Environmental Effects) Regulations (1988). SI 1988:1199.
(6) Lowe, P. (1989). Development of a sludge incineration strategy for a UK water authority. Conference on drainage and waste management into 1990s, Dundee, Scotland, 9-11 May.
(7) Technische Anleitung zur Reinhaltung der Luft (1986). GMBI S95 ber S202.
(8) Waste Management Paper No. 23 (1981). Special wastes ... a technical memorandum provides guidance on their definition. HMSO.
(9) Department of the Environment (1989). Dioxins in the environment. Pollution Paper No. 27. DoE.
(10) Lowe, P. (1988). Incineration of sewage sludge - a re-appraisal. J. Inst. Water and Environ. Management, *2* (4), 416-422.
(11) Atkins, W. S. (1985). Incineration of sewage sludge: technical and economic review. Water Research Centre.
(12) Coombs, C. R. (1986). Sewage sludge dewatering by belt filter press - a survey of capital and operating costs. Water Research Centre.
(13) Institute of Chemical Engineers (1981). Model form of conditions of contract for process plants (lump sum).

DISCUSSION SESSION 3

QUESTION: C. ROWLANDS.
Could Mr Hill and Mr Blakey comment on the degree of settlement that is caused by co-disposal?

ANSWER: C. P. HILL.
We have noticed that within 12 months of co-disposal with baled waste we are getting settlement of about 75-100 mm within a depth of 3m of waste which is quite significant. With loose waste and sludge it is not as pronounced, the amount of settlement is perhaps 50 mm maximum.

ANSWER: N. C. BLAKEY.
It is very much a differential settlement with baled wastes where we have pockets of sludge within the bales, those zones are more prone to settlement than other areas of the fill. It would seem that the best way of operating would be to spread sludge over relatively large areas within domestic wastes to alleviate such problems.

QUESTION: D BEKER, RIVM, NL.
What is the basic idea behind adding sewage sludge or coal ash to municipal solid waste? Why is methane production starting earlier compared with only municipal solid waste?

ANSWER: N. C. BLAKEY.
It is a two-phase process where the initial stage in the landfill is the production of the volatile acids. If that is allowed to proceed too rapidly, which may happen where you get ingress of a lot of water, conditions that are conducive for production of these acids, the pH can often drop below the optimum at which methane generating bacteria can digest the acids. If the pH drops much below 5.5, methane production almost ceases and you get nothing but strong acid. What tends to happen with the addition of sludges, coal ash and other alkaline materials is they act like buffering zones to prevent the significant decrease in pH caused by the production of acid material and it does seem to promote methanogenic conditions more rapidly in domestic waste.

QUESTION: R. D. DAVIS, WATER RESEARCH CENTRE.
Is there a difference in gas production between digested and undigested sludge? Is there an effect of introducing a large number of methanogenic bacteria when you put digested sludge into the landfill?

ANSWER: N. C. BLAKEY.
We found some minor effects but I think the overall effect really is that the sludge is actually acting as an inert material rather than an active material. We are obviously adding extra bacteria into the system and that will assist in the overall production of methane but the predominant effect must surely be the buffering effect that the sludge gives.

QUESTION: K. M. PANTER, THAMESGRO LAND MANAGEMENT.
One of the reasons people are looking at landfill at the moment is the increasing costs of other forms of disposal but there may also be a rise in landfill airspace prices as well. A general reaction against sewage sludge is its non-compactibility, it is generally argued that it occupies the space that it is delivered at. Would you say there are any effects of the combination of sludge and refuse giving greater compactibility overall and greater final settlement, and greater amounts of landfill being put in the same airspace overall?

ANSWER: N. C. BLAKEY.
In one trial we actually disposed of the same quantity of domestic refuse in the control cell as a cell containing limed sludge at a ratio of 1:5 sludge/refuse. The sludge seems to have disappeared into the voids achieving very much better compaction.

ANSWER: C. P. HILL.
As far as the compactibility is concerned, with normal controlled waste with steel-wheeled compactors we can achieve 1 t/m^3. The problem of adding the sludge is that in order to prevent contamination of the plant you have got to put quite a large layer of loose waste on top which limits the effectiveness of the landfill compactor. Eventually you will find that the landfill will assume an ultimate density which you would get anyway but it is just a matter of the time factor involved.

QUESTION: L. E. DUVOORT-VAN-ENGERS, RIVM, NL.
Did you find any influence of sludge treated with polyelectrolytes in either the leachate quality or the gas quality?

ANSWER: N. C. BLAKEY.
We found very similar effects on gas production with the polyelectrolyte sludge in the field trial experiments as we were able to generate in laboratory conditions with other types of sludge, with gas production starting very soon after the deposition of the waste materials.

QUESTION: C. D. BAYES, WATER RESEARCH CENTRE.
Have you detected any change in the water balance in landfills as a result of co-disposal and how might that have affected the actual leachate loads being generated as opposed to the leachate concentrations quoted?

ANSWER: N. C. BLAKEY.
In general, although we are adding a lot more water to the system by putting significant amounts of sludge into refuse, there does not seem to be an amazing change in landfill water balance, although obviously what is likely to happen is you will achieve field capacity possibly more rapidly. All that it is likely to do is to shorten the timescale for the initial production of leachate. As far as the quality is concerned, at one site the ammonia level did increase quite significantly and we have put that down to the introduction of sludge to the waste.

QUESTION: P. HYDES, SOUTHERN WATER.
Do you know what effects there are on pathogens in codisposal?

ANSWER: R. COSSU.
The fact is that we have very few data at all on pathogens and viruses in landfill. There are some results in some US EPA manual which shows that pathogen survival is unlikely in landfill.

QUESTION: J. MANN, THAMES WATER.
Did you monitor the temperature in the layers of the sludge? I am thinking in particular in terms of its relationship with composting.

ANSWER: N. C. BLAKEY.
There is considerable data on temperature from the laboratory scale experimental work. We have not monitored it in the field but I think you will find there are not any vast differences in temperature. One normally achieves 30-40°C in the initial phases of degradation but the effect of adding the sludge made no difference.

QUESTION: L. E. DUVOORT-VAN-ENGERS, RIVM, NL.
There is a tendency to decrease the amount of urban waste being landfilled in the Netherlands because of the policy of recycling, making compost of carbon waste and food waste etc. Will that policy influence both the quality of the leachate and gas in the landfill?

ANSWER: R. COSSU.
Recycling is a trend but I think that it is not affecting the quality of refuse because it is such a very low percentage. 99% of the waste in the world is

disposed of in landfill and of this only 20% is properly disposed of in sanitary landfill. I think that we need not worry if some portion of food waste is transformed into compost.

QUESTION: R. C. FROST, WATER RESEARCH CENTRE.
Could Mr Lowe describe the method adopted at Esholt to preheat the stack gases to avoid visible plume formation, and could he comment on the stack gas preheating if one needs to pre-dry the sludge fed to the incinerator? I ask this because energy balance modelling which we have carried out indicates that where you do need to pre-dry the sludge feed, then the spare heat available for stack gas preheating may be only just sufficient.

ANSWER: P. LOWE.
At Esholt the suppression of the plume is quite effective. We are heating up about the same quantity of fresh air and injecting it into the stack, so that has the double benefit of heating up the whole of the gases emitted plus a dilution effect to lower the vapour pressure. The heat balances show that there is plenty of surplus heat for that and we estimate about a megawatt. I would say that over a range of plants a megawatt of heat is probably typical of the amount that you would need.

COMMENT: L. E. DUVOORT-VAN-ENGERS, RIVM, NL.
In the Netherlands a new directive has been accepted recently in which the emission rate from incinerators, compared to the German TA Luft directive, has been reduced by 1/6th based on dust. This value must be met for sewage sludge incinerators as well and companies have guaranteed to our ministry that these values can easily be reached.

QUESTION: C. B. POWLESLAND, WATER RESEARCH CENTRE.
Could Mr Lowe comment on the out-turn capital cost of the Esholt incinerator which would perhaps give us an idea of how the route compared to, say, landfilling or agriculture over a 20 year period?

ANSWER: P. LOWE.
The out-turn price has been about £4.5m for Esholt and our target price for the Blackburn Meadows plant is £6m. The difference in cost, although they are dealing with roughly the same amount of sewage sludge, is due to the building already at Esholt whereas at Blackburn Meadows it was a green field site and so I would think that this is probably a better indicator of the current capital costs of sewage sludge incinerators for about 15,000 tds/yr that we are building at the present time. Our target operating costs between £40-£45 goes up at Blackburn Meadows to round about £50-£55/tds.

Those are very comparable costs with some of our other route options. Incineration was seen as an essential option in order to protect the agricultural land outlets in Yorkshire. We wanted to maintain the agricultural recycling programme but we could see that there were certain sludges that actually would never really fit in to that recycling operation. So this was not just a case of going for the cheapest option *per se* for Esholt or for Blackburn Meadows, but it was an option that enabled the other routes to be protected.

QUESTION: R. P. EARTHY, WESSEX WATER.
Could Mr Lowe comment on what sort of public opposition there has been from the people in Yorkshire bearing in mind the concerns in other parts of Europe?

ANSWER: P. LOWE.
We have taken a very positive attitude towards the public in the question of incineration. As far as the Esholt and Blackburn Meadows plants were concerned, we had incinerators on the sites previously and that eased the way as far as the public acceptability was concerned. We have a third plant that is under consideration at the present time and we are currently preparing a detailed environmental impact statement for the local planners. This is in quite a sensitive area at Huddersfield where you have got quite a number of industrial emissions. We see this as being one of the stages that you must go through now if you want to build an incineration plant, and we have tried to take the public with us by having open days at Esholt to explain how the emission standards will operate.

QUESTION: R. COSSU.
What does Mr Lowe think about co-disposal of sludge and municipal solid waste in incinerators? One of the processes which has been proposed often is to use the steam produced during the municipal solid waste incineration for drying the sludge and then feeding the sludge into the furnace together with the municipal solid waste.

ANSWER: P. LOWE.
We did look at this when we were doing the initial planning. The problem is that there is much more municipal refuse than sewage sludge and the result is that the major investment would have to be in the municipal refuse incinerator, about £30m-40m as opposed to the £5-6m for a sludge incinerator.

SESSION 4
Other Uses

Economics and marketing of urban sludge composts in the EEC

J-L. MARTEL

Agro Developpement SA, 2 rue Stephenson,
78181 St. Quentin, Yuelines Cedex, France.

SUMMARY

The position of sludge compost processing in Europe is described and compared to the situation in North America. The various composts made from urban sludges are outlined and three different analytic and agronomic profiles of sludge composts are defined. The regular procedure in France for distributing and marketing composts is explained; sludge composts are not well defined under present regulations. A technical and commercial approach was employed for a survey of six composting sites operating in France. The importance of the commercial approach is outlined by the example of a French city of 15,000 population equivalent, and an economic comparison between a controlled sludge landspreading organisation and a sludge composting organisation including compost commercialisation is reported. Nine sludge composting plants operating in Europe are described briefly; four plants are located in south Europe (one in Spain and three in Italy), and five plants operate in north Europe (one in the United Kingdom, one in the Netherlands and three in the Federal Republic of Germany).

From the north to the south of Europe, composting aims remain approximately the same with different responses to environmental constraints. In France, sludge composting promoters have to face intermediary conditions of climate, soils and environmental pressure in their attempt to build cost-effective composting facilities.

1. INTRODUCTION

This paper presents some of the numerous data collected by Agro Developpment on behalf of various bodies such as French public agencies, the Commission of the European Communities and French municipal authorities, in relation to quality and marketing of urban sludge and composts.

At the present time and in terms of quantity, sludge composting remains relatively marginal in the European Community; 40-50 plants produce in total 175,000-200,000 tonnes of compost per year. In comparison, in the United States, 100 plants are in operation and another 100 are planned; 500,000 tonnes of compost per year are produced by only 36 facilities.

In the Federal Republic of Germany, 30 units of less than 15,000 population equivalent (pe) produce together 40,000 tonnes of compost per year. Great difficulties are met in developing new operations and in marketing composts containing relatively low concentrations of heavy metals.

In France, seven sites produce 20,000-25,000 tonnes of compost. Selling sludge compost remains a difficult job because of numerous factors, such as the fear of heavy metals, the poor quality of some products, and the absence of specialised marketing teams.

In contrast, one Dutch private firm and two Spanish companies produce large quantities of sludge compost; 30,000 tonnes per year in Madrid from digested sludge from three waste water treatment plants (wwtp) and 100,000 tonnes in the Amsterdam suburbs from urban and industrial sludges from four provinces.

2. TYPES OF COMPOSTS PRODUCED FROM URBAN SLUDGES

It is quite difficult to classify composts because of the wide variations in appearance and composition. The composting processes applied to raw or digested sludges are very different; they depend on the quantity of dry matter to be treated each day and, even more, on the reaction of the environment to odour.

Plants can use sophisticated equipment in sensitive urban environments, such as:
a) Enclosed buildings with processing by windrow turning or with forced ventilation;
b) Negative ventilation applied to static piles (Beltsville process), the effluent gas being filtered through a pile of cured compost or barks, or even treated in a deodorising tower (scrubbing system);
c) Vertical or horizontal in-vessel composting systems (BAV, Weiss, Siloda, Royer, Fairfield, American Biotech, Purac, Buhler ...).

Other plants treating small quantities of sludge or operating on isolated sites can use the following equipment:
a) Windrow composting with front loaders;
b) Windrow composting with specialised compost turners;
c) Static pile composting with forced ventilation;
d) Filtration of liquid sludge through a bulking agent like straw, then mixing and composting.

Fig. 1. Location of sludge composting units in the European Economic Community.

Because of the diversity of products currently on the market, only general descriptions of their agronomic qualities can be given. Five products tested or marketed in France, Spain and in the United States are described in Table 1.

Table 1. General descriptions of five sludge composts.

Location Technical parameters	Nantes Nord (France) Agro-compost	Blois (France)	Brametot (France)	Washington DC (USA) Compro	Madrid Valdemingomez (Spain)
Sludge treatment	Primary sedimentation	Anaerobic digestion	Anaerobic digestion	Anaerobic digestion and lime	Anaerobic digestion
Sludge quantity t DM/day	13	2	-	80	30
Sludge quality					
DM %	28.5	25	26.5	18	25
VS %	77	55	55	55	-
Total nitrogen % [1]	3.0	3.8	3.0	-	-
Compost quality					
DM %	50	45	61	64	70
OM % [2]	39	27	27	42	34.5
Total nitrogen % [2]	1.0	0.7	0.7	0.9	1.5
Composting process	BAV 550cm	Windrow C 400	Siloda	Aerated static pile	Windrow after sun drying
	50 days	21 days	14 days	21 days	42 days
Bulking agent	Sawdust	Sawdust	Pretreated refuse	Woodchips 80% recycled	Pellets of sludge

[1] Dry weight basis.
[2] Wet weight basis.

With respect to the organic matter content of compost, in Blois digested sludge produces a compost containing less organic matter than one produced in Nantes from raw sludge, but other parameters should also be considered; the quality and composition of the bulking agent, the dry matter content of the dewatered sludge, the process of composting, and the possibility of screening.

In fact, three qualities of sludge products on the market can be defined:
a) A fine texture compost or "French compost" made with sawdust (BAV process or windrow composting). This product is of good colour, relatively marketable, mainly used as organic amendment, but difficult to mature (200 mg O_2/kg DM per hour after 6 months) and to use in a substrate;
b) A rather coarse compost made with woodchips or barks by static pile composting. This product is easier to mature (35 mg O_2/kg DM per hour after 1 month) and to use in a substrate;
c) Special composts resembling "dried sludge" made without a bulking agent (Madrid compost) or with the minimum of bulking agents (Canterbury compost); they are nearly organic fertilisers.

3. SITUATION IN FRANCE

3.1 Legislative approach

General legislation on waste control does not specifically mention composts. Special regulations place them into three categories:
a) Organic fertilisers (with a minimum content of N, P and K);
b) Organic amendment (soil conditioner);
c) Plant medium.

The law on fertilisers, laid down on 13 July 1979, organises the control of fertilising products with standardisation for importation, marketing, and free distribution, but it does not cover already normalised products (AFNOR standard), products with labels, waste included in the regulations concerning waste water, waste from classified plants, and raw organic products from the farm. The Decree No. 80-472 of 16 June 1980 established the Commission for Fertilisers and Plant Media, the Standards Committee, and the Commission for Toxicity Analysis. The following illustrates the procedures involved in launching a product:

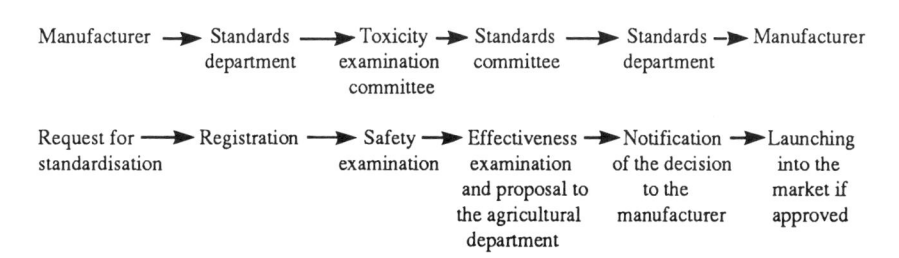

Manufacturer → Standards department → Toxicity examination committee → Standards committee → Standards department → Manufacturer

Request for standardisation → Registration → Safety examination → Effectiveness examination and proposal to the agricultural department → Notification of the decision to the manufacturer → Launching into the market if approved

3.2 Production control

Composition. The Decree of 8 December 1982 states that self-inspection is to be carried out by the seller. The seller must check that the correct constituents given on the label are within specified limits. The seller must also check the safety of the directions for use stipulated by the Standards Committee.
Labelling. The compulsory and optional indications are stated by decree.
Responsibility. The company launching the product onto the market is responsible for law suits as far as the quality of the product is concerned. Standardisation is not a guarantee. As far as a spreading service is concerned, an insurance can be taken out.

3.3 Technical and commercial approach

The composting process allows organic by-products to be treated but does not necessarily lead to a commercial product. To obtain a marketable product, technical, legal and commercial criteria must be respected. Regulation provides a technical context for the product but a commercial strategy must be created to transform it into a marketable product. In France, sludge compost is sold at between 50 and 500 FF per tonne under different packaging; bulk or in bags.

Desirable properties of compost are:
a) High content of organic matter;
b) Significant content of organic nitrogen with slow mineralisation;
c) Pleasing physical appearance with no undesirable substances such as glass or plastics and a good colour from brown to black.

Psychological aspects should be mentioned since the customer, before ordering, compares sludge compost generally to well-known products such as manure. Therefore, it is important for the seller to provide precise information based on laboratory tests and experimental results.

A good market survey must identify the competitive products and each potential market with its specific requirements. To enter the market, prices must be fixed and distribution channels chosen taking into account commercial investments, advertising and packaging. Each market is different and until now no real commercial advertising campaign for sludge composts in the EEC has been defined. The existing information sources (agricultural advisers) must be taken advantage of and followed up with personal research.

Information concerning six French operating sites is summarised in Table 2.

Table 2. Survey of six operating sites in France.

Location	Capacity (pe) Production of compost (t)	Bulking agent	Process	Production cost F/t	Selling price F/t		
Pau	30,000 pe 1,200 t	Woodchips	Forced ventilation	500	Free		
Bormes les Mimosas	35,000 pe 4,000 t	Refuse	Windrow	600	0-15 mm 0-25 mm	35 15	
Chateaurenard	3,600 pe 200 t	Straw	Forced ventilation	275	Free		
Sin le Noble	30,000 pe 11,000 t	Refuse	Windrow	300	32		
Blois	50,000 pe 2,000 t	Sawdust	Windrow	250	110		
Cernay, Thann Mossch, Guebwiller	64,000 pe 14,000 t	Refuse	Forced ventilation	380	0-7 mm 7-15 mm 15-40 mm	80 40 Free	

3.4 Sludge landspreading or sludge composting?

For large cities, the alternative to composting as regards sludge disposal has been controlled and uncontrolled landspreading or landfilling. In the past few years, landfills have become increasingly rare and access fees increasingly high. Moreover, controlled landspreading operations, as required by local authorities and by farmers receiving sludge, commit municipal authorities to important investment programmes, involving storage facilities, mixing, pumping and landspreading equipment, sometimes without any real commitment from the local agricultural lobby. A highly sensitive local environment can lead to a cost-effective sludge composting organisation. The main results of an economic comparison between four organisations, applied to an annual production of 2,000 t of sticky sludge, are presented in Table 3.

The fourth organisation, using windrow composting, appears to have the least expensive mode of treatment because of the commitment of a private company to buy all the compost production at a good price.

Table 3. Economic comparison of landspreading and composting for sludge from a 15,000 pe sewage works.

Production:	2,000 wet tonnes (wt) per year at 14% DM.
Financial conditions:	Government grant of 50% of total investment; Banker's loan at rate of 10% over 10 years.

Commercial conditions:	10 FF/wt of sludge paid by farmers. 60 FF/wt of crude compost paid by private company engaged on all the production.

	Landspreading		Composting	
Organisation	1	2	3	4
Capacity of storage	900 m^3	1,150 m^3		
Composting process			Forced ventilation	Windrow composting
Landspreading by	Contractor	Producer		
Compost production			1,300 wt/yr	2,300 wt/yr
Total investment (FF)	1,200,000	1,700,000	1,370,000	1,430,000
City investment (FF)	600,000	850,000	685,000	715,000
Financial cost (FF)(1)	37,500	53,500	43,000	45,000
Depreciation (FF)(2)	42,500	80,000	70,500	64,000
Operating costs (FF)(3)	213,000	211,000	227,000	225,000
Total costs (FF) (4) (4) = (1) + (2) + (3)	293,000	344,500	340,000	334,000
Revenue (FF)(5)	20,000	20,000	79,800	138,000
Balance (FF)(4) - (5)	273,000	324,500	260,200	196,000
Cost FF/t DM	1,092	1,298	1,040	784

4. PRESENTATION OF NINE SLUDGE COMPOSTING PLANTS IN FIVE COUNTRIES OF THE EEC

Individual visits to a representative selection of sludge composting plants operating in Europe were carried out during autumn 1987 for the Commission of the European Communities.

Four plants were located in southern Europe (one in Spain and three in Italy) and five units were visited in northern Europe (one in the United Kingdom, one in the Netherlands and three in the Federal Republic of Germany).

A short extract of the main characteristics and a comparison between the nine units are shown in Table 4 (southern Europe) and Table 5 (northern Europe).

4.1 Southern Europe

Italy: Two of the three units are located on the Adriatic Coast in a tourist area which is mainly why the municipal authorities of Senigallia and Pesaro chose a relatively expensive sludge composting process (in-vessel system). Mature compost is sold at 25,000 lire (20 US$) per m^3 in Senigallia and is free to municipal public parks services in Pesaro. At the third plant, Schio, a very good quality compost is manufactured from industrial sludge and organic material of biostabilised household refuse. Marketing this compost is rather difficult because there are many dairy farms near Schio and the plant is 100 km away from the nearest areas where organic matter is in demand.

Spain: Only the regions of Catalonia and Madrid seem to be developing important programmes of urban waste recycling with sludge and garbage. The composting process used anaerobic digested sludge from two waste water treatment plants (la China and Sur). It is rather extensive and works only from May to September in Madrid. During winter, sludge is stored in small piles on site.

Composting operations are cost-effective; no bulking agent is used except pellets of dried sludge recycled by compost screening, and only two workers produce 10,000 tonnes of manufactured compost. Private contractors produce and sell compost without receiving financial assistance from the district.

4.2 Northern Europe

Federal Republic of Germany: The three organisations described in Table 5 show that German municipalities have to face strict regulations and lack of space for landfilling. Sludge composting is becoming more and more important. Special care is given to levels of heavy metals.

The Netherlands: Halweg's capacity is so great that it resembles some plants operating in the USA. In fact, this unit is quite unique; different wastes entering the treatment centre come from four provinces and a private company operates the composting and sells compost at a low price (5-10 US$/wet tonne). Operating costs are paid mainly by industrial and urban sludge producers; up to 170 US$/t DM.

United Kingdom: All the household refuse composting plants (twelve in 1975) have been closed down over the last few years; operating costs were too high and there was no market for compost. Sludge composting has started recently because of odour problems during winter storage and increasing difficulties in landspreading. The Water Research Centre at Medmenham is working on minimal co-composting sludge/refuse for practical landfilling, and co-composting sludge/straw for marketing.

Table 4. Comparison of four sludge composting plants in southern Europe.

Composting plant	Sludge type	Bulking agent	Composting process	Compost product	Marketing
Senigallia, Italy (municipal)	Raw sludge digested in summer. Seasonal variations	Sawdust 1,500 m^3/yr 8 US$/m^3	In-vessel BAV	1,500 - 2,000 m^3/yr 30% OM (organic matter)	Municipal services. 20 US$/m^3. Bought by vegetable growers
Pesaro, Italy (Aquagest)	Raw sludge 20% ds. Seasonal variations	Sawdust 1,500 m^3/yr 4US$/m^3	In-vessel BAV	2,500 m^3/yr 30% OM	No selling, free for municipal needs
Schio, Italy (Lanerossi)	Digested sludge 20-25% ds	Organic fraction screened refuse	Windrow (5% sludge by weight)	7,000 t 27% OM	Difficult. 9 US$/t
Valdemin-gomez, Spain (Vertresa)	Digested sludge 25% ds 30 tds/day	Screened pellets of sludge	Drying then windrow (May-Sept)	10,000 t/yr	Easy. 25 US$/t. Valencia specialist growers

Table 5. Comparison of five sludge composting plants in northern Europe.

Composting plant	Sludge nature	Bulking agent	Composting process	Compost production	Marketing
Horn Bad Meinberg, W. Germany	Prolonged aeration sludge 25-32% ds	Sawdust 9 US$/m³ 600 m³/yr	In-vessel Gebruder Weiss/Kneer	1,200 m³/yr	Municipal. 18 US$/t in bulk, 66 US$/t in bags
Lemgo, W. Germany (Lippe c.)	Sludge from 9 STW 25% ds	Organic from shredded screened refuse 60,000t	Static pile screened after curing 10 and 18mm	20,000 t/yr 25% OM	7-16 US$/t in bulk, 70 US$/t in 35 kg bags
Hilchenbach, W. Germany	Prolonged aeration sludge 25% ds 100,000 pe	Pine sawdust	In-vessel BAV	1,000 t/yr	Municipal. 16 US$/t
Canterbury, UK (Southern Water)	Limed raw sludge 30% ds 10 tds/day	Cereal straw 40 US$/t	Static pile Rutgers strategy	4,000 t/yr 24% OM	First year of production trials
Halweg, Netherlands (Rutte Recycling)	Urban and industrial sludge 90,000 tds	Woodchips and woodbark	Static pile under cover, screening	100,000 t 20% OM	5-10 US$/t Sold on national scale

5. CONCLUSIONS

From the north to the south of Europe, composting targets remain appoximately the same with different reactions to environmental constraints. Organisations concerned with sludge composting try to build an economic sludge disposal system by producing a stabilised product which is easy to store and market.

Contrasting situations are seen in Halweg (near Amsterdam) and in Valdemingomez (suburbs of Madrid). In each case, private companies produce and sell compost in very different ways. In the Netherlands, environmental constraints are very important and a composting unit is considered as a treatment centre where sludge can be treated with a high access fee; up to 1,000 FF/t DM (165 US$/t). In Spain more flexible regulations and an important need for organic matter allow the emergence of cost-effective composting units.

Standardisation of regulations among different countries of the European Community and optimisation of composting processes developed in Europe will make sludge recycling possible inside a developing organic matter market. However, in the course of the 1990s, local economic distortions will certainly appear as organic by-products are increasingly imported into France, and French manufacturers will have to face this challenge by increasing the quality of their products and of their services.

Compost - a sewage sludge resource for the future

P. J. MATTHEWS[1] and D. J. BORDER[2]

[1]Anglian Water, Chivers Way, Histon, Cambridge, CB4 4ZY, UK.
[2]Hensby Biotech Ltd., Woodhurst, Huntingdon,
Cambridgeshire, PE17 3BS, UK.

SUMMARY

Cereal production in the UK results in the production of about 13 million tonnes of straw per year. Straw has many uses but these still leave a surplus of about 7 million tonnes each year. This surplus is presently burned or chopped and incorporated directly into soil. Straw burning is now considered to be antisocial and often dangerous, while direct incorporation of straw is at best an inefficient use of the material. It has been found that much of this excess waste straw is appropriate for co-composting with liquid, untreated sewage sludge to produce a range of composts suitable for use in agriculture, land reclamation and horticulture. It has further been found that modifications of the process developed to accomplish this co-composting can allow the production of useful composts from a wide variety of other waste materials both in the UK and overseas.

1. INTRODUCTION

Existing methods of composting sewage sludge normally use dewatered sludge cake as a starting material. Such composting methods have been successfully used in providing a means of sludge disposal or in solving specific problems such as the removal of smells which occur in association with stored sludge cake.

The novelty of the process described in this paper (patent applied for) is that liquid, untreated sewage sludge is used for composting and not sludge cake. Instead of the sludge being dewatered prior to composting, cereal straw is used to absorb the liquid component. This technique has some advantages over previous methods both in terms of the quality of the product produced and the general applicability of the method in composting other waste materials.

The potential problems of the heavy metal content of the resultant composts and the possible presence of pathogens in the final material have been examined carefully.

The process has been developed over a period of three years. The approach has always been strictly commercial, in that the process has been designed not just to dispose safely of straw and liquid sludge but also to produce compost which can be sold at a profit. To achieve this end the technique has been kept as simple and as cost-effective as possible.

The process has also been deliberately designed to be capable of composting a wide variety of waste materials other than cereal straw and liquid sewage sludge. The applicability of the process therefore extends far beyond the UK and the sewage treatment industry.

2. THE BASIC COMPOSTING METHOD

Organic material naturally decomposes over a period of time under the influence of the bacteria, actinomycetes and fungi which occur in soils. In nature the process is a slow one. However, if suitable material is gathered together into relatively large masses, and if appropriate conditions of compression, aeration and moisture are met, considerable heat is generated within the masses by thermophilic micro-organisms. The temperatures within the masses rise and the rate of decomposition increases rapidly. The technique of composting is, in essence, the manipulation of the physical and chemical environment within the masses to encourage the proliferation of the correct micro-organisms to bring about this accelerated decomposition.

In its simplest form, the process under consideration involves the use of just two components: untreated liquid sewage sludge and wheat straw (1). The sludge provides most of the water, nitrogen and phosphorus needed for microbial activity. The straw is used to absorb the liquid component of the sludge and also to provide the main source of carbon and potassium. The straw also acts as a bulking agent, keeping the mix open in structure and thereby allowing a relatively free movement of air.

The wheat straw used in the process is stored in the form of 500 kg compressed Hesston bales. In order to ensure even composting the bales are broken up thoroughly into loose straw before use. Typical analyses for the straw are shown in Table 1.

The liquid sewage sludge is delivered to the compost production site in tankers and is pumped into large storage tanks. The sludge normally contains about 4% dry solids. Typical analysis figures are given in Table 1.

The sludge and straw are intimately mixed to provide as uniform a starting material as possible. Each tonne of straw can normally absorb 3 tonnes of liquid sludge without significant liquid run off. This initial mixing and absorption process takes 5 - 7 days. The temperature of the mix rises rapidly to in excess of 60°C.

The mix is then formed into compressed, vertically-sided windrows. These are 2 metres wide, 2 metres high and about 70 metres long. Each windrow weighs about 150 tonnes at this stage. It is emphasised that these compressed windrows behave quite differently to the loose windrows commonly encountered in sludge cake composting.

Table 1. Typical analysis values of raw materials, and compost after one week of windrow composting.

		Straw	Sludge	Compost
pH		6.9	6.03	7.6
Dry solids	%	90	4	27
Volatile matter	% ds	96	79	80
Carbon:Nitrogen		48	5	<20
N	% ds	0.7	4.6	3.0
P (P_2O_5)	% ds	0.5	3.4	2.0
K (K_2O)	% ds	1.0	0.4	1.0
Zn	mg/kg ds	20	390	185
Cu	mg/kg ds	6	250	50
Ni	mg/kg ds	<0.1	16	7
Cd	mg/kg ds	0.05	5.0	1.0
Cr	mg/kg ds	<0.01	22	6
Pb	mg/kg ds	<0.01	56	14
Hg	mg/kg ds	0.03	0.4	0.1
Salmonellae	mpn/100g	ND	14,000	ND

ND = None detectable

The compressed mix acts as its own heat insulator, and thermophilic micro-organisms and chemical reactions in the mix cause the compost temperature to rise rapidly. Temperatures in excess of 80°C are routinely obtained through the bulk of the windrows. These temperatures are maintained over a 7 day period.

The temperature of the windrows decreases from the inside to the outside; the outer 10 to 15 cm is at, or only just above, ambient temperature. It is important to make certain that all of the mix is exposed to the required high temperatures and to ensure that the composting process is as uniform as possible. In order to accomplish this, specialised turning equipment is used to turn the windrows "inside out". This turning is carried out at regular intervals over a week long period. The internal temperature of the composting material drops during the turning to about 40°C but recovers within a short time of turning being completed. Various materials such as additional sources of nutrients can be added to the windrows during the turning stage. The typical smell of sludge disappears during this stage.

No forced air ventilation is carried out. Aeration is accomplished by taking advantage of the "chimney effect". Hot air rising through the compost draws in fresh air through the sides of the windrow. All of the windrow, with the exception of a region at the centre bottom, is kept aerobic by this means.

At the end of this 7 day period of windrow composting the compost is suitable for agricultural applications. Typical analysis figures for the compost at this stage are shown in Table 1.

The first sludge/straw composting facility is run by Hensby Biotech Ltd at Huntingdon, Cambridgeshire. The site has the potential of producing about 50,000 tonnes of this agricultural compost per year. The compost is marketed under the name of "Natgro".

This windrow composting process can be continued for a further period of time to provide a drier, more friable material with a lower carbon:nitrogen ratio and lower pH. The material is suitable for many horticultural applications. Alternatively, the agricultural grade compost can be taken through large pasteurisation tunnels, each tunnel holding up to 100 tonnes of compost. Within these tunnels a forced air system controls temperature to better than ±1°C. Compost temperatures are kept at 49°-50° C for a period of 5 days or more. Such temperatures have been found to provide the optimum composting regime. The process is kept fully aerobic. Both the extended windrow method and the tunnel method produce similar material.

3. MODIFICATIONS TO THE BASIC COMPOSTING METHOD

The basic method utilises untreated liquid sewage sludge and wheat straw only. These are not the only materials which can be composted by this method.

Although wheat straw has been found to be ideal as a source of carbon and as a bulking agent, other straws can also be used. Straw from rice, barley, beans and rape can all be utilised successfully with suitable modifications to the technique. Other waste materials such as sugar cane bagasse can also be used. The main requirement of such materials is that they should be capable of absorbing a significant amount of water without losing their structure.

The nitrogen sources used can be untreated liquid, untreated or digested sludges, dewatered sludge cake, animal manures, animal slurries or other relatively high nitrogen containing wastes from agriculture or industry.

The water required for the process can be applied either directly from a mains or river supply or in the form of one of the sludges or slurries mentioned above.

The physical and chemical properties of these various waste materials differ considerably, and these differences must be taken into account when the materials are used in making up a composting mix. A computer programme has been developed which takes account of the physical properties and chemical analyses of potentially useful waste materials and uses them to calculate an ideal mix to provide the best conditions to encourage aerobic composting.

4. QUALITY CONTROL PROCEDURES

The safe use of potentially hazardous materials such as sludges and slurries requires that extensive quality control procedures are employed in the composting process.

These quality control procedures operate at three levels:
a) Procedures to ensure efficient composting;
b) Procedures to ensure the effective pasteurisation of any human, animal or plant pathogens which may occur in any of the starting materials;
c) Procedures to control levels of heavy metals in the resultant composts.

4.1 Composting quality control

The chemical analysis of each batch of starting material is required in order to exert effective control over the composting process. The moisture, pH, N, P, K, Ca, Mg, ash and carbon:nitrogen ratio of each batch is determined. These figures are then used to calculate the correct proportions of the initial mix of materials.

At appropriate intervals throughout the compost process samples are taken for further chemical analysis. Compost temperatures are also recorded. The results of these tests provide sufficient data to enable a controlled and reproducible composting process to be carried out. If necessary, adjustments to the mix, the compression of the windrow, or the moisture level of the compost can be made.

4.2 Pasteurisation quality control

The pathogenic organisms and micro-organisms normally found in sewage sludge can be effectively destroyed in a number of ways: heat (particularly moist heat), microbial competition, chemical inhibition and destruction of nutrients. All of these methods are utilised in the composting process to ensure effective pathogen kill.

The thermal death point of an organism is used as an indication of the temperature regime required to ensure the destruction of a particular population. This thermal death point can vary according to the stage of the life cycle of that organism. For example, the spores of some fungi are particularly resistant to death by the use of elevated temperatures. An indication of the temperatures required to kill typical human pathogens, along with the time required, is found in Table 2.

Table 2. Thermal death points of typical human pathogens.

Organism	Temperature	Time
Salmonella typhosa	55-60°C	30 min.
Salmonella spp.	55°C	60 min.
Escherichia coli	55°C	60 min.
Shigella spp.	55°C	60 min.
Streptococcus pyogenes	54°C	10 min.
Taenia saginata	71°C	5 min.

The high compost temperatures, in excess of 80°C, and the time the compost is exposed to these temperatures during the 7 day windrow composting stage of the process ensure that a very effective pathogen kill takes place. Each batch of compost is routinely tested for the presence of salmonella and the results are received before any compost is allowed to leave the production site. No finished compost has ever been found to contain salmonellae.

4.3 Heavy metal quality control

The sludge utilised in the manufacture of the compost is taken from selected sewage works. Each of the works is selected for use on the basis of its geographical proximity to the production site and the historical chemical analyses of the sludges produced. Only sludges from those works whose analyses routinely indicate low levels of heavy metals are used.

Sludges are taken from works on a rotational basis in order to reduce the chance of "spikes" of heavy metals in a particular batch of sludge significantly influencing the analysis of the final compost. Deliveries of sludges are mixed in the storage tanks to ensure a forced dilution of any particular batch of sludge by batches from other works.

The stored sludge is tested at regular intervals for the levels of heavy metals present. This provides an extra barrier against sludge containing more than the set limits of heavy metals entering the manufacturing process. The compost itself is also tested for heavy metals after one week of composting. Results are received before any compost leaves the production site. Typical heavy metal analysis figures for the compost are given in Table 1.

The way in which the compost is produced results in approximately 85% of the dry matter of the compost being of straw origin and only 15% of sludge origin. This provides a very useful dilution of the heavy metal content in the final compost. This feature of the process is one which provides a considerable advantage over some alternative methods. The heavy metal content of the final compost is very low, well within the limits recently discussed for sludge-derived composts (2), and also well within limits set by UK groups associated with the production of organic and conservation grade food.

5. COMPOST APPLICATIONS

A number of trials have been carried out to determine the usefulness and commercial value of the various types of composts produced by the above process (3, 4, 5). Initial results indicate that many successful applications can be found for this type of material, either on its own or in association with other substances.

The agricultural type of compost can be used in bulk as a mulch, as a tree back-fill or as a soil ameliorant to improve the structure of soils together with their water retention properties, fertility and nutrient content. It can also be used in combination with a number of types of subsoils or damaged topsoils to produce a topsoil equivalent to satisfy BSI standards and the needs of professional landscapers. It has been accepted by a major food producing collective as being an acceptable organic input.

The further composted, or horticultural grade of compost, suitably fortified with readily available nitrogen, has found applications as a partial or total substitute for peat-based composts. Other organic waste materials from the UK and overseas are being utilised to improve the physical characteristics of the horticultural grade compost (6).

6. CONCLUSION

The aerobic windrow composting system which has been developed provides an alternative environmentally acceptable method of sludge and waste straw disposal. The method can be adapted to handle a wide variety of potentially polluting or otherwise difficult waste materials. In addition, the method produces, as end products, a variety of composts with important applications in agriculture and horticulture. This factor allows the process to be carried out as a successful commercial operation.

Note: The opinions expressed in this paper are the authors' and are not, in any way, attributable to Anglian Water.

REFERENCES

(1) Border, D. J., C. Coombes and M. Shellens (1988). Composting straw with untreated liquid sludge. Biocycle, *29*, 54-55.
(2) Zucconi, F. and M. de Bertoldi (1987). Specifications for solid waste compost. Biocycle, *28*, 56-61.
(3) Smith, S. R., J. E. Hall and P. Hadley (1989). Composting sewage sludge wastes in relation to their suitability for use as fertiliser materials for vegetable crop production. (in press)
(4) Border, D. J. (1989). The properties and uses of composts derived from co-composted cereal straw and sludges. (in press)
(5) Vaidyanathan, L. V. (1988). ADAS preliminary report on straw-based compost made with liquid sewage sludge. (unpublished report)
(6) St Lawrence, L. and D. J. Border. (1989). The horticultural use of a renewable waste material with properties superior to peat. (in preparation)

The manufacture of a quality assured growing medium by amending soil with sewage sludge

K. M. PANTER[1] and J. E. HAWKINS[2]

[1]Thamesgro Land Management Ltd., Slough WPC Works,
Wood Lane, Slough, Berkshire, SL1 9EB, UK.
[2]Thames Water, Ryemeads WPC Works, Stanstead Abbots,
Ware, Herts., SG12 8JY, UK.

SUMMARY

Dried sewage sludge was used for many years in the London landscape market. Tightening environmental controls and malpractice by contractors diminished its use. Thamesgro Land Management Limited (TLM) was set up to market sludge in the land reclamation and landscape business in 1986. It has been demonstrated that there is a market for a soil growing medium based on blended, screened sludge and soil, currently running at 100,000 m^3 per year. To satisfy the need for environmental control and user satisfaction, soil specifications for various applications have been drawn up based on best advice. A quality assurance programme has been instigated at TLM soil production sites. Graded soils are consistently being manufactured at sites in the London area. Current site potential will deal with a sludge from a population equivalent of about one million people.

1. INTRODUCTION

Historically, London area sewage works disposed of a large proportion of their digested sludge in dry form through haulage contractors. The contractors used sludge for soil improvement and occasionally sold it directly as a growing medium. There were problems with sludge being used inappropriately. With the introduction of guidelines (ADAS 10 and STC 6), the emphasis changed to agricultural disposal of lagooned sludges over a wide geographical area, through operations such as Cinagro, Hydig and Thamesgro Organic Soil Treatment. During this period dried sludge continued to be made and occasionally marketed on an opportunistic basis. Because of the continuing, if sporadic, demand for dried sludge and concern about the environmental control of the product in the market place, a separate company was set up in 1986. The aims of Thamesgro Land Management

Limited (TLM) were to encourage the use of sludge in land reclamation and landscaping, to control the end use of the material in compliance with known guidelines and to make a profit.

The use of sewage sludge to improve soils is well documented. Coker *et al.* (1) showed that a single application of dried sewage sludge had a permanent effect on the fertility and structure of soil.

An infertile calcareous clay subsoil was much more able to respond to further additions of nitrogen fertiliser five years after additions of sewage sludge than a control soil. A reduced bulk density was an indirect measure of structure formation. No harmful effects were noted, despite soil concentrations which were in excess of STC 20 guidelines for phytotoxic metals. Digested sludge has been found to be particularly effective in improving soil aggregate stability (2).

The simple principle for "growing soil" derived from this paper, is that additions in excess of 100 tds/ha of sewage sludge solids give a satisfactory growing medium. Other work confirms this; trials monitored by Liverpool University at Stockley Park, Heathrow, showed that sewage sludge gave better results for a wide range of landscape shrubs, trees, plants and grass than other organic amendments. As a result of these trials, 90,000 m^3 of air dried sludge is being supplied by TLM and used to produce soil for a golf course and amenity areas on this prestigious development.

Two other factors which are important when using sludge to improve soil are to ensure that sludge is incorporated in a dry form (to prevent anaerobic effects), and that it is mixed thoroughly.

The best way to get an intimate mix of substrate soil and sludge is to mix, shred and screen soil, subsoil and sludge using a mechanical screen. The screened mixture has the appearance of fine structured topsoil, but has no immediate structure. It is nascent soil which is very amenable to pedogenesis when it is laid out in the right environmental conditions (temperature and moisture). A soil created in this way was used by the Greater London Council to reclaim land at Harefield, Middlesex.

A soil structure was formed and grassland established quickly. Earthworms which were seeded onto one part of the field did not appear to have an effect on the structuring of the soil. This process was subsequently used by a contractor to produce a topsoil substitute which was marketed west of London.

As a result of the success of the principles of soil creation and the knowledge that there was a market for recycled soils in the London area, it was decided to enter the topsoil market with a sludge/soil growing medium.

2. THE TOPSOIL MARKET

The market is very fragmented in the London area, with very few key companies. An estimate of market size in the London area is about 400,000 m^3 per year. TLM is currently selling 100,000 m^3 per year, and is aiming for 150,000 m^3 per year, with a sludge: soil mix of 1:5; this would account for 30,000 m^3 of air dried sludge per year, which is equivalent to the sludge production from a population of one million.

The major opportunity identified in the market was the need for topsoil to meet a high specification. In a paper discussing topsoil quality (3) Bloomfield *et al.* showed that, of 44 so-called topsoils available on the market in the Liverpool/Manchester area, half were much below the standards to be expected of topsoil and one failed to grow anything. They concluded that topsoil can be a doubtful commodity when bought-in specially and should only be used when its quality has been critically assessed. A similar situation exists in London; because of green belt policy, no greenfield sites are being stripped and so many of the soils on the market are recycled materials. Furthermore, even when soils are original topspit topsoil, this does not mean that they conform to the highest specifications required by landscape professionals. A leading soil laboratory has indicated that probably only about 5% of soils analysed meet the specification required for landscaping purposes on one or more parameters.

Bearing in mind that these represent soils being presented at the top end of the market, the conclusion here is that either virtually all soils are poor quality or the specification used may be too demanding. TLM's experience, being closely involved with major users of soils, is that both conclusions apply. Landscape contractors complain that many of the soils they use cause them aftercare problems, including lack of fertility, lack of moisture-holding capacity, perennial weeds, compaction and occasionally contamination. They also complain that many soils which give satisfactory results fail the highest specification on certain parameters, most often conductivity, one of the potentially toxic elements (PTEs), or a deficiency in one of the major plant nutrients. The contractors, therefore, find themselves caught between practice and theory.

Therefore, it was decided to examine the parameters involved in current specifications and to produce a standard set of specifications for recycled soils. The manufacturing process was revised to ensure that products of consistent quality were produced.

3. SOIL SPECIFICATIONS

The British standard for soil (BS 3882:1965) concentrates on the physical appearance of soil; it does not quantify chemical parameters. Thus, whilst soils may have a sandy loam appearance, they may also have a wide range of acidity, nutrients and contaminants. Because of this, users or specifiers of soil have produced their own quantified specifications.

The latest to be published is that of Voelker and Dinsdale (4), which is widely accepted by many of the leading landscape architects as representing a high level specification defining a good soil for general landscaping purposes.

Thamesgro has designed three specifications:

Grade 1. A high level specification very much in line with that of Voelker and Dinsdale. This defines a premium product which can be used without any constraints on most landscape planting schemes involving non-food chain use.

Grade 2. A high level specification with some relaxation on some of the parameters which, based on evidence currently available, does not affect materially the success of landscaping projects.

Grade 3. A lower level specification for a soil which has enough fertility to establish growth in verges, rough grass areas and reclamation sites.

The specifications are summarised in Table 1.

4. SPECIFICATION RATIONALE

4.1 Texture of fine earth

The specifications for Grade 1 and 2 have been set to achieve a product of loamy texture (except silt loams), as this provides the best combination of properties for healthy plant growth and bulk handling. The limits for clay, silt and sand are similar to those described by Hodgson (5).

4.2 Stone content

Earth stripped from long term permanent pasture has a low content of stones, due to the build up of a large humose layer. Such soil is now rarely offered to the London market. In the London and Lea Valley area, garden and agricultural soils quite commonly have 40% stones by weight.

Grade 1 and 2 soils are screened below 20 mm. Soils with 10% stones of 2-20 mm appear virtually stone free. TLM does not receive complaints on soils less than 25% stones by dry weight.

Table 1. Thamesgro growing media specifications.

	Grade 1	Grade 2	Grade 3
Clay content %	5-35	5-35	-
Silt content %	<65	<75	-
Sand content %	<85	<85	-
Stone size mm	<20	<20	-
Stone content %[1]	<20	<25	-
Organic matter %[1]	5-20	>5.0	>4.0
pH	6.0-7.5	6.0-7.8	6.0-8.5
Conductivity mmho/cm	<3.0	<4.0	<5.0
Zinc mg/kg[1]	<300	<600	<900
Copper mg/kg[1]	<130	<190	<250
Nickel mg/kg[1]	<70	<110	<150
Boron mg/kg[2]	<3.0	<4.5	<6.0
Chromium mg/kg[1]	<600	<800	<1000
Cadmium mg/kg[1]	<15	<15	<15
Lead mg/kg[1]	<2000	<2000	<2000
Mercury mg/kg[1]	<20	<20	<20
Nitrogen %[1]	>0.2	>0.2	>0.1
Phosphorus mg/kg[3]	>50	>50	>25
Potassium mg/kg[3]	>150	>100	>100
Magnesium mg/kg[3]	>80	>80	>40

Notes:
Grade 1: For Zn, Cu and Ni, where soil pH is maintained >7.0, concentrations may be increased up to 450, 200 and 110 mg/kg respectively.
Grades 1 and 2: Product to be free of perennial weeds (couch grass and convolvulus), and to be free of objectionable odour.

[1] Dry matter basis.
[2] Water soluble.
[3] Extractable.

Stones have two major effects on the suitability of soil. Oversize stones can damage grass cutting machinery and interfere with cultivation, and stones effectively dilute organic matter and nutrients; this is less important if those parameters are present in satisfactory quantities, guaranteeing nutrient status and moisture-holding capacity.

4.3 Organic matter

Sludge used in Thamesgro soils has been air dried and then stored for at least six months. As a consequence, a high level of aerobic stabilisation of the original anaerobic product occurs. Organic matter, expressed as loss on ignition, usually falls from between 50-60% of dry matter to below 30%. However, to allow for any further degradation of organic matter whilst soil humus forms, the soils are manufactured with a higher than normal organic matter. This also results in soils with good moisture retention capacity and nutrient reserves.

Sludge is a good source of organic matter, as it has a high organic nitrogen content and low C:N ratio. Adding organic matter with a high C:N ratio (such as peat) can lead to depletion of soil nitrogen to the detriment of plant growth. Also, peats can be very unstable, and can degrade biologically at a rapid rate.

4.4 Soil reaction - pH

The pH range has a minimum of pH 6.0 to prevent any excess availability of PTEs occurring; otherwise pH values are in the normal range.

4.5 Conductivity

Conductivity is an indirect method of measuring the total amount of soluble salts in soils. The word salt in this context is used to cover a wide range of chemical salts, such as calcium sulphate (gypsum), ammonium nitrate (Nitram etc.) and sodium chloride (sea salt), some of which are beneficial and some of which, especially sodium chloride, are harmful in excess. Some of the soils TLM manufactures have a higher level of conductivity than average soils, but in practice and in seedling trials no adverse effects have been observed. This also applies to high conductivity compost in seedling trials.

Because of the difference in these observations and the conductivity limits currently being promulgated, a literature search was commissioned by Thames Water, and carried out by the Soil Survey and Land Research Centre, Silsoe in March 1989.

The main findings of this report are summarised thus:
a) The availability of reliable data on conductivity is scant;
b) Methodology and terminology are confused;
c) Soils of equal conductivity may have totally different salts present. If present as sodium chloride, toxic effects are most likely. Many UK soils and composts have high levels of calcium, sulphate, nitrate and

phosphate which, whilst giving conductivities above normal, are not toxic at the range of concentrations normally observed;
d) There is evidence to show that a wide range of plants can be affected by the presence of salt (probably as sodium chloride);
e) However, the greatest cause of death in amenity planting is induced osmotic stress (i.e. the soil water is too concentrated to allow water to pass into the roots). This is normally caused by inadequate watering during the early stages of growth.

TLM has subsequently carried out a survey of successful and award winning planting schemes in London and, once again, found a wide range of conductivities.

The US Soil Salinity Laboratory recommends that conductivity analyses above 3.0 mmho/cm should be supplemented by a sodium analysis (6). If this figure is low (less than 15% exchangeable sodium expressed as a percentage of total exchangeable cations), then the conductivity figure is of much less significance. A wide range of TLM soils have been tested and found to have low sodium levels, but high calcium, sulphate and nitrate levels.

4.6 Potentially toxic elements

4.6.1 Phytotoxic elements These include zinc, copper, nickel and boron. Grade 1 soil specification uses the same levels as Voelker and Dinsdale, namely those published in ICRCL 59/83 (7). It must be borne in mind that these values were derived from those used in a Code of Practice for most sensitive agricultural crops. Plants such as lettuce and spinach, with a high water uptake and rapid growth, are particularly prone to phytotoxic effects due to PTEs. However, the majority of plants can tolerate much higher levels. Therefore, the PTE levels quoted in Grade 1 represent trigger concentrations for sensitive agricultural crops. The levels in Grade 3 are based on a WRc draft Code of Practice for restoration (8). The PTE levels in the Grade 2 specification are set at a midway point between 1 and 3.

4.6.2 Zootoxic elements Since the plants, shrubs, trees or grass grown on TLM soils are not intended for human or animal consumption, it was decided that it is not necessary to set tight limits for zootoxic elements in the specifications. Nevertheless, limits for cadmium, lead and mercury have been included to provide the necessary protection of the environment in accordance with ICRCL 59/83 (7).

4.7 Weeds

All soils, unless heat treated, contain weed seeds and the more fertile the soil the quicker the weeds will re-establish. Weeds from germinating seeds are to be expected, and rapid cultivation, planting and mulching are standard practice in landscape schemes as measures to keep weed growth to a minimum. More troublesome are stolons of weeds, such as couch grass and convolvulus, as they need specialist spraying.

4.8 Pathogens

All sludge used has undergone anaerobic digestion, drying and storage at a Thames Water works, and is equivalent to USEPA PFRP status (Process to Further Reduce Pathogens). It is soil-like in appearance.

5. QUALITY ASSURANCE OF THE MANUFACTURING PROCESS

At its simplest, the Thamesgro process entails selecting texturally appropriate soils and subsoils, blending them with stabilised sludge followed by shredding and screening. Because of the variability of soil substitute feedstock and the problems of site management, it was decided to adopt a quality assurance programme in accordance with BS 5750 Part II. The programme also covers storage and haulage of soils, which can have a profound effect upon soils (9).

Quality assurance is a management technique for ensuring that a manufactured product has the required characteristics and properties, as well as no undesirable ones. In TLM's case, it also serves to ensure that customers get value for money, and that TLM can provide evidence that a given batch of product is of the required quality. To achieve these objectives, a quality system has been set up, which includes the following features:
a) Specifications for raw materials and products;
b) Procedures for sampling raw materials and products;
c) Procedures for the production process at each site where manufacturing takes place, together with details of site layout, storage areas and site security;
d) Procedures for each test and analytical method employed to monitor the quality of raw materials and products;
e) A description of the system for keeping records on the quantities and quality of each batch of product that has been set up by TLM;
This permits audit of the production process at any time;

f) A statement of the conditions of sale;
g) A general statement of the training needs of TLM employees;
h) Procedures for customer service and remedial action;
i) An account of the arrangement for internal quality audit and document control, particularly for altering written procedures when changes have been made in the production process, product specification etc.

To assist with the implementation of the quality assurance scheme, a full-time soil chemist was appointed in January 1989. This has resulted in the improvement of on-site processing.

Table 2 shows success in achieving selected parameters over the period January-June 1989, at three sites in the London area. The results of this exercise have demonstrated that it is possible to select and predict the quality of the final product. It is possible to produce consistently a Grade 2 product and, progressively, Grade 1 soils. Having done this, TLM are now actively promoting the standard product (Grade 2), and can supply a premium product (Grade 1) on request.

Manufacture is carried out at three sites around London, which give a good geographical coverage of the market. The process is, however, very capital intensive, involving a feed 360° excavator, screens and stockpilers, mechanical shovel, and a dumptruck at each site. A typical site layout is shown in Figure 1. The sites require planning and licensing, because of the importation of subsoils, and therefore careful siting is important. The operation is somewhat seasonal, with a midsummer lull when stocks are built up for the autumn season.

Overall, Thamesgro Land Management is now well-established in the topsoil market, and is aiming to be a market leader through quality control and volume sales.

ACKNOWLEDGEMENTS

R. Finch and P. Shortis for all the sampling and analysis. T. Evans and P. Loveland for helpful suggestions concerning specifications. M. Brophy for proving it could be done. The views of the authors do not necessarily represent the views of Thames Water.

K. M. Panter and J. E. Hawkins

Table 2. Thamesgro graded soil products. Summary chart results January to June 1989.

Location	pH	Cd	Cr	Metals mg/kg DM Ni	Cu	Zn	Pb	%DM	%VM	Conductivity µmhos MAFF No. 24	Avery and Bascombe	% Stoniness dry weight basis 2-12mm	12-20mm	720mm	Total stone	Overall Specification
Manor Farm product 7 samples	6.7-7.4	3.0-5.1	31-100	17-67	24-65	70-188	26-72	74.5-86.1	3.7-5.7	260-710	810-1270	1.9-3.9	0.8-3.8	0-3.4	4.7-10.2	Grade 1
Mean	70	4.2	60	35	49	144	45	82.6	4.8	463	946	3.1	2.7	0.5	6.5	Organic matter needs a little enhancement
Specification	1	1	1	1	1	1	1	1	2	-	1	-	-	1	1	
Beddington product 9 samples	6.4-7.1	3.4-15.6	69-152	26-76	73-280	225-754	146-377	67.4-84.6	5.0-15.2	557-1410	1980-3720	8.2-12.3	4.7-5.1	0-9.8	9.5-29.3	Grade 2
Mean	6.7	7.8	86	39	138	422	232	78.3	9.0	876	2528	10.3	4.9	1.1	15.2	Constraints restricting Grade 1 Zn and Cu
Specification	1	1	1	1	2	2	1	1	1	-	1	-	-	1	1	
Lower Hall Lane product 5 samples	7.0-7.3	3.9-5.0	61-164	55-77	126-274	310-396	139-224	64.5-71.7	7.2-11.6	1890-2723	3100-4180	12.7-19.4		NIL	13.4-28.9	Grade 2
Mean	7.2	4.4	66	66	189	346	160	72.5	9.9	2210	3456	15.2		NIL	19.6	Constraints restricting Grade 1 Cu, Zn and conductivity
Specification	1	1	1	1	1+	1+	1	1	1	-	2	-	-	1	1	

+ Using 50% limit enhancement for soils >pH 7.0

R Finch and P Shortis June 1989

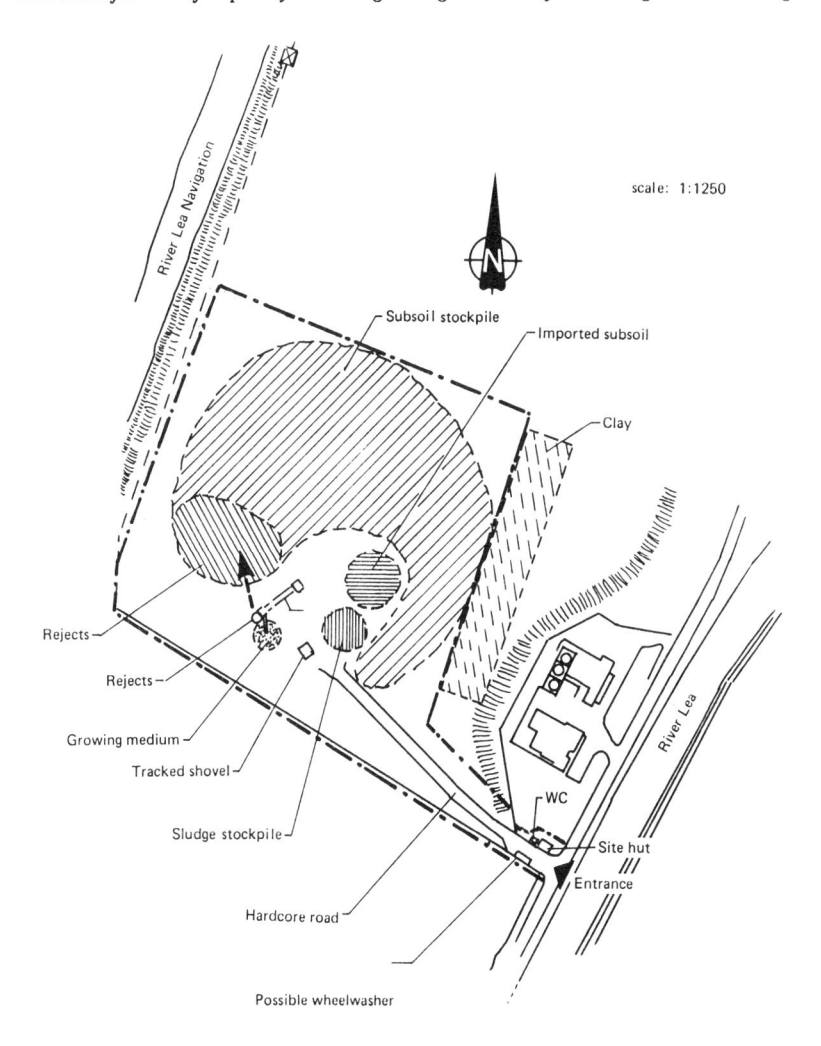

Fig. 1. Typical site layout.

REFERENCES

(1) Coker, E. G., R. D. Davis, J. E. Hall and C. H. Carlton-Smith (1989). Field experiments on the use of consolidated sewage sludge for land reclamation: effects on crop yield and composition and soil conditions, 1976 -1981. Technical Report 183. Water Research Centre, Medmenham, UK.

(2) Pagliai, M., G. Guidi, M. La Marca, M. Giachetti and G. Lucamante (1981). Effects of sewage sludges and composts on soil porosity and aggregation. J. Environmental Quality, *10*, 556-561.

(3) Bloomfield, H. E., J. F. Handling and A. D. Bradshaw (1981). Topsoil quality. Landscape Design, *135*, 32-34.

(4) Voelker, R. and M. Dinsdale (1989). A guide to specifying topsoil. Landscape Design, *178* and *179*.

(5) Hodgson, J. M. (1976). Soil survey field handbook. Technical Monograph No. 5, Rothamsted Experimental Station, Harpenden.

(6) Wild, A. (Ed.) (1988). Russell's soil conditions and plant growth, Longmans, 11th Edition.

(7) Department of the Environment (1983). Guidance on the assessment and redevelopment of contaminated land. Inter-departmental Committee on the Redevelopment of Contaminated Land, ICRL 59/83.

(8) Hall, J. E. (1989). The use of sewage sludge in land restoration. Draft code of practice. Report PRS 1783-M. Water Research Centre, Medmenham, UK.

(9) Hunter, F. and J. A. Carrie (1956). Structural changes during bulk soil storage. J. Soil Science, *7*, 75-80.

Energy from sludge

R. C. FROST and A. M. BRUCE

Water Research Centre, PO Box 85, Frankland Road, Blagrove, Swindon, Wiltshire, SN5 8YR, UK.

SUMMARY

A number of methods for the treatment and disposal of sewage sludge which seek to make use of the energy content of sludge are surveyed in this paper. Included for comparative purposes are others which do not. The bases for comparison of the selected routes are their net primary energy efficiency (ratio of net primary energy saved to energy content of raw sludge) and their conjectured contribution of "greenhouse effect" gases to the atmosphere. These matters are put into perspective through a comparison with the total combustion of fossil fuels in the UK. A value of about 35% seems to be representative of the maximum net energy efficiency achievable by the routes surveyed, anaerobic digestion being the common factor in the four most efficient (of 16) routes. Waste heat generated by a number of the routes could be used to produce hot water. If this hot water were put to use, energy efficiency gains of between 15% and 45% might be achieved. Measured in terms of their net contribution to the "greenhouse effect", routes involving dispersal of sludge to sea appear to be favoured whilst the disposal of sludge cake to landfill is not. The ranking of these routes is determined by the nature of the disposal outlets whilst that of other routes is heavily influenced by their energy efficiency.

1. INTRODUCTION

Following the energy supply crises and steep price rises of the 1970s, concern about the safety and costs of nuclear power and the "greenhouse effect" has put the issue of energy once again to the forefront of public consideration. Measures to reduce energy consumption and the use of renewable sources consequently attract a high degree of interest by the media, general public and politicians. If against this issue are juxtaposed the constraints which increasingly are being placed on the use of current methods for sewage sludge treatment and disposal, it is easy to appreciate the great interest that the prospects for recovering energy from sludge can generate.

Offered as a contribution to the continuing debate on the use and disposal of sewage sludge, this paper has two aims. The first of these is to identify the potential for energy recovery from sludge which might be achieved in practice. The second is to rank a number of treatment and disposal routes - in not all of which is energy recovery sought - in terms of their net contribution to the "greenhouse effect". No route should or is likely to be adopted on the basis of a single issue and, in the context of the subject addressed here, it is important to reflect also on the national role which the use of energy from sludge could play, and on non-energy related environmental impacts, operational aspects, and costs. This paper attempts to place the potential for energy recovery from sludge into perspective by comparing it with data for *per capita* energy consumption and waste. It is outside the scope of this paper, however, to offer more than passing mention to the other issues indicated above. Finally, some thoughts and pointers for the future are offered.

2. CALORIFIC VALUE OF SLUDGE

The basic raw material from which energy may be obtained is the solid content of the raw sludge produced at a works, though in general this will not be the form in which it is disposed. It is relevant, therefore, to consider briefly the factors which might affect the calorific value and mass production of raw sludge and some typical values.

The heating value or calorific value (CV) of unit mass of dry raw sludge solids will be a function of both the fraction of dry solids which is combustible and the composition of that material. In turn these will depend on a number of factors such as:
a) Diet;
b) Food preparation and waste disposal practice;
c) Sanitary habits and practice;
d) Nature and volume of trades wastes discharged to sewer;
e) Design and operation of sewerage system (separate/combined, retention time);
f) Design and operation of sewage treatment works.

Taken as a whole these factors are likely to give rise to national, local and temporal variations in calorific value but the available information is inadequate to quantify these in any detailed sense.

Results from a recent unpublished Japanese survey of 17 polyelectrolyte conditioned raw sludges confirm that net CV (H1 GJ/tds) correlates strongly with combustible solids content (VS% in the range 40-80):
$H1 = 0.217 * VS - 0.181$.

Energy from sludge

The low value of the second constant suggests that the CV of the combustible solids varies only slightly, having a value of 21.5 GJ/t at a VS of 75%. Indeed, the earlier work of Fisher and Swanwick (1) indicated a constant value of 21.4 GJ/t VS. Such correlations of data from a number of different works, however, do not reveal the variations in CV that may be encountered at an individual works. Data presented in Figure 1 for a medium-sized primary sedimentation and filtration works sampled over a three-month period one winter indicate that the variations may be significant. They also suggest that, on a dry ash-free basis, the calorific value of an anaerobically-digested sludge can be about 2 GJ/t less than that of the feed raw sludge. Some degree of correlation between the dry ash-free CV and VS content is apparent in Figure 1. However, samples collected over a seven-day period at another works fail to confirm this observation. Some 80% of the data lay within ±2.5% of the mean value of 21.3 GJ/t VS and all measured values lay within ±7.5%.

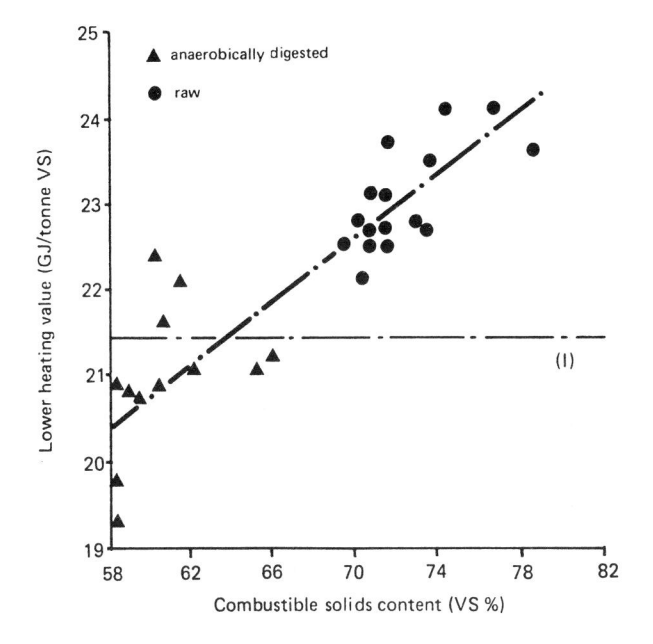

Fig. 1. Data showing the variation in calorific value of the combustible sludge solids at a works.

Based on the above, and other data (2-4) which indicate gross CVs for sludges of UK origin ranging up to 27 GJ/t VS, the CV for raw sludge adopted in subsequent analysis is 25 GJ/t VS gross, equivalent to 23.25 GJ/t VS net (the ratio of net to gross CV is normally 0.93 for sewage sludges).

3. ENERGY FROM SLUDGE IN PERSPECTIVE

In arriving at any investment decision the issues involved need to be considered with a due sense of proportion. This can be made more difficult if the issues involved are emotive and attract media attention. Whilst everyone should welcome in principle any scheme for sludge treatment and disposal which makes good use of its energy, and minimises the contribution to the "greenhouse effect", due regard must also be paid to the replication potential within the industry and the proportional savings at a national level. Any energy savings must also be balanced against potentially adverse environmental impacts which might result and associated costs. In the absence of a good case being vigorously promoted, however, it is not difficult to imagine a scenario for the future in which an ill-informed general public and media pressurise governments and operators to abandon or not implement routes which might be thought to be energy inefficient or to exacerbate the "greenhouse effect" - and, indeed, to abandon some routes for other reasons though their abandonment might in fact result in an increased emission of "greenhouse" gases.

With the above partly in mind, a diagram illustrating the direct and indirect UK energy consumption and waste on a *per capita* basis has been constructed, drawing on a number of sources (5-10); see Figure 2. The units of all energy flows shown are GJ/head per year and are on a primary energy basis. *Per capita* production of municipal solid waste (MSW) as received and sewage sludge solids is taken to be 300 and 30 kg/year respectively, the sludge production relating to a works providing primary and secondary treatment. A raw sludge ash content of 25% is assumed.

In this analysis energy and energy products are consumed either directly through individuals exercising some measure of choice or indirectly through membership of society and participation in the national economy. Thus, whilst indirect *per capita* consumption of energy by individuals exceeds direct usage by a factor of 1.3, it may be seen that the total direct consumption outweighs the theoretical energy potential in sludge by a factor of about 130! Even the recommended dietary intake exceeds sludge generation by seven-fold on an energy basis, whilst if all the energy content of raw sludge could be harnessed it may be equated on a *per capita* basis to the power consumed by a 100W bulb operating for 1 hour a day. On any rational basis, therefore, a national programme to conserve energy and adopt other measures to combat the "greenhouse effect" would place the use of energy from sludge quite low down in its order of priority.

Narrowing the field of vision to consider the water industry as a user of energy a somewhat different picture emerges. Sludge can be seen to represent a potential energy source equivalent to about 80% of the current

Energy from sludge

primary (fossil fuel) energy consumption of the water undertakings (65% if the equivalent power generated from nuclear and hydro sources is included). On this sectorial basis the widespread and efficient use of energy from sludge, in particular at the large works, can be significant and worthwhile. Recovery of energy from the waste sludge may be regarded as complementary to the achievement of operational savings in energy consumption. The operational consumption of energy is predominantly accounted for by the electricity required to fulfil the function of water distribution, with other functions individually accounting for small fractions of the whole. The potential for operational savings is not addressed here but references (11-16) may be consulted for further information.

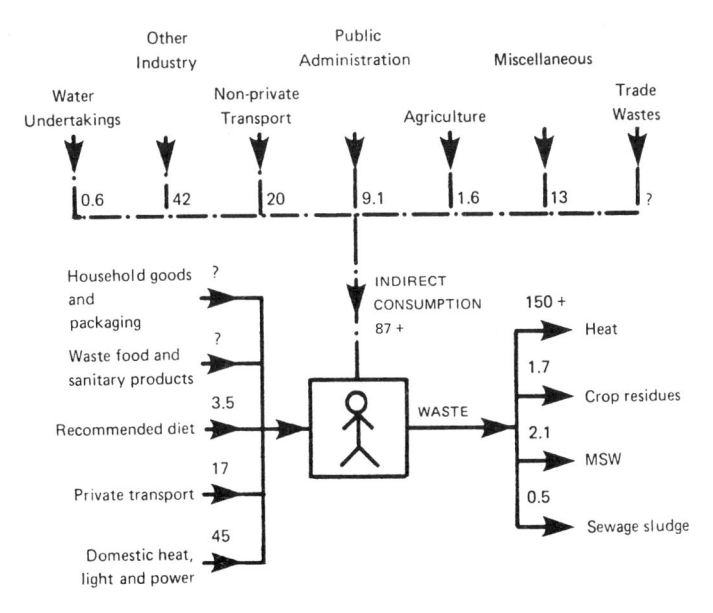

Fig. 2. *Per capita* energy consumption and waste production for the UK *ca*. 1987. (All flow data units are GJ primary energy/head per year. Power generated from nuclear and hydroelectric installations is excluded.)

4. METHODS FOR EXPLOITING ENERGY IN SLUDGE

At their point of use fuels may be gaseous, liquid, or solid and sewage sludge and its derivatives can be exploited in any of these forms, in principle. For example, fuel gas can be generated through anaerobic digestion or the application of high temperature gasification technology, whilst oils can be produced using low temperature pyrolysis

and liquefaction techniques (17) and both cake and dried sludge may be regarded as solid fuel. The uses to which these energy forms can be put are numerous but may be classified as:
a) Process and space heating;
b) Power generation;
c) Vehicular fuel;
d) Pasteurisation and disinfection of sludge;
e) Sludge conditioning (low and high temperature);
f) Total oxidation and volume reduction of sludge.

In its initial state raw sludge is a thin suspension of solids and of little value as a fuel; much has to be done if its energy potential is to be exploited. The techniques which may be applied to develop this potential are numerous, as are their combinations. However, within any route it is possible to

Table 1. Activities fundamental to the exploitation of sludge energy.

Activity	Examples
1. Feed preparation	- gravity thickening - mechanical dewatering - thermal drying
2. Processing	- mesophilic anaerobic digestion ⟶ digester gas - pyrolysis ⟶ oil
3. Use of fuel	- combined heat and power (CHP) - submerged combustion ⟶ pasteurisation - thermophilic aerobic digestion ⟶ pasteurisation - combustion of cake sludge ⟶ power/hot water - combustion of dry sludge ⟶ process heat/power/hot water - substitution of pyrolysis oil for conventional fuels
4. Internal energy recycling	- digester gas combustion ⟶ digester heating/sludge drying - heat recovery from CHP and dryer vapours ⟶ digester heating - heat recovery from incinerator gases ⟶ preheat combustion air and stack gases and pre-dry cake feed - combustion of pyrolysis products ⟶ drying and pyrolyser

identify up to four classes of activity which describe the purpose of these technologies; see Table 1. Of course the combined effects of needing to recycle energy within a route, inefficiencies in conversion, and the consumption of power and other imported forms of energy, may result in the net primary energy yield of a route being substantially less than the energy content of the feed sludge.

Sixteen routes for the treatment and disposal of sewage sludge have been selected for analysis in this paper; see Table 2. These comprise a spectrum of technologies and disposal outlets and include routes which are at the development stage, in addition to those which are in common practice. The rationale for the coding of these routes will be made clear in a subsequent section considering contributions to the "greenhouse effect". Not all of the selected routes purport to use the energy content of sludge but they have been included for comparative purposes. Further comment on some of the less usual or developed routes is offered below.

4.1 Drying

Route C (Figure 3) incorporates mechanical dewatering followed by indirect thermal drying of the sludge cake, the heat for the dryer being provided by a digester-gas fired boiler and the vapours being used for digester heating via an indirect hot water circuit. Indirect drying is a fully established technology for sludge treatment, especially in situations where sludge solids contents of about 40-60% ds are adequate, but also where sludge solids contents of above 80% ds are to be obtained, as in Route C. Examples of locations where such drying plants are in operation include Nice, Brugges, Nancy and Toulouse. It is envisaged that the dry sludge produced in this route would be used as a substitute fuel in cement and brick manufacture or co-incinerated with municipal solid waste (MSW) in a plant equipped with power generation facilities. In principle this route ought to be competitive in cost terms with those described below, as the burden of capital investment in combustion and gas cleaning plant would have to be borne by these facilities irrespective of whether sludge was burnt as a fuel or not. The marginal additional costs that would be incurred might be expected to be less than if sludge was processed in a dedicated plant as described below. A further benefit of this route could be the fixation of heavy metals in the solid matrices of cement, concrete and bricks and the implementation of a more integrated approach to solid waste disposal.

Table 2. Summary route details.

Route	Outlets	Principal operations
A	Sea	Sh
B	Sea	Th/MAD/CHP/Sh
C	Solid fuel	Th/MAD/MP/Dry/Ro
D	Agriculture	Th/MAD/CHP/Ro
E	Agriculture	Th/MAD/CHP/MP/Ro
F	Liquid fuel + ash	Th/MAD/CHP/MP/Dry/Pyr/CC/Ro
G	Ash	Th/MAD/CHP/Cen/PD/Inc/HR/Ro
H	Liquid fuel + ash	MP/Dry/Pyr/CC/Ro
I	Gaseous fuel + agriculture	Th/MAD/Scr/Ro
J	Ash	MP/Inc/HR/PG/Ro
K	Agriculture	Ro
L	Agriculture	MP/Ro
M	Agriculture	MAD/Ro
N	Landfill	Th/MAD/CHP/MP/Ro
O	Agriculture	Th/TAD/Ro
P	Landfill	MP/Ro

Note:
All routes start with raw sludge produced at 4% ds; ash content = 25%;
C H O = 56.8.30 wt% (daf); gross CV = 25 GJ/t (daf); necessary storage is assumed for all routes.

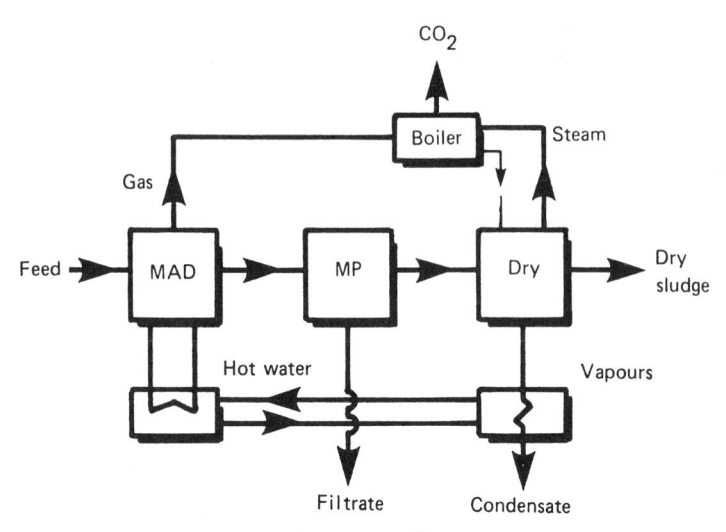

Fig. 3. Flow diagram for Route C.

Energy from sludge

Key to Table 2.

Sh = Shipping.
Th = Gravity thickening to 6% ds.
MAD = Mesophilic anaerobic digestion, 40% conversion of volatiles to gas at a yield of 1085 m^3/t VS converted, the gas comprising 65% CH_4 by volume, and the CV of the digested solids being 23 GJ/t (daf) gross.
CHP = Combined heat and power with 28% of net heating value of the gas converted to power and 50% to hot water.
Dry = Indirect thermal drying to 85% ds Route C and 95% ds in Routes F and H.
MP = Membrane pressing of raw and digested sludge to 30 and 28% ds respectively.
Ro = Road transport, tanker or truck.
CC = Combustion of char, liquor, gas and support oil followed by gas cleaning.
Pyr = Thermal conversion of sludge at 450°C to oil, char, gas and liquor.
Cen = Centrifugation to 24% ds.
PD = Indirect thermal drying to 40% ds.
HR = Heat recovery to ensure autothermic operation and stack gas pre-heating to 125°C.
Inc = Fluidised bed incineration at 850°C with gas cleaning by electrostatic precipitation and wet scrubbing.
Scr = Surplus gas scrubbing, compression and pipeline transmission to end-user.
PG = Power generation.
TAD = Thermophilic aerobic digestion using air.
Ash = Ash from sludge and product combustion transported to landfill.

4.2 Pyrolysis

Raw and anaerobically digested sludges are pyrolysed at a temperature of about 450°C in Routes H and F, respectively; see Figure 4. Both routes provide for membrane pressing and indirect thermal drying of the sludge before thermal conversion in the pyrolyser to oil, char, liquor and gas. Estimated product yields are given in Table 3. The heat for drying and thermal conversion would be provided by a char burner in which liquor and gas would also be burnt. The conversion process is currently undergoing development at the Wastewater Technology Centre in Canada using a 40 kg ds/hour reactor scaled up from a continuous bench-scale model. Planning for a demonstration plant is understood to be well advanced but no final decisions have been made. Research which WRc has commissioned and undertaken has indicated that the product oil might be marketable as a residual fuel oil subject to an appropriate price discount on conventional fuels, to it satisfactorily passing trial burn tests, and to its odour being rendered acceptable. However, much development work probably has to be undergone still on product upgrading to a diesel fuel, to widen the potential cost differential over incineration, and on product testing, before this route is likely to be implemented on a fully commercial basis (17).

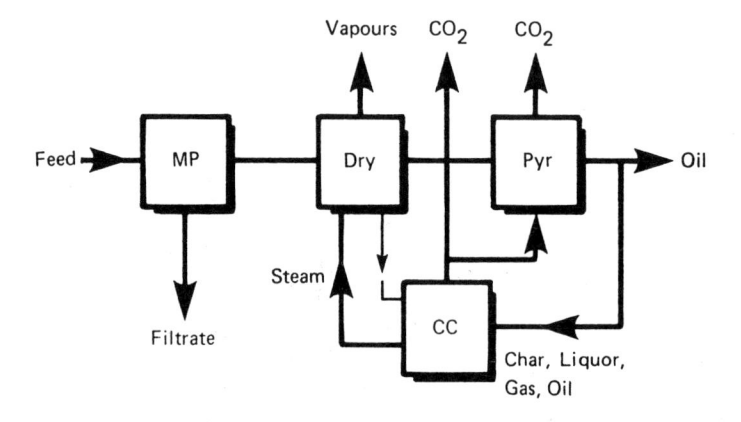

Fig. 4. Simplified flow diagram for dewatering and downstream processes in Routes F and H.

Table 3. Estimated yields for low temperature pyrolysis.

Product	Mass (t/t raw ds)		Energy (GJ/GJ raw solids)[1]	
	H Raw	F Anaerobically digested	H Raw	F Anaerobically digested
Digester gas	0.000	0.375[2]	0.000	0.435[2]
Oil	0.346	0.163	0.599	0.286
Char	0.446	0.377	0.351	0.228
Liquor	0.131	0.101	0.019	0.014
Gas	0.077	0.059	0.030	0.024

[1] On the basis of lower heating values.
[2] Apparent imbalances result from water being a reactant in anaerobic digestion process at the assumed performance.

4.3 Incineration

Incineration of sewage sludge in plants of modern design to meet strict emission standards (18, 19) is represented by Routes G and J, anaerobic digestion being included in Route G. Autothermic operation could be ensured for both routes but the manner of achieving this is different. In Route J raw sludge is dewatered to an autothermic cake solids content without the need to pre-dry, and the surplus heat in the flue gases is used to generate power. In Route G, however, the digested sludge cake is partially dried using heat recovered from the furnace gases before feeding to the incinerator and there is minimal surplus heat available. Route G is illustrated in Figure 5. Historically regarded as an expensive route, the costs of incineration have been reappraised, and for large works they can look attractive (4,18).

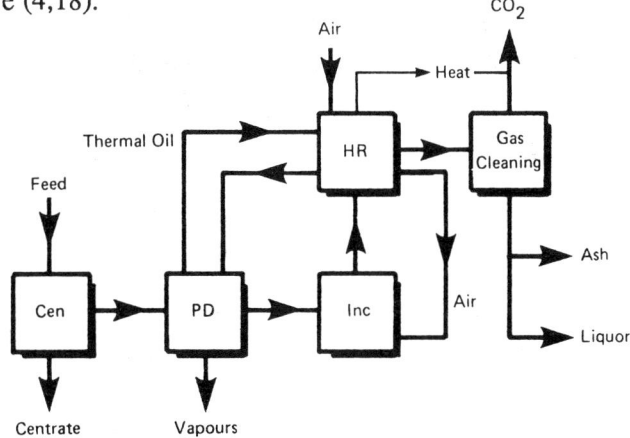

Fig. 5. Simplified flow diagram of Route G downstream of anaerobic digestion.

4.4 Thermophilic aerobic digestion

The above routes, especially perhaps those involving pyrolysis and incineration, will be most suited to large works. In contrast, Route O is most suited for works serving populations of up to 10-20,000 where there is a requirement to stabilise and perhaps pasteurise sludge before disposal to agricultural land. This route incorporates thermophilic aerobic digestion in which raw sludge is aerobically digested at a temperature of about 55°C for a period of about 10 days. Sludge in the reactor is continuously withdrawn, air injected via a venturi, and the aerated sludge returned to the reactor vessel. Heat to sustain the process is primarily derived from the oxidation reactions occurring. The process has the benefits of robustness and versatility but suffers from the disadvantage - in the context of this paper - of a substantial power consumption. A power consumption of 25 kW, appropriate for a plant serving a population of 20,000, has been assumed for this exercise but the specific power consumption can be appreciably higher for smaller sized plants.

5. CONTRIBUTIONS TO THE "GREENHOUSE EFFECT"

The net contribution of "greenhouse effect" gases to the atmosphere by a route can be expressed as the sum of three components A, B, and C as follows:

Net contribution (kg CO_2 equivalent/head per year) = A + B - C
where:

A = Direct emissions to the atmosphere of CO_2 and contributory gases (as CO_2) resulting from the final disposal of raw sludge and its derivatives.

B = Primary fossil fuel energy (expressed as CO_2 equivalent) consumed directly or indirectly (as electricity and as capital energy) in sludge processing and disposal operations.

C = Primary fossil fuel energy (expressed as CO_2 equivalent) which would have been consumed in the economy but is conserved through the use of raw sludge and its derivatives as substitute fuel.

5.1 Component A

The component A is clearly a function of the outlets which are adopted for the safe and final disposal of a sludge and the assumptions which are made concerning each outlet. For routes where the sludge or its derivatives are burnt there is no uncertainty; all of the carbon in the sludge proceeding to that outlet will be emitted to the atmosphere as carbon dioxide at the time of combustion. However, estimating the emissions resulting from the use of other disposal outlets (agricultural land, landfill, and the sea) involves some judgement and is less clearly defined. Assumptions have to be made concerning the extent to which the organic carbon is converted to carbon dioxide and methane, or "fixed", and the partitioning of these gases between the land, sea and air over a meaningful timescale. Furthermore, not

Energy from sludge

only must account be taken of the enhanced absorption by methane of infra-red radiation but also of its decay in the atmosphere. It is understood that methane is about 30 times more effective an absorber than CO_2 on a molar basis (20) and that its life in the atmosphere is approximately seven years (21). Other assumptions made here and their bases are stated below.

Whilst initially resulting in a build-up of organic carbon and humic material in a soil, it is assumed that the organic content of sludge applied to agricultural land would in a few years after deposition be completely oxidised to carbon dioxide, and that this gas would be emitted to the atmosphere.

The organic carbon content of sludge discharged to a dispersal site at sea is similarly assumed to be largely oxidised to carbon dioxide, with the remainder deposited in the sediments. However, the opportunity for the CO_2 derived from sludge to cross the boundary from the sea to the atmosphere is judged to be minimal and the rate of transfer would be correspondingly slow. Zero emissions to the atmosphere are therefore assumed.

Modelling the fate of organic carbon deposited in a landfill is more complex than in the above two cases. Since the environment will be anaerobic, some of the carbon may be expected to be effectively non-biodegradable or "fixed" whilst the remainder will migrate from the site as leachate and as landfill gas. Assumptions have to be made, therefore, concerning not only the relative sizes of these pathways but also whether or not the leachate and landfill gas are intercepted and, if so, what is done with them. For the case involving the deposition of anaerobically digested sludge it is assumed that 50% of the organic carbon becomes "fixed", that 25% is recovered in the leachate (a sealed site is assumed) which is treated on the surface with the result that this carbon is emitted as CO_2 to the atmosphere, and that the remaining 25% is released to the atmosphere as landfill gas, comprising 70 vol.% CH_4. For the case where raw sludge is deposited the emissions are modelled as a two-stage process. As a first stage it is assumed that 40% of the volatile solids content of the raw sludge is converted to gas which has the normal composition of digester gas (65 vol.% CH_4), and that this gas is released to the atmosphere in an uncontrolled manner, whilst the emissions from the second stage are assumed to be identical to those estimated for the deposition of anaerobically digested sludge. Whichever type of sludge is deposited the greater infra-red absorption capacity of methane has to be weighed against its transient nature in the atmosphere. A linear rate of oxidation to CO_2 and water over seven years is assumed and an average over a characteristic period of the effective contribution of CO_2 equivalent has been calculated. This parameter varies with the characteristic period chosen, as shown in Table 4, and for this exercise a period of fourteen years is considered to be reasonable.

Table 4. Dependence of effective contribution of atmospheric CO_2 equivalent from landfill, with time-scale.

Characteristic period (years)	Average effective contribution over the period (kg CO_2 equivalent/t C deposited)	
	Raw	Anaerobically digested
7	22,000	11,200
14	12,300	6,500
28	7,400	4,200
56	5,000	3,000
>100	2,600	1,800

Based on the assumptions above, the calculated direct emissions (A) of CO_2 and equivalent gases to the atmosphere from the sixteen routes of Table 2 are listed in Table 5. In general, these emission rates should be regarded as quasi-steady state values over a medium-term time scale.

Table 5. Direct emissions to the atmosphere arising from the final disposal of sludge (component A).

Route	Emission rate (kg CO_2 equivalent/head per year)
A	0
B	18
C - M	46
N	68
O	46
P	155

5.2 Components B and C

The second and third components of the equation defining a route's net contribution of "greenhouse gases" to the atmosphere can be amalgamated and the combined term calculated from an analysis of the energy balance across the route together with estimates of the consumption of electricity and diesel fuel (for transport operations). In principle, the energy invested in the capital plant used in a route should also be allowed for, but other analyses (16, 22, 23) indicate that this can be a minor item compared with the direct consumption and generation of energy, and this item is therefore disregarded here. The values used in this study for the primary energy equivalents of different fuels and energy sources are listed in Table 6 (6).

Table 6. Primary energy equivalents of different fuels and energy sources[1].

Fuel	Primary energy equivalent (GJ/GJ)
Electricity	3.5
Coal (solid fuel)	1.03
Gas	1.06
Fuel oil	1.1
Derv	1.1
Hot water	1.8

[1] Calculated as the mean of the values for gas and oil divided by an assumed efficiency for water heating of 0.6.

These calculations have been made for the sixteen routes and the results are presented in Figure 6. The left-hand scale of the vertical axis illustrates the net saving in primary fossil fuel energy (C - B), expressed in kg CO_2/head per year, achieved through use of the indicated route (a normalised factor of 73.7 kg CO_2/GJ which is representative for the UK as a whole is applied). The right-hand scale, however, represents the primary energy efficiency of a route which is given by the net saving divided by the lower heating value of raw sludge expressed as a percentage. The solid elements of the bar chart indicate the basic level of energy efficiency which might realistically be achieved, whilst the cross-hatching suggests the additional efficiencies which might be gained if the hot water which might also be produced in some routes could be usefully applied. It should be noted that in determining these energy balances seasonal influences have not been taken into account, an average ambient and feed sludge temperature of 10°C being assumed. In practice, the potential quantity of surplus hot water would vary over the course of a year for those routes incorporating CHP, and all routes capable of generating surplus hot water might find marketing in the summer months a difficult proposition.

Route O, which provides thermophilic aerobic digestion, stands out as the least efficient route on an energy basis but, otherwise, routes seem to have basic efficiencies close to zero (equivalent to a neutral contribution of "greenhouse effect" gases), or cluster in the range 15 to 35%. However, if it is assumed that hot water could be used for purposes other than sludge processing then some significant improvements in efficiency levels to as high as 80% would appear to be possible, with a correspondingly beneficial impact on the net release of "greenhouse" gases to the atmosphere.

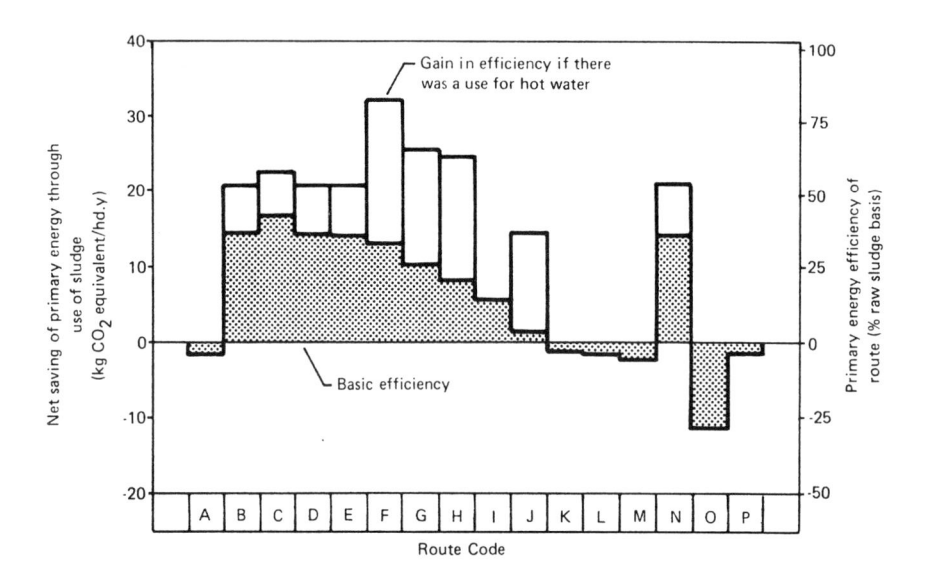

Fig. 6. Net primary energy savings and energy efficiencies of sixteen routes.

5.3 Net contribution

Subtraction of the values provided in Figure 6 from the equivalent data in Table 5 gives the net contribution of "greenhouse effect" gases to the atmosphere on a *per capita* basis; see Figure 7. These data may be compared with the emission of 11,100 kg CO_2/head per year associated with the 150+ GJ/head per day waste heat depicted in Figure 2 and a *per capita* respiration rate of about 280 kg CO_2/year. One conclusion that clearly emerges from this analysis is that a route's contribution to the "greenhouse effect" does not necessarily reflect its energy efficiency. For example, Routes A and P have approximately equal energy efficiencies of about - 3% yet represent extremes separated by a factor of about 100 in terms of "greenhouse gas" release. The final disposal outlet selected for a sludge, in fact, seems to be the major determining factor, with energy efficiency as a moderating influence.

Energy from sludge

Sea dispersal of sludge is clearly beneficial in this context whereas disposal of sludge cake to landfill, especially raw sludge cake, is not. Other conventional routes involving anaerobic digestion, CHP and disposal to agricultural land, D and E, rank highly. Ranking third is Route C, involving the thermal drying of sludge and its use as a solid fuel, a route which has not been implemented in the UK although the technology exists, whereas the higher technology thermal processing routes involving incineration and pyrolysis (Routes F, G, H, J) form a cluster towards the middle of the rankings. The export of digester gas surplus to digester heating does not rank highly at I. Forming another cluster of routes with low rankings (L, M and O) are the disposal to agricultural land of raw sludge or of anaerobically digested sludge in the absence of CHP.

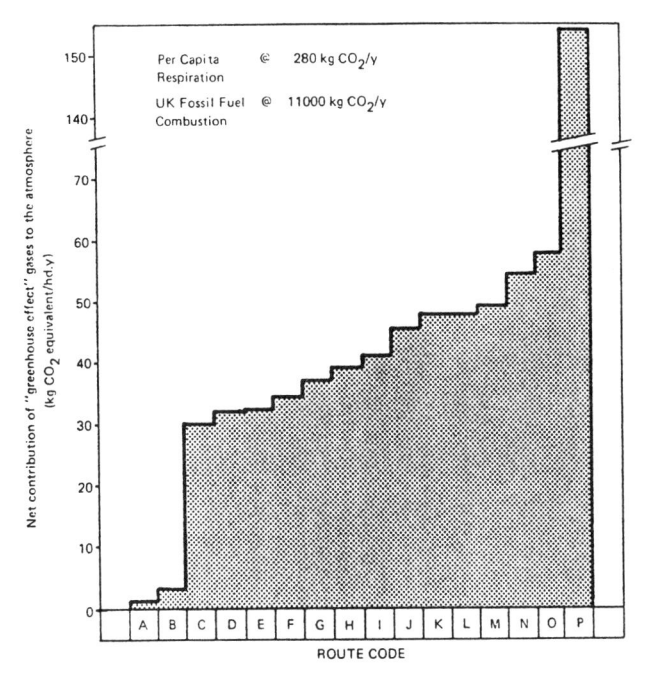

Fig. 7. Ranking of sixteen routes by their conjectured contributions to the "greenhouse effect".

6. THE FUTURE

The issue of energy and the "greenhouse effect" will be with us for the foreseeable future and all sectors of society and industry will be expected to play their part in measures to reduce long-term global problems. The water industry may be seen to have only a minor contribution to make in any

national strategy, and current practices with regard to sewage sludge treatment and disposal already seem to represent (albeit fortuitously) a fair balance of opposing factors. Thus, in the UK:
a) Sea dispersal is used for a substantial fraction of the sludge produced at treatment works and for sewage flows arising from coastal and estuarial communities, i.e. minimal "greenhouse effect";
b) Some 51% of raw sludge produced is anaerobically digested before disposal, the figure rising to 69% if works serving populations in excess of 100,000 only are considered. Most large works use digester gas for energy production.

Amongst the measures which might be adopted increasingly in future are:
a) Increased use of CHP for smaller works;
b) Sludge drying using digester gas and use of the dried sludge as a fuel;
c) More effective use of waste heat from sludge thermal processing (e.g. for district heating).

However, all energy-saving schemes will continue to need close economic assessment to justify their use unless, of course, other priorities prevail.

REFERENCES

(1) Fisher, W. J. and J. D. Swanwick (1971). High temperature treatment of sewage sludges. J. Inst. Water Pollution Control, 70 (4), 355-373.
(2) Gale, R. S. (1972). The sludge treatment and disposal problem. Symposium on incineration of refuse and sludge, University of Southampton.
(3) Department of the Environment/National Water Council (1978). Report of the sub-committee on the disposal of sewage sludge by incineration. Department of the Environment.
(4) Lowe, P. and R. C. Frost (1989). Developments in sewage sludge incineration. Paper presented at the Sewage Sludge Treatment and Disposal seminar at Lake Balton, Hungarian Ministry for Environment and Water Management, Budapest.
(5) Department of Energy (1988). Digest of energy statistics for 1988.
(6) Gilchrist, A. R. and P. Taylor (1989). Personal communication from ETSU, Harwell.
(7) Anon. Nutrition. In: Encyclopaedia Britannica, Vol. 25.
(8) Woodfield, M. (1987). The environmental impact of refuse incineration in the UK. Report prepared for the Department of the Environment, Warren Spring Laboratory, p 108.

(9) Water Authorities Association (1988). Water facts '88. Water Authorities Association.
(10) Bruce, A. M. and J. M. Hill (1985). Survey of sewage sludge production and of production and use of digester gas at sewage works in the United Kingdom. Report 363-S. Water Research Centre, Stevenage, UK.
(11) Cameron, R. (1985). Pump scheduling in water supply. WRc Technical report TR 232, p 71. Water Research Centre, Stevenage, UK.
(12) Thomas, V. J. and N. J. Slater (1988). Optimisation of mechanical surface aeration - final report. Report 802-S. Water Research Centre, Stevenage, UK.
(13) Hobson, J. A. (1987). Sewage pumping - An energy saving manual. Technical report TR 250, p 70. Water Research Centre, Stevenage, UK.
(14) Rachwal, A. J. and B. J. E. Hurley (1982). Energy savings in wastewater treatment processes. Paper presented to IPHE, 17 June 1982, Imperial College, London.
(15) Clough, G. F. G. (1979).The efficient use of energy in sewage disposal. J. Inst. Water Pollution Control, 78 (2),156-165.
(16) Thompson, L. H. and A. A. Larkin (1976). Energy in the used water industry. Public Works Congress.
(17) Frost, R. C. and W. J. Fisher (1989). Production of oil from sewage sludge - a reassessment. Report ER 375. Water Research Centre, Swindon, UK.
(18) Frost, R. C. and K. A. Bull (1988). Review of sewage sludge incineration, Vol. 1 Systems design and costs. Report 792-S. Water Research Centre, Stevenage, UK.
(19) Bull, K. A. and R. C. Frost (1988). Review of sewage sludge incineration, Vol. 3 Legislation and the control of emissions and residues. Report 781-S. Water Research Centre, Stevenage, UK.
(20) Bower, J. (1989). Personal communication from Warren Spring Laboratory.
(21) Marsh, R. (1989). Personal communication from Climatic Research Unit, University of East Anglia.
(22) Bartlett, D. C. (1984). Digester gas - an energy resource. In: Sewage sludge stabilisation and disinfection (A. M. Bruce, Ed.) Chap. 17. Ellis Horwood.
(23) Bayley, R. W. (1979). Contribution to paper given by G. F. G. Clough, in J. Inst. Water Pollution Control, 78 (2),156-165.

Resource recovery through unconventional uses of sludge

M. D. WEBBER

Environment Canada, Wastewater Technology Centre, PO Box 5050, Burlington, Ontario, L7R 4A6, Canada.

SUMMARY

Resource recovery through unconventional uses of sewage sludge, with emphasis on information published since 1986, is presented. Several uses including land reclamation, forest fertilisation, composting (except for vermicomposting) and sludge use for energy recovery were not addressed because they are the subjects of other papers in these proceedings. There are numerous potential unconventional uses for sewage sludge and their costs, benefits and acceptability are likely to be highly dependent upon local factors. In most cases, costs are ill-defined or highly variable, and benefits are strongly dependent upon fluctuating product prices. Efforts to shape public opinion are likely to play a key role in acceptance of these uses.

1. INTRODUCTION

Agricultural utilisation, landfilling, incineration and ocean disposal are the "conventional" municipal sludge disposal methods practised by communities in the developed world. However, they are subject to a variety of influences including economics, public opinion and environmental restrictions, and the future of some of these options is uncertain. In an effort to ensure the availability of options, sludge disposal in many communities is under continuous review and a variety of "unconventional" uses are being considered.

The unconventional uses of sludge involve a wide range of utilisation/disposal alternatives. They include:
a) Land reclamation, forest fertilisation and compost production and use;
b) Using sludge as a feedstuff for animals, poultry, fish, earthworms and aquatic detritivores;
c) Extracting valuable constituents such as protein, lipid, metals and phosphate from sludge;
d) Treating sludge to produce valuable commodities such as building

bricks, construction aggregate, and gaseous, liquid and solid fuels.

The category a) alternatives may be considered similar to conventional agricultural utilisation because sludge ultimately is applied to land. The category c) alternatives are related to conventional incineration because they involve heat treatment of sludge.

Information concerning unconventional uses of sludge was reviewed in 1986 (1, 2). However, three years have passed since those reviews were completed and the purpose of this paper was to update selected sections of the previous information. Several methods including land reclamation, forest fertilisation, composting (except for vermicomposting) and sludge use for energy recovery were not addressed in this paper because they are covered elsewhere in these proceedings. The information databases searched during preparation of this paper were as follows: Agricola, Agris International, Aqualine, Biosis, CAB Abstracts, Dissertation Information Service, Engineering Index, Enviroline, Pollution Abstracts, Water Resources Abstracts, an International Development Research Centre (Canadian) database and the 1986 to 1989 literature review issues of the Journal of the Water Pollution Control Federation (USA).

2. SLUDGE AS A FEEDSTUFF

With the exception of a few studies involving fish, there are almost no recent studies of sewage sludge as a feedstuff. Comprehensive reviews of this topic (3, 4, 5) indicate that a considerable amount of study was conducted during the 1970s and early 1980s but very little has been conducted since. Although most of the information available predates the previous reviews of unconventional sludge uses (1, 2), a résumé of information follows.

2.1 Sludge composition

The composition of sewage sludge varies depending upon the nature of the wastewater and the treatment process used. However, activated sewage sludge is similar in composition to yeast and bacteria (3). It contains large amounts of crude protein, small amounts of crude fibre (carbohydrate) and fat and a wide range of minerals and vitamins (Table 1). In addition, it may contain highly variable amounts of ash.

Activated sewage sludge contains practically all the essential amino acids and compares well with fish and soya bean meal as a protein source (3). About 70-80% of the nitrogen in sludge occurs as amino acids with the remainder present as nucleic acids, ploymeric hexosamines and ammonia.

Table 1. Approximate resource content of co-settled primary and activated sludge.

Constituent	Representative concentration (kg/tds)	Nominal market value of constituent (£/kg)	Nominal gross value of resources (£/tds)
Crude protein	320	0.2 (low grade feedstuff)	64
Fat	150	0.2	30
Total nitrogen	52	0.25 (as fertiliser)	13
Total phosphorus	16	0.6 (as fertiliser)	9.6
Potassium	3.5	0.15 (as fertiliser)	0.5
Mineral oil	31	0.17	5.3
Fibre	120	0.01	1.2
Starch	9	-	-
Sugar	4	-	-
Vitamin B_{12}	0.0025	4,200	10
Cadmium[1]	0.016	1.2	0.02
Copper[1]	0.69	1	0.7
Lead[1]	0.57	0.3	0.2
Nickel[1]	0.13	4	0.5
Zinc[1]	1.4	0.6	0.8
Chromium[1]	0.78	4.1	3.2
Mercury[1]	0.003	8.5	0.03
Tin[1]	0.098	9.5	0.9
Silver[1]	0.017	150	2.6
Gold[1]	0.001	8,470	8.5
Energy content (MJ/tds)	18,000	0.0020 to 0.0042 (£/MJ as primary fuel: coal, gas, oil)	36-76

[1] Concentration covers a wide range.

The carbohydrates occur mainly as polysaccharides, with small amounts of free sugars, and as cellulose and hemicelluloses associated with lignin. Approximately one third of the fat is not saponifiable. The micronutrient, toxic and non-essential element concentrations in activated sewage sludge generally are much larger than in conventional feedstuffs.

There is concern about the potential for toxicity to animals and fish, and the build-up of toxic elements in their tissues when sludge is included as a diet ingredient. Consequently, studies of sludge as a feedstuff have included consideration of the effects of toxic elements in addition to assessment of its value as a nutrient and energy source.

2.2 Animals

In a previous review (2) prepared by this author, it was concluded that "Although sludges contain protein, lipid and vitamin B_{12} which are of interest as feed supplements, they contain low levels of digestible nutrients and metabolisable energy and there is increasing evidence that their main effect is to act as a diet diluent. This being so they may serve a useful purpose in maintenance diets, but are unsuitable as a constituent of high energy diets".

Recent information (3) concerning the animal feed value of sewage sludge appears to be consistent with the above conclusion. For example, rats fed diets containing up to 25% sewage sludge gained energy value but absorbed very low amounts of sludge protein. Moreover, the growth of third generation weanling rats was reduced as a result of feeding a diet containing 30% activated sludge to three generations of rats. Sewage sludge reduced the dry matter, nitrogen and energy digestibility of swine diets, although, the inclusion of small amounts of sludge (<5% dry weight in the diet) resulted in no adverse effects on the animals. No differences in nitrogen retention by sheep were found between conventional diets and diets containing 18% of the nitrogen as dried sludge. Nitrogen retention and biological values for beef cattle were lower for a ration containing dried sludge than for the control containing soya bean oil meal.

Studies conducted in the USA with cattle, sheep and swine indicated that low metal concentration sludges did not increase heavy metal concentrations in animal tissues, but high Cd sludges increased Cd in the liver and kidneys (2, 6). The most consistent potential problem arising from sludge ingestion was thought to be a reduced Cu concentration in the liver, resulting from interference by Zn, Cd, Fe and possibly Mo. Ingestion of sludge rich in Fe induced Cu deficiency in cattle.

2.3 Poultry

A limited number of sludge feeding studies with poultry indicated generally positive results. For example, a diet containing 50% of the protein as sewage sludge fed during one year resulted in increased weight gain and egg production compared with a control diet containing no sludge (3). Diets containing <7% ds of dried Chicago sludge showed no significant effect on the daily feed intake, feed efficiency, body weight, egg production and egg weight of Leghorn hens (7). Similarly, there was no effect of diets containing 6% ds of dried Chicago sludge on mineral accumulation in the muscle tissue of broiler chickens, although there were increased levels of Cd in their livers and kidneys. Recent studies to evaluate the replacement of 5% and 10% of the groundnut cake by dried poultry droppings and sludge in broiler poultry rations indicated significantly improved performance for the 10% sludge treatment (8). There were no significant differences in cooked meat texture, tenderness, and juiciness between the control and substituted diets, but taste scores for flavour indicated a preference for meat from broilers on the unsubstituted control diet, while meat from the 10% poultry droppings and sludge diets was least palatable (9).

2.4 Fish

There has been long and extensive use of sewage and animal wastes in fish culture, particularly in the Far East (4, 10). Traditionally, common carp, *Cyprinus carpio* and *Labeo jordani* were cultured and the productivity of ponds was a function of the number of people and farm animals in their watersheds. Human and animal wastes represented the major source of fish food and frequently latrines were sited directly over or adjacent to fish ponds. There is also a tradition of fish culture using human and animal wastes in Israel and Eastern and Central Europe.

Where sewage has been used for fish culture, it has generally proved satisfactory only after treatment or dilution. The readily degradable organic matter in sewage results in severe oxygen depletion and fish kills, when it is present in water in excessive amounts. Juvenile tilapia *Sarotherodon mossambica* and a polyculture (several species) of carp have been cultured successfully in secondary effluent from an activated sludge sewage treatment system and a mixture of effluents from both the activated sludge and high-rate biological filter systems (11, 12). Similarly, carp have been cultured in the terminal ponds of a cascade system at a sewage farm near Moscow, and primary effluent diluted 1:4 with water has been sprayed onto carp ponds at the Munich sewage treatment works (4).

Only limited study of sewage sludge utilisation in fish culture has been reported. Five week studies of sludge as a replacement for white fish meal and wheat midlings, at up to 33% of total feed, for rainbow trout indicated either no effect or an increased gross body protein content and decreased fat content (4). Goldfish *Carasius auratus* grew equally well on a control diet and one containing 30% sludge, but catfish *Ictalurus punctatus* did not. Increased levels of metals (Cr, Fe, Ni, Pb and Zn) were reported in trout fed a diet containing 30% activated sludge for a 10-week period, although the increases were small compared with the metal enrichment in the experimental diet.

Several potential problems with sewage sludge use in fish diets have been identified (4) as follows:
a) Activated sludge is deficient in sulphur containing amino acids, particularly methionine, the concentration of which is about 25% of that in white fish meal;
b) Heavy metals from sludge may accumulate causing either physiological damage in fish or unacceptably high concentrations in the human diet;
c) Activated sludge contains high concentrations of non-α-amino N, principally in the form of nucleic acids which may disturb fat, carbohydrate and uracil metabolism in fish;
d) Sludges contain parasites and pathogens, particularly *Clostridia* spp., which might either damage the fish or be transferred to man;
e) Sludges may impair the quality (i.e. flavour and texture) of fish for human consumption.

2.5 Culture of earthworms

Earthworms play an important role in processing organic wastes in natural systems and can also be used to process organic wastes in artificial systems. The products of the artificial systems are an inoffensive compost-like residue, which exhibits favourable soil amendment properties, and worm biomass which is rich in protein. Earthworm culture for the production of residue is referred to as vermicomposting and the product vermicompost. Earthworm culture for the production of worm biomass is referred to as vermiculture.

2.5.1 Vermicompost Vermicomposting of sewage and farm wastes is the subject of recent reviews (13, 14). The decomposition of organic matter during vermicomposting results from the combined action of worms and microorganisms. The burrowing action of the worm maintains aerobic conditions in the material and the digestive processes of the gut cause physical and chemical breakdown of the organic matter. A portion of the organic matter is converted into worm biomass and respiration products.

The excreted material, or castings, have a greatly increased surface area which promotes vigorous microbiological activity and rapid decomposition and stabilisation. Also, due to the large surface area, drying of the castings may proceed up to twice as fast as for non-vermicomposted material.

The worm-worked material contains nutrients that are more available to plants than in the original waste, and exhibits increased moisture-holding capacity and porosity. Plant growth trials conducted in the UK with worm-processed animal wastes have shown them to be suitable growing media (14). A wide range of plants were grown successfully in undiluted vermicompost and in a number of mixes including 3:1 and 1:1 ratios of vermicompost to peat, pine bark and loam soil. Frequently, germination was more rapid and growth was better with vermicompost than with recommended growing media. The earthworm *Eisenia foetida* (Savigny) has been identified as the species most suitable for vermicomposting since it feeds on decaying organic matter and naturally colonises the upper litter layers of the soil, manure piles and compost heaps (13). Much information about its biology and preferred physicochemical conditions has been established through extensive laboratory trials conducted in the USA.

There are reports of vermicomposts being marketed in Italy as organic based pot-plant additives at a very high price (15) and in 1987 the author toured a commercial sludge vermicompost operation near Valencia, in Spain. Despite extensive investigations into sludge vermicomposting in the USA, there was only one operational vermicomposting project in that country in 1988 (16). This compares with one operational and two pilot projects in 1987 (17) and one operational and one pilot project in 1986 (18). The total numbers of operational sludge composting projects in the USA during those years were 115, 107 and 89, respectively. In the UK, vermiculture has been commercialised by the formation of a company called British Earthworm Technology (BET). This company, which has subsequently ceased trading, operated profit-sharing contracts with farmers and industrial organisations to grow worms in their organic wastes (14). BET then processed and marketed the compost and protein. In addition, the potential for using vermiculture to reduce odour associated with unstabilised sludge cake was assessed. However, it was concluded that "whilst it is technically possible to vermicompost most types of sludge, the process is too slow to be operationally or economically viable.". Thus, although sludge vermicomposting is possible, it has not gained favour in the UK.

2.5.2 Worm biomass Earthworms consist of 60-70% protein, 7-10% fat, 8-20% carbohydrate and 2-3% minerals, with a gross energy of 4,000 kcal/kg (14). The nutrient spectrum of worm tissue is excellent, at least as good as meat or fish meal. It is rich in the essential feed amino acid lysine and in the sulphur containing amino acids. It is rich in vitamins, particularly niacine, riboflavin and vitamin B_{12}, which are valuable in animal feeds. Chickens, pigs, rats and mice have been shown to grow as well or better on worm meal as on commercial diets. Variable results with fish were related to palatability which was affected by the method of preparation. Edwards et al. (14) reported that the chemical composition of earthworm species differs little and is relatively unaffected by the type of waste upon which the worms feed, although Hartenstein et al. (19) reported that the problematic heavy metals Cd, Ni, Pb, Zn and Cu can accumulate in earthworm tissues. They recommended that "since these metals pose a problem for organisms in general, careful thought must be exercised in the use of earthworms in sludge management.".

2.6 Culture of aquatic detritivores

Sludge use for the intensive production of detritivore invertebrates for fish food has been suggested and the most recent information available on this subject is contained in a review of minor uses of sludge, published in 1984 (20). In brief, it was suggested that the potential advantages of detritivores are high rates of growth, high fecundity, a high protein content and tolerance of low concentrations of dissolved oxygen and high stocking densities. Three species with population doubling times less than one month that were suggested for further study were *Acellus aquaticus*, *Chironomous riparius* and *Tubifex tubifex*. Encouraging results were obtained in initial laboratory studies but a need for more work on the fate of heavy metals and other sludge contaminants in the systems was recognised.

3. RECOVERY OF VALUABLE SLUDGE CONSTITUENTS

3.1 Extraction of fat and protein

Sewage sludge contains appreciable amounts of both fat and protein (Table 1) and the potential for extracting these constituents has been assessed. A thorough investigation of this option, carried out in the UK in the late 1970s by Thames Water Authority and Unilever Ltd., was reviewed in (1). In that work, the extraction of fat was optimised because of its potential for non-edible uses such as soap manufacture. Financial analysis indicated that the viability of the process was critically dependent upon the market price of tallow, although other factors such as losing some digester gas production were also significant. The process was not considered financially viable and was not implemented by Unilever Ltd. in the early 1980s because of future projected prices of alternative feedstocks. More recent information concerning fat and protein extraction from sewage sludge was not found.

3.2 Extraction of vitamins

Activated sludge contains a wide range of vitamins and is a rich source of vitamin B_{12} in a form that is used efficiently by chicks on a diet low in animal protein (3). Vitamin B_{12} extraction from sludge would probably not be economic but sludge might be used as a protein supplement and the vitamin considered as an added advantage. Study in the UK indicated that vitamin B_{12} extraction from sewage sludge would be uneconomic unless it offered some advantage in addition to vitamin production (2).

3.3 Extraction of metals and phosphorus

Sewage sludge contains a few percent ds of phosphorus and smaller amounts of a wide variety of metals (Table 1). Information concerning the extraction of these constituents from sludge has been reviewed (1, 2, 20) and a more recent publication on this topic was not found. Canadian and USA studies conducted about 1980 indicated that approximately 80% of the P, Fe, Al, Zn, Mg, Ni, Cd, Cr, Cu, Pb and Mn could be extracted from sludge following acidification to pH 1.5-2. However, it was concluded that the process would be uneconomic, largely because of the cost of chemicals.

A later Canadian study indicated that a few sewage sludges and sludge incinerator ashes exhibited high concentrations of Au and Ag as compared to normal concentrations in the earth's crust (1). Assuming values of $500 (Cdn)/ounce Au and $12/ounce for Ag and a segregation roasting/flotation concentration process, it was calculated that recovery of these metals could be economic for the large capacity Ashbridges Bay wastewater treatment plant in Toronto.

Recently proposed US EPA regulations would severely restrict land application of sludge due to metals, particularly copper. Land application is frequently the most economic sludge disposal option and there is considerable interest in developing techniques to improve sludge quality. One possibility that has received consideration is non-destructive solvent extraction of metals (R. Bastian, US EPA Washington DC, personal communication). Details concerning the nature and cost of this technique were not obtained.

4. PRODUCTION OF VALUABLE COMMODITIES

4.1 Brick production

Firing the raw materials in brick and cement manufacture can be assisted by the presence of organic matter which reduces the fuel requirement.

The findings of investigations conducted in the Netherlands and the USA in which sewage sludge was used as a clay replacement in brick manufacture were reviewed in (1) and (2). Briefly, the results in the two countries were similar and bricks with less than 40% sludge by volume in the wet mix met the appropriate technical standards. There were no discernible differences in the appearance, texture or smell between the sludge-amended and regular bricks. Suggested advantages of sludge-amended manufacture included economic use of clay, water and energy and a light-weight product with improved water absorption and transfer properties. "Biobricks" have been used by the Washington Suburban Sanitary Commission in the USA to construct several buildings, including a picnic pavilion (21).

An investigation of biobrick manufacture, conducted in Singapore (22), indicated that the maximum percentages of dried sludge and sludge incinerator ash that can be mixed with clay for brick manufacture are 40% and 50%, respectively. The texture and finish on the surface of bricks made with dried sludge was rather poor and water absorption increased with increasing sludge addition. Bricks made with sludge ash had lower water absorption values and would be expected to exhibit greater durability than those made with dried sludge. The compressive strength of bricks was 87.2 N/mm for 0% sludge, 37.9 N/mm for 40% dried sludge and 69.4 N/mm for 50% sludge ash. For bricks with 10% dried sludge and 10% sludge ash respectively, the compressive strengths were 30% lower than, and about equal to that of the control brick.

4.2 Cement production

Dried sludge (70-75% ds) has been proposed for use in cement production (1). It was suggested that a practical limit on sludge use as a fuel supplement in cement kilns is 15% of the total heat input and that the total chlorine input should not exceed 0.02% with respect to the final product. Although no tests had been performed, adverse effects of sludge on cement clinker quality were anticipated.

The feasibility of using sludge incinerator ash as a filler in concrete has been studied (23). There was a slight improvement in workability of the concrete with increasing ash content up to 20%, although there were no significant effects on setting times, shrinkage and water absorption. The

28-day compressive strength of the concrete decreased with increasing ash content. However, with 10% replacement of cement by ash, the compressive strength was reduced only about 10% and was well within the normally required range.

4.3 Construction aggregate and landfill cover production

A variation on the N-Viro Soil process, described later, is currently being used in Wilmington, DE. Sewage sludge, kiln dust and fly-ash are mixed to prepare an aggregate which is used as landfill cover and by the Army Corps of Engineers in dyke construction along the Delaware River (S. Robinson, N-Viro Energy Systems, personal communication).

There is also considerable interest in the USA in treating sewage sludge by the Chemfix process and using the product as landfill cover. Currently, Los Angeles CA is employing this technique and the states of Florida, North Carolina, New Jersey and Washington have expressed interest in it (R. Bastian, US EPA Washington DC, personal communication).

4.4 Fibre production

A proprietary process has been described in which pulverised solid domestic refuse and sludge are mixed in a 10:1 ratio by dry weight and digested anaerobically to produce methane. The residue remaining after digestion is screened and used as fibre in the board industry (20). The methane production is more than sufficient to provide the whole energy requirement for the process. A plant capable of processing 7 tonnes of refuse and 700 kg ds of sludge per day has been built and operated successfully in the UK. Activated sludge with a solids content of 2-7% was used.

A Japanese patent describes a process in which sludge is one of the components of a mixture which is compressed and heated to produce wallboard (20).

4.5 Activated carbon production

Activated carbon of varying degrees of activity has been prepared from sludges and has potential applications in sewage treatment (20). Activated carbon assists in sewage treatment through adsorption of organics from mixed liquor and effluent. It is unlikely that a process designed specifically for this purpose would be economic, since temperatures of several hundred degrees Celsius are required to activate the carbon. However, it is possible that the "char" resulting from a sludge pyrolysis process might be effective and affordable.

4.6 Methane production

Sludge gas resulting from anaerobic digestion has been suggested as a feedstock for the organic chemical industry (20). Production of methanol and cyanide both of which require methane are two possible uses. However, it would be necessary to determine whether sludge gas could be competitive with methane of fossil origin and whether the removal of carbon dioxide, water vapour and the traces of hydrogen sulphide and organic vapours would be necessary. Proximity of sludge digestion plants to organic chemical works would be a prerequisite.

Currently, sludge gas is used to keep digesters at the operating temperature, to run dual-fuel engines and to heat wastewater treatment plant buildings. Excess gas production which usually occurs during summer is burned to avoid explosion hazard. It is suggested that through better insulation and more conservative use, the amount of sludge gas available for industrial use might be increased substantially.

Use of sludge gas as a vehicle fuel was recently revived by the Anglian Water Authority in the UK with a break-even point, compared with the cost of gasoline, at 40,000 m^3 of methane per year (20).

4.7 Carbon dioxide production

Carbon dioxide, which comprises about 33% by volume and 60% by weight of sludge gas, is usually regarded as a useless constituent (20). However, it has many applications, for example, in the drinks industry, in fire extinguishers and in the manufacture of "dry ice". The author is not aware of any beneficial use of carbon dioxide resulting from sludge digestion.

Carbon dioxide from composting has been proposed as an additive to glasshouse atmospheres to increase rates of crop growth, particularly as crops can rapidly exhaust the natural supply (20).

5. SLUDGE USE FOR PRODUCTION OF SOIL AMENDMENTS

5.1 Production of black earth

"Black earth", also referred to as "black soil", is gaining increasing interest in the Netherlands as an alternative processing and use option for sewage sludge (24, 25). It is prepared for the commercial market by mixing sludge and sandy soil, followed by composting and maturation as required to provide a desirable soil amendment material. The composition is varied depending upon the end use.

Black earth is used for various green belt applications such as establishing vegetation covers on landfills and construction areas, establishing horticultural plantings and for forestry. Moreover, there is interest in using it as an agricultural soil amendment and regulations to protect soil and crop quality are being developed. Organic fertilisers, mainly for use in agriculture, are prepared by adding chemical nutrients to black earth.

The cost to prepare black earth varies with the size of the operation and the amount of processing required. This in turn is dependent largely upon the degree of stabilisation and moisture content of the sewage sludge (24). However, cost calculations assuming an 80,000 person equivalent wastewater treatment plant indicated that black earth production and use would be much more economic in the Netherlands than sludge disposal by incineration (25).

5.2 Production of soil conditioner/fertiliser containing free lime

In the USA, sewage sludge applied to land must be treated by a Process to Significantly Reduce Pathogens (PSRP) (26). However, further public health protection is required when sludge is applied to land where crops for direct human consumption are grown less than 18 months after waste application. In these instances, the sludge must be treated by a Process to Further Reduce Pathogens (PFRP). One such process, approved in January 1988 by the US Environmental Protection Agency (EPA), is "Advanced Alkaline Stabilisation with Subsequent Accelerated Drying (AASSAD)".

The AASSAD process involves admixing a strongly alkaline material with dewatered sludge followed by aeration (27). It combines extended alkaline disinfection, exothermic reaction from quicklime, Portland Cement kiln dust or other alkaline material and drying by aeration or mechanical process. It is completed within seven days and is much less expensive than alternative PFRP processes. The product is a nearly odourless, granular, organic soil conditioner and fertiliser.

In 1985, N-Viro Energy Systems in the USA patented a process based on AASSAD which involves adding kiln dust, a strongly alkaline byproduct of the Portland Cement industry, to sewage sludge (28). The product, called N-Viro Soil, costs approximately $90 US per dry tonne to produce, and is a soil conditioner and balanced NPK organic fertiliser. It combines the nutrient, organic matter and liming resources of the two materials. The kiln dust contributes K, S and Ca to the mixture. Because N-Viro Soil results from a PFRP process it is suitable for both agricultural and urban use. There is considerable interest in the N-Viro Soil process and several USA cities including Portland ME, Toledo OH, Jupiter FL and Oklahoma City OK have adopted it or will adopt it shortly (S. Robinson, N-Viro Energy

Systems, personal communication).

6. CONCLUSIONS

This review has shown that there are numerous unconventional uses for sewage sludge. Some, such as worm biomass, aquatic detritivore production and extraction of vitamin B_{12}, are theoretically possible whereas others, such as use in animal feedstuffs and in brick manufacture, have been demonstrated at pilot or full scale. The costs, benefits and acceptability of all such uses are likely to be highly dependent upon local factors. In most cases, costs are ill-defined or highly variable, and benefits are strongly dependent upon fluctuating product prices. Efforts to shape opinion will play a key role in public acceptance of these uses.

REFERENCES

(1) Frost, R. C. and H. W. Campbell (1986). Alternative uses of sewage sludge. In: Processing and use of organic sludge and liquid agricultural wastes. (P. L'Hermite, Ed.), pp. 94-109. D. Reidel, Dordrecht.
(2) Webber M. D., L.E. Duvoort-van Engers and S. Berglund (1986). Future developments in sludge disposal strategies. In: Factors influencing sludge utilisation practices in Europe (R. D. Davis, Ed.), pp. 103-116, D. Reidel, Dordrecht.
(3) Vriens, L., R. Nihoul and H. Verachtert (1989). Activated sludge as animal feed: a review. Biological Wastes, 27, 161-207.
(4) Edwards, R. W. and J. W. Densem (1980). Fish from sewage. In: Applied Biology, Vol. 5 (T. H. Croaker, Ed.), pp. 221-270. Academic Press.
(5) Sludge - Health risks of land application (1980). (G. Bitton, B. L. Damron, G. T. Edds and J. M. Davidson, Eds.), pp. 367. Ann. Arbor Science.
(6) Chaney, R. L., R. J. F. Bruins, D. E. Baker, R. F. Korcak, J. E. Smith and D. Cole (1987). Transfer of sludge-applied trace elements to the food chain. In: Land application of sludge (A. L. Page *et al.*, Eds.), pp. 67-99. Lewis Publishers.
(7) Edds, G.T., O. Osuna and C. F. Simpson (1980). Health effects of sewage sludge for plant production or direct feeding to cattle, swine, poultry or animal tissue to mice. In: Sludge - health risks of land application, (G. Bitton *et al.*, Eds.), pp 311-325. Ann. Arbor. Science.

(8) Ologhobo, A. D. and S. O. Oyewole (1987). Replacement of groundnut meal by dried poultry droppings (DPD) and dried activated sewage sludge (DASS) in diets for broilers. Biological Wastes, *21*, 275-281.

(9) Ologhobo, A. D. (1988). The effects of dried poultry droppings (DFD) and dried activated sewage sludge (DASS) on broiler carcass quality. Biological Wastes, *23*, 99-105.

(10) Kaur, K., G. K. Seghal and H. S. Seghal (1987). Efficacy of biogas slurry in carp, *Cyprinus carpio* Var. *communis* (Linn.), culture - effects on survival and growth. Biological Wastes, *22*, 139-146.

(11) Sin, A. W. and M. T. L. Chiu (1987). The culture of tilapia (*Sarotherodon mossambica*, in secondary effluents of a pilot sewage treatment plant. Resources and Conservation, *13*, 217-229.

(12) Sin, A. W. and M. T. L. Chiu (1987). The culture of silver carp, bighead, grass carp and common carp in secondary effluents of a pilot sewage treatment plant. Resources and Conservation, *13*, 231-246.

(13) Daw, A. P. (1989). Vermicomposting of sewage sludge. Report PRU 1922-M. Water Research Centre, Medmenham, UK.

(14) Edwards, C. A, I. Burrows, K. E. Fletcher and B. A. Jones (1986). The use of earthworms for composting farm waste. In: Composting of agricultural and other wastes (J. K. R. Gasser, Ed.), pp. 229-242. D. Reidel, Dordrecht.

(15) Fieldson, R. S. (1985). The economic feasibility of earthworm culture on animal wastes. In: Applied biology (T. H. Croaker, Ed.), Vol. 5, pp. 243-254. Academic Press.

(16) Goldstein, N. (1988). Steady growth for sludge composting. Biocycle, *29*, (10), 27-36.

(17) Goldstein, N. (1987). Sludge composting on the rise. Biocycle, *28*, (10), 24-29.

(18) Goldstein, N. (1986). Sewage sludge composting maintains momentum. Biocycle, *27*, (10), 21-26.

(19) Hartenstein, R., E. F. Neuhauser and J. Collier (1980). Accumulation of heavy metals in the earthworm *Eisenia foetida*. J. Environ. Qual. *9*, 23-26.

(20) Montgomery, H. A. C. (1984). Minor uses of sewage sludge. Report 747-M., Water Research Centre, Medmenham, UK.

(21) Anon. (1986). Constructing with "biobricks." Biocycle, *27*, (2), 45.

(22) Tay, J. H. (1987). Bricks manufactured from sludge. J. Environ. Eng., *113*, 278-284.

(23) Tay, J. H. (1987). Sludge ash filler for Portland Cement concrete. J. Environ. Eng., *113*, 345-351.

(24) van den Berg, J. J. (1986). An alternative application for sewage sludge: black earth. In: Processing and use of organic sludge and liquid agricultural wastes (P. L'Hermite, Ed.), pp.523-531. D. Reidel, Dordrecht.

(25) Ten Wolde, J. G. (1986). Processing organic wastes to black soil and organic fertilisers. In: New developments in processing of sludges and slurries (A. M. Bruce, P. l'Hermite and P. J. Newman, Eds.), pp. 59-66, D. Reidel, Dordrecht.

(26) US EPA (1979). Criteria for classification of solid waste disposal facilities and practices. Federal Register, *44*, (179), 53438-53464.

(27) Farrell, J. B. (1987). Outline of a presentation to the national conference on alkaline treatment and utilisation of municipal waste water sludge, August 26 and 27, Toledo, Ohio.

(28) Slakter, A. (1987). Toledo puts its sludge out to pasture. Chemical Week, September 2, 1987.

The development of a sludge treatment and disposal strategy

C. POWLESLAND

Water Research Centre, Henley Road, Medmenham,
PO Box 16, Marlow, Buckinghamshire, SL7 2HD, UK.

SUMMARY

The development of a sludge treatment and disposal strategy involves the identification of existing and likely future constraints, the establishment of objectives for the study, collection of data, the selection and evaluation of alternative options and the selection of the preferred option.

It is important that constraints on the treatment and disposal operation are quantified as far as possible during the data collection phase of the study. Alternative treatment and disposal options which overcome these constraints can then be identified. In selecting suitable options for future evaluation, the treatment and disposal operation should be treated as an integrated whole and not as separate processes. The evaluation of these options should consider the environmental impact, cost and security of the operation. In this way the most environmentally acceptable, secure and cost-effective treatment and disposal strategy can be identified.

1. INTRODUCTION

Sewage sludge disposal typically accounts for approximately half the cost of sewage treatment and a significant proportion of the pollutants arising from sewage are present in the sludge. It is, therefore, regarded by water undertakings as a major problem. There are two possible ways of tackling this problem; the first involves seeking to reduce the quantity of sludge produced. However, although processes such as 'LOSLUJ' (1), have shown promise there seems little prospect of being able to reduce sludge production in a cost-effective manner in the near future.

The alternative is to carry out the disposal of sludge as effectively as possible. This paper seeks to illustrate, by drawing on a hypothetical example, the general principles which could be employed in developing an environmentally acceptable and cost-effective disposal strategy. The principal features of the study area are shown in Figure 1.

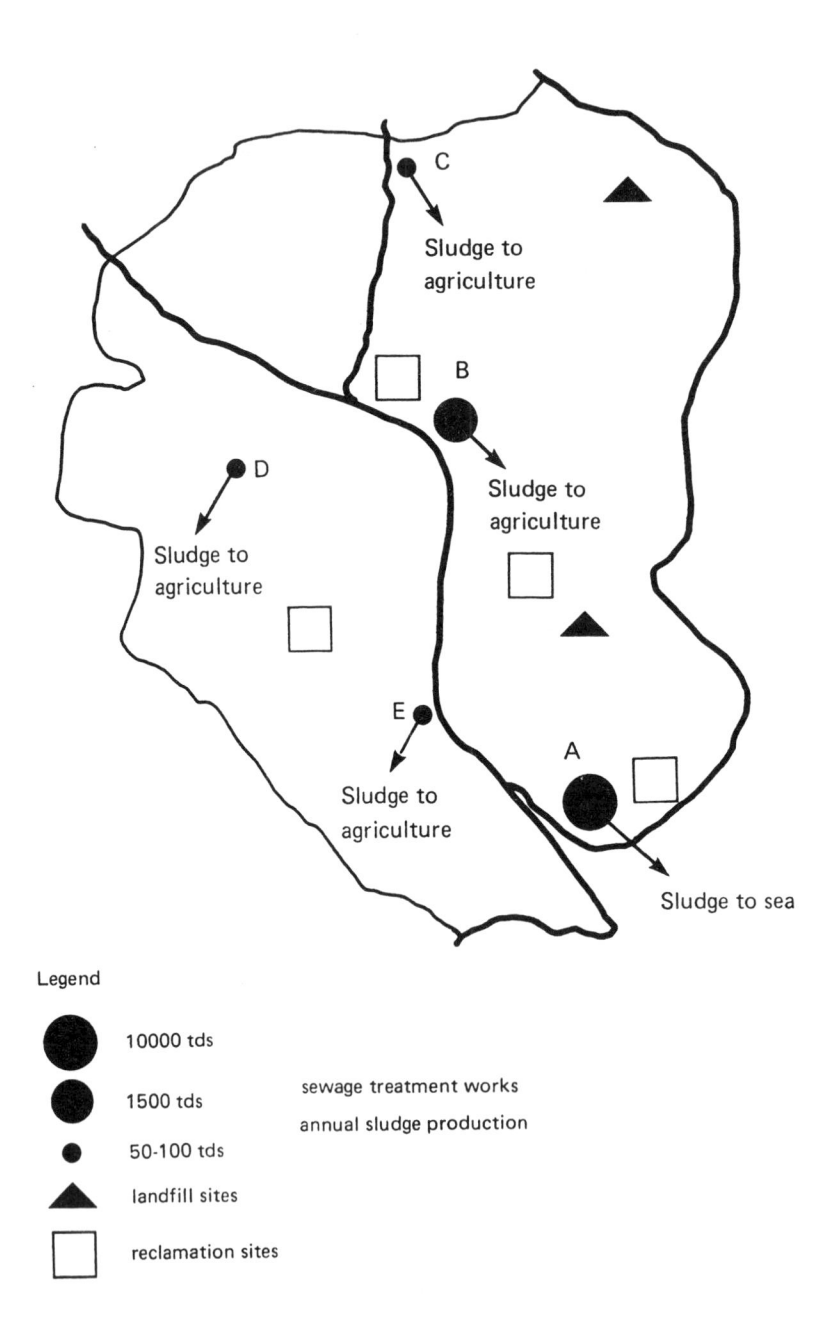

Fig. 1. Study area and principal sewage treatment works.

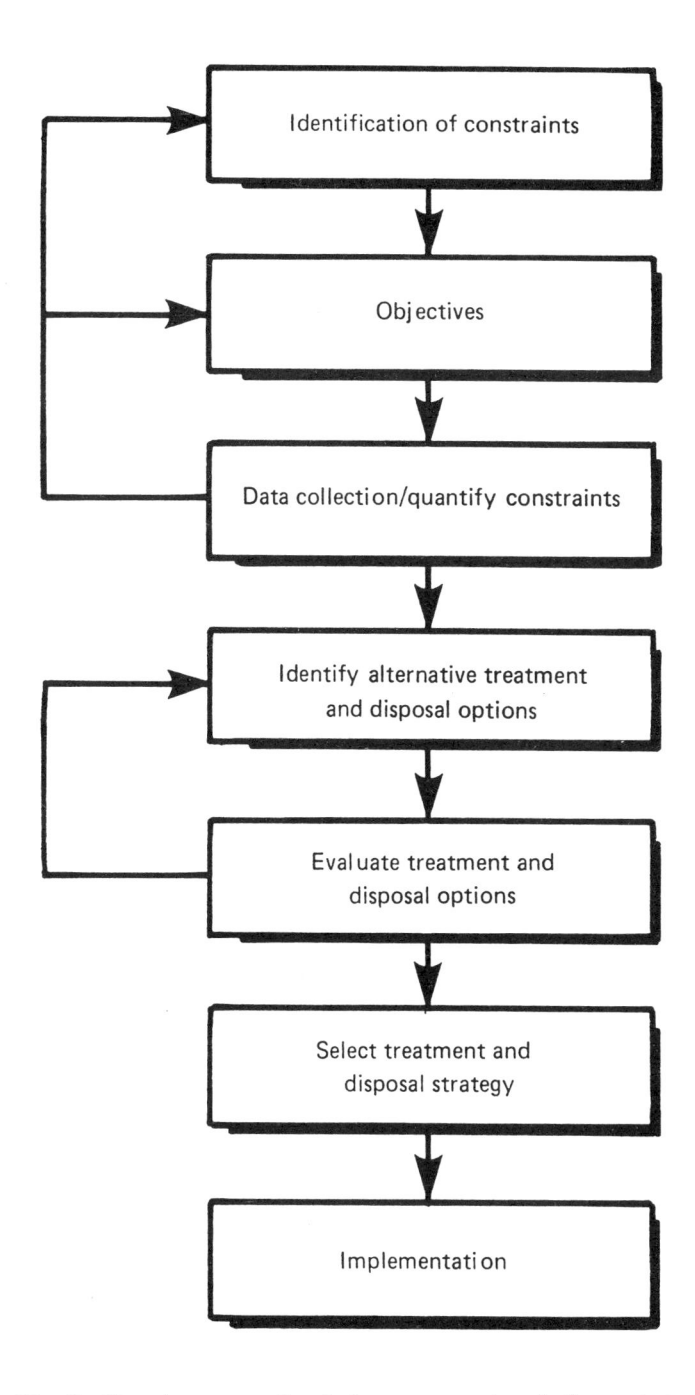

Fig. 2. Development of a sludge treatment and disposal strategy.

2. OUTLINE OF STRATEGY DEVELOPMENT

The development of a sludge treatment and disposal strategy falls into the series of steps illustrated in Figure 2 and involves the identification of existing and likely future constraints, the establishment of objectives for the study, the collection of data, the identification and evaluation of alternative options and the final selection of the preferred option. Whilst the objectives of disposal strategies may vary, it is likely that these steps will be a common feature of these studies.

3. IDENTIFICATION OF CONSTRAINTS

The initial identification of constraints provides the starting point for the study and leads to the definition of the objectives for the assessment.

Constraints on the disposal operation could be financial, environmental, operational or legislative and should be quantified during the data collection phase of the study. For the purposes of this paper it might be considered that environmental pressures on the sea disposal outlet have led to the requirement for a reappraisal of the disposal strategy in the area.

4. STRATEGY OBJECTIVES

It is important that the objectives for the study are clearly stated as these will define the scope of the study and provide the benchmark against which the success, or otherwise, of the final strategy will be judged.

Bearing in mind the principal constraint on the disposal operation in the example study area, the objective of the study would be the development of an environmentally acceptable and cost-effective sludge disposal strategy which will accommodate the sludge from Works 'A' if the sea disposal route is no longer available.

5. DATA COLLECTION AND QUANTIFICATION OF CONSTRAINTS

5.1 Study area

It is necessary to include in the study area other nearby works whose normal operation may conflict with or impinge on the options being considered for the principal works in the study.

5.2 Sludge production, sludge quality and costs of existing treatment and disposal operation

For each of the works in the study area data should be collected on the current operation as this provides the control against which the effectiveness of alternative options can be assessed. It will also provide much of the basic data required in developing and assessing other treatment and disposal options.

A typical data collection exercise would seek to collect data on sludge production, quality, treatment facilities and the costs of treatment and disposal together with information on potential constraints on the operation of particular works; for example, poor access, odour problems or lack of space for future development.

It is important to ensure that the data relate to the same period and that cost information is considered consistently across different processes or disposal outlets. Particular attention should be paid to the establishment of transport costs since they are a major element in any disposal strategy.

5.3 Agriculture

5.3.1 Assessment of land availability The availability of agricultural land for sludge disposal is governed by:
a) Physical features such as local topography, cropping patterns, weather, ease of access and soil type;
b) Environmental considerations such as proximity of habitation, aquifer protection areas, protection of surface water quality and soil metal concentrations;
c) The farmer, who will decide whether or not to make the land available to the utility;
d) In the longer term the availability of land may be affected by changes in farming practice or environmental legislation;
e) Sludge quality.

In order to estimate the area of agricultural land available, a number of approaches are available:
a) A field survey to assess the suitability of agricultural land and direct questioning of farmers. This approach would be likely to provide the most reliable information, however, for large areas it is often impractical and expensive;
b) An assessment based on existing utility records of fields which have been used at some time in the past. This method is likely to be of use when assessing the implications of different options over the short term. However, there is no guarantee that these fields will remain available and the

method gives no indication of land potentially available to the utility;
c) An alternative approach is to assess the land availablity on a more theoretical basis based on published information. For example, in the UK the Ministry of Agriculture, Fisheries and Food (MAFF) parish returns can be used to identify the total agricultural area under a variety of different crops for each parish in the study area. This 100% availability can then be corrected for the presence of aquifer protection zones, the effects of topography, soil type and the proportion of farmers willing to accept sludge. Since soil type and topography are closely linked, suitability of land for sludge disposal can be assessed to some extent by consideration of soil maps for the area concerned. Information on the proportion of farmers prepared to accept sludge could be obtained by means of a market survey or as a subjective assessment based on local experience.

The advantage of method (c) is that it is relatively quick and easy and enables the potential land availability to be assessed. However, it assumes that the agricultural land is evenly distributed over the parish area and between soil types, which is unlikely to be the case. It is important, therefore, that some type of sample survey is carried out in order to validate the results of the study.

In addition to estimating the total area of agricultural land, the disposal operation will be constrained by the seasonal distribution of land availability. The pattern of land availability can often be inferred from existing sludge to land records. It is important to check that seasonal fluctuations in the quantity of sludge disposed to land are due to variations in land availability or trafficability and are not compounded by outside influences such as changes in sludge production, or operating practice.

If there are few available seasonal data on sludge to land operations then the distribution can be estimated in a more theoretical manner. If it is assumed that sludge can only be applied to arable land in the period between harvesting and planting the next crop, then for each crop type a window of availability can be built up using typical harvesting and planting dates for the area concerned (2). The seasonal availability of pasture land can be estimated in a similar manner by looking at grazing and harvesting patterns of grassland in the study area. Information of this type is usually available from local agricultural (ADAS) advisors. The theoretical seasonal availability of arable and pasture land in the case study area is shown in Figure 3.

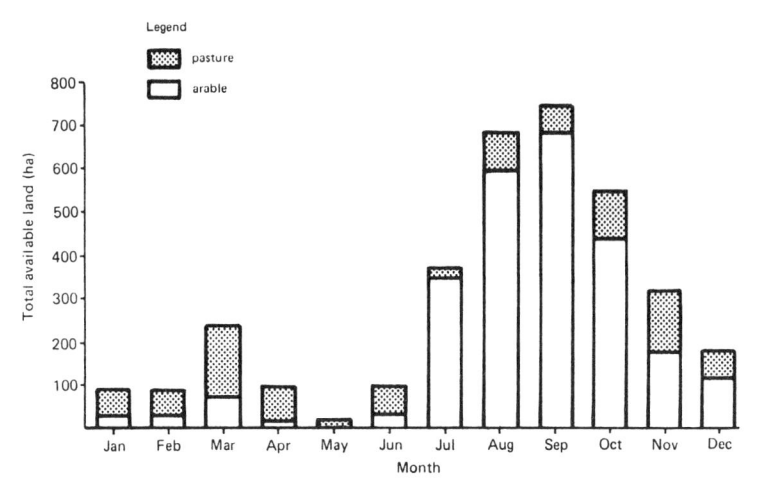

Fig. 3. Theoretical seasonal availability of arable and pasture land for sludge disposal.

5.4 Landfill

In England and Wales landfill sites are licenced by the local waste disposal authorities, which are usually County Councils in England and District Councils in Wales.

The availability of landfill sites for sludge disposal is dependent on:
a) The licencing authority approving the disposal of the material to a particular site;
b) The approval of the National Rivers Authority with regard to potential ground or surface water pollution;
c) The site operator being willing and able to accept sludge;
d) Sludge quality.

The views of all these organisations need to be canvassed in order to build up a realistic picture of site availability in the study area. For sites where sludge disposal is acceptable, data on the site life, capacity for sludge disposal and tipping charges are required in order to be able to assess the option objectively.

5.5 Land restoration

In the UK, potentially large quantities of sludge may be utilised in restoration schemes carried out by local authorities and other organisations such as CEGB, British Coal and other members of the mineral extraction industry (3). However, the availability of sites is strongly dependent on the individual restoration agency concerned and is often related to the timing of other site operations and/or the availability of suitable funding. Consequently, the use of sludge in land reclamation requires a substantial degree of forward planning to enable sufficient sludge to be available at short notice, and good information on the future availability of sites.

Apart from the difficulties in matching sludge availability with demand, the operation could be constrained by:

a) The Collection and Disposal of Waste Regulations (1988), which requires that the use of sludge in restoration schemes is notified to the disposal authority and if sludge is to be stored on site prior to use a waste disposal licence is obtained. It should also be noted that in order to obtain a waste disposal licence, planning permission for the site to accept waste must have been granted;

b) The provisions of the 1986 EC Directive on the use of sewage sludge in agriculture, if the site is to be restored for agricultural use. However, where the land is to be restored for amenity use, metal concentrations suggested by the Department of the Environment for the redevelopment of derelict land may be more appropriate (4);

c) Local concerns over odour, ground or surface water pollution and environmental health.

However, many of these environmental constraints can be overcome if the operation is carried out according to recently published guidelines on the use of sludge in land restoration (5).

In assessing the potential for sludge utilisation in land restoration schemes it is important to contact likely site owners to establish the area and location of sites potentially available, the local planning and waste disposal authorities and ADAS for their views on the use of sludge in this manner.

5.6 Sea disposal

The quantity and quality of sludge disposed of to sea are limited by the conditions of the licence issued by MAFF under the terms of the Food and Environment Protection Act 1975 Part II.

In 1985 the Commission of the European Communities introduced a proposal on the dumping of waste at sea which, if adopted, would impose tighter controls on the quantity and quality of sludge disposed of to this outlet. In addition, the Ministerial Declaration from the Second International North Sea conference, whilst preserving the disposal of sewage sludge to sea, requires urgent action to reduce the concentration of contaminants disposed of to the North Sea compared with 1987 levels. It is likely, therefore, that in the future this outlet will come under increasing environmental pressure.

If sea disposal is already carried out by the utility then information on the licence conditions and costs of the operation should be obtained.

5.7 Incineration

Following the introduction of new technology incorporating a more integrated approach to incineration, there has been an increased interest in this disposal method. For the purposes of a strategy study it is important to consider the potential location for the plant and its likely impact on the surrounding environment together with the costs of the operation.

Although potentially this outlet offers the utility the greatest control over the disposal operation, increasing public concerns over incinerator emissions will require the highest possible standards to be met. Opinions on the possible location of plant should be sought from HMIP and from the local planning authority.

It must also be remembered that incineration is not a technique for total destruction of sludge and that a significant quantity of residual ash will require disposal. Information should be sought, therefore, on the availability of suitable landfills and charges for the disposal of this material.

5.8 Other outlets

Information on the potential availability and costs of other disposal outlets, for example forestry, which may be locally significant should be obtained.

6. IDENTIFICATION OF ALTERNATIVE TREATMENT AND DISPOSAL OPTIONS

6.1 Options available

It is important that the treatment and disposal of sludge is considered as an integrated operation and that sludge of appropriate quality is produced to match the requirements of the disposal outlet available. If there are no constraints on the operation then there are a large number of possible treatments and options. The principal options at a particular works are illustrated in Figure 4.

C. Powlesland

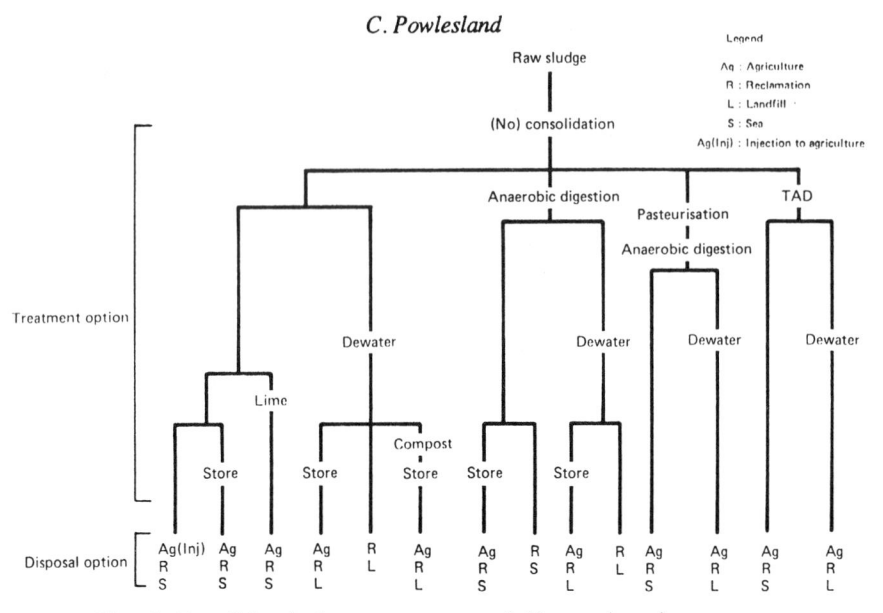

Fig. 4. Possible sludge treatment and disposal options.

6.2 Selection of options

6.2.1 Selection of disposal outlets From the information collected earlier, a range of possible disposal outlets should be chosen. It could be suggested that a suitable outlet would have the following characteristics:
a) Sufficient capacity to enable the outlet to make a significant contribution to the strategy, either locally, or as a major outlet for large quantities of sludge;
b) Long term viability;
c) Environmentally acceptable.

The outlets chosen would normally include agriculture, landfill, land restoration and, for larger works, incineration. It is important to include as many outlets as possible and not to prejudge the outcome of the study.

The availability of disposal outlets in the example study area is shown in Figure 5. From Figure 5 it can be seen that agriculture and landfill have sufficient capacity and likely long or medium term availability, whereas for land restoration it is difficult to identify suitable capacity beyond five years hence. The availability of sea disposal illustrates the hypothetical constraint imposed on this outlet. It is also likely that incineration would be possible at this works since it is a coastal site with little surrounding habitation.

6.2.2 Selection of treatment methods The selection of sludge treatment processes at each works could be considered in a number of stages:

a) Selection of processes which will provide sludge of sufficient quality to meet the requirements of the available disposal outlets;
b) Identification of appropriate processes for the sludge quantity and quality encountered at the works;
c) Review of the processes chosen in the light of operational constraints at the works; for example, availability of space for new plant, odour, traffic nuisance or other local problems.

As a result of this selection procedure it may become apparent that further development of treatment facilities at a particular works is inappropriate, in which case sludge transfer to another works must be considered.

In this example it has been assumed that there are few constraints on the treatment operation. The treatment and disposal options selected for further evaluation are shown in Table 1.

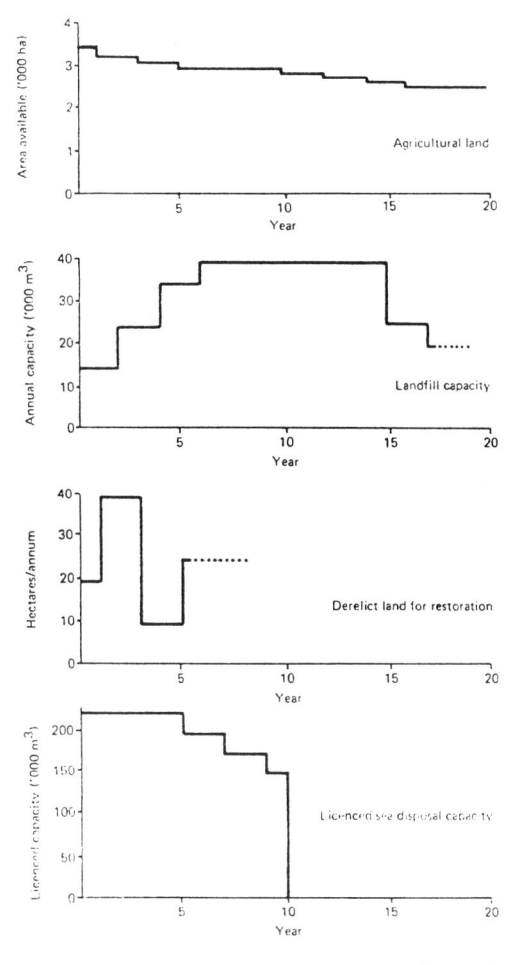

Fig. 5. Capacity of sludge disposal outlets in study area.

7. EVALUATION OF OPTIONS

7.1 Costs

The assessment of costs is usually carried out by an economic appraisal of the different options using a discounted cash flow technique. It is beyond the scope of this paper to discuss this method in detail and reference should be made to standard texts (6,7).

The evaluation of the trial options in this study was carried out using WRc's sludge treatment and disposal model (WISDOM). The results of the economic evaluation are shown in Table 1.

Table 1. Economic evaluation of sludge treatment and disposal options (Net present cost £M).

Option	Capital	Operating	Total
Current sea disposal operation	-	11.4	11.4
C/D-CHP/DW/A	1.2	6.8	8.0
C/D-CHP/A	1.2	7.6	8.8
C/Inj	0.2	9.0	9.2
C/DW/L	0.4	9.0	9.4
Incineration	2.3	10.1	12.4

Where:
C = Sludge consolidation.
D-CHP = Anaerobic digestion and combined heat and power plant.
DW = Dewatering by belt press.
A = Disposal to agricultural land by surface spreading.
Inj = Disposal to agricultural land by injection.
L = Disposal to landfill.

The costs of the current operation shown in Table 1 assume that the sea disposal outlet is available throughout the evaluation period and provide a theoretical control against which to judge the alternative options. Table 1 indicates that from the options evaluated, anaerobic digestion with energy recovery through a combined heat and power plant and disposal of sludge cake to local agricultural land offers the most cost-effective solution to the envisaged constraints. However, the assessment should not be based on economic aspects alone but should also involve a detailed consideration of the environmental effects and security of operation of the different options.

7.2 Environmental considerations

The environmental impact of each of the major options under consideration should be evaluated. It is likely that these will be based largely on subjective assessments but where possible quantitative data should be used. For example, in the case considered here, the sludge metal concentrations when sludge is applied at agriculturally acceptable rates give rise to annual metal loadings which approach, or in some cases exceed, the 10 year average application rates given in the EC sludge to land Directive. Table 1 indicates there are significant financial benefits to disposing of sludge to agriculture compared with the next option of tipping to landfill. It could, therefore, be argued that some of these "benefits" could be spent in (say) improved trade effluent control which would enable a sludge of acceptable quality to be produced. The costs and benefits of this approach should be evaluated and the options re-evaluated in the light of this appraisal.

In addition, the possible effects of accidents or spillages should also be considered; what is the likelihood of these events and what will be their impact on the environment and the future of the operation? In this example, even with better trade effluent control, the potential risk of contamination from accidental discharges to sewer is still considered to be significant. Other factors which should be taken into account include potential ground or surface water pollution, number of vehicle movements, odour and visual effects resulting from the operation. Care should be taken to ensure that the options being considered do not simply move the problems from one portion of the environment, say water, to another, for example air.

7.3 Security of operation

The security of the treatment and disposal operation is dependent to a large extent on the number of factors over which the utility is able to exert control. The principal factors which influence the security of these outlets are summarised in Figure 6. It is difficult to quantify the effects of many of the factors shown in Figure 6 but it may be possible to rank the options in order of increasing security based on a subjective assessment of the risks. For example, an option requiring the cooperation of a single farmer will be inherently more risky than one where sludge is disposed of to many farms.

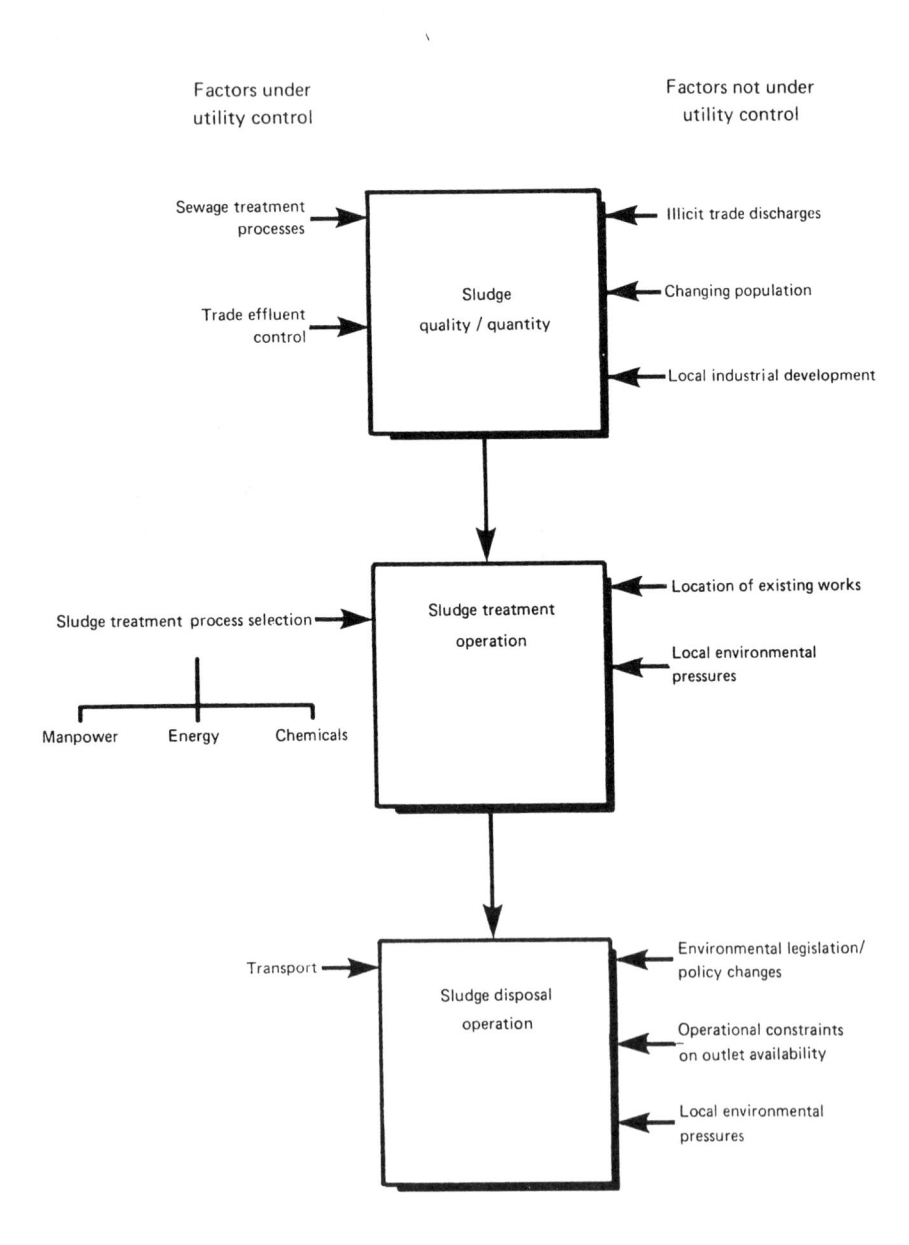

Fig. 6. Factors affecting the security of sludge treatment and disposal operations.

7.4 Sensitivity of options

Much of the information that has been used in the appraisals has been estimated from various sources. The results of the appraisal will be only as good as the information on which it is based. The options should be examined to identify whether they depend, for example, largely on a particular item of plant or disposal outlet. If so, what percentage change in costs, availability of outlet or contaminant level would be required for the rank order of the options to change? If a certain option is found to be particularly sensitive to change then the assumptions, costs and timing of actions that go to make up this option should be re-examined in detail.

In the case of the example options considered here it can be shown that benefits from CHP amount to approximately £780,000 in net present cost terms over the evaluation period. From Table 1 it can be seen that if these benefits were not realised there would be no change in the least cost option but soil injection of raw sludge would become more economic than a liquid digested sludge to land operation. It can also be shown that for incineration to be more cost-effective than landfill the cost of transport and disposal of sludge would have to increase by approximately £16/m^3 or nearly 1.5 times the original estimated value. From these two examples it could be suggested that the potential benefits from CHP should be examined in more detail, but it is unlikely that landfill costs will exceed those for incineration. It should be noted that a more rigorous analysis would also investigate the effects of changing incineration costs on the order of options.

8. IDENTIFICATION OF STRATEGY

The selection of the final strategy needs to take into consideration the economics, environmental acceptability, security and sensitivity of the different options. There can be no single answer which is correct in all circumstances; all cases should be considered individually.

Table 1 indicates that in the example given here, the disposal of anaerobically digested sludge to agriculture offers the most economic disposal outlet. However, the data suggest that contaminant levels in the sludge are significantly greater than those acceptable for agricultural disposal and that even with improved trade effluent control there would always be a significant risk of accidental contamination of the sludge. Therefore, it can be suggested that in this case the disposal of raw sludge to landfill offers the most environmentally acceptable and cost-effective alternative for sludge disposal should the sea disposal outlet no longer be available.

REFERENCES

(1) Hoyland G. and D. Roland (1984). Biological filtration of finely screened sewage. Technical Report 198. Water Research Centre, Medmenham, UK.

(2) Arnold, J. (1979). Yorkshire Water Authority's operational code of practice for the recycling of sewage sludge to agricultural land. In: Utilisation of sewage sludge on land; papers and proceedings (A. M. Bruce, Ed.), pp. 431-448. WRc Conference, November 1979.

(3) Hall J. E., A. P. Daw and C. D. Bayes (1986). The use of sludge in land reclamation. A review of experiences and assessment of potential. Report ER 1346-M. Water Research Centre, Medmenham, UK.

(4) Department of the Environment (1983). Guidance on the assessment and redevelopment of contaminated land. Inter-departmental Committee on the Redevelopment of Contaminated Land, ICRL 59/83.

(5) Hall J. E. (1989). The use of sewage sludge in land restoration. Draft code of practice. Report PRS 1783-M/2. Water Research Centre, Medmenham, UK.

(6) Rees, R. (1973). The economics of investment analysis. Civil Service College Occasional Paper No. 17. HMSO.

(7) HMSO (1976). Waste management paper No. 1, Reclamation, treatment and disposal of wastes. Ch. 11. HMSO.

DISCUSSION SESSION 4

QUESTION: C. ROWLANDS, SEVERN TRENT WATER.
In considering the energetics of agricultural disposal of sludge, did Dr Frost take into account the reduction in inorganic fertiliser that would be required when you beneficially use the sludge in agriculture? I would have thought there might be quite a large saving in terms of both pollution in producing the inorganic fertilisers and also the energy consumed in production and spreading.

ANSWER: R. C. FROST.
We did not take that into account.

QUESTION: R. J. UNWIN, MINISTRY OF AGRICULTURE.
Is it correct to attribute any greenhouse effect to carbon dioxide coming from sludge when a lot of that carbon has initially been fixed from the atmosphere before going into the food chain?

ANSWER: R. C. FROST.
It is an internal recycle effectively, whether the sludge is either combusted or whether it is oxidised in agriculture or what have you, then that is a good point. Where methane is produced and released to the atmosphere, then that would be a net effect. If the carbon dioxide emissions are actually locked into the sea sediments then it is a net benefit.

COMMENT: B. LIVESEY, ICI FERTILISERS.
With regard to Dr Powlesland's graph showing the time when agricultural land was available with the great proportion of land being vacant in the autumn. Agriculture is currently being asked to curtail the amount of nitrogen it uses in the autumn. The time when the maximum amount of land is available for sludge application is the time of minimum plant uptake. Since plants need nitrogen and phosphorus in the spring not in the autumn, that fact that the land is available does not indicate it is the right time.

COMMENT: B. METCALFE, YORKSHIRE WATER.
It is important to look at all the environmental impacts of everything that we do and look at all the possible routes. One possibility that I think we have got to look at as far as environmental impact is concerned, and the possiblity of accidental discharges, is the separation of industrial discharges, at least until after primary treatment, especially those that are likely to give environmental impact implications for the sludge. It would not always be easy but in some cases where the industrial discharges are close to the sewage

works, separate treatment may give us a more secure disposal route to agriculture or elsewhere.

QUESTION: J. J. VAN DEN BERG, GRONTMIJ, NL.
Can Dr Powlesland explain why the operating cost of current sea disposal operation is so expensive compared with incineration?

ANSWER: C. B. POWLESLAND.
That is the cost of the current operation and it is based on an average sea disposal cost for the UK. It is often quite an expensive operation when you have to pay for ships and the movement of a lot of liquid sludge.

QUESTION: P. V. HORSMAN, GREENPEACE.
I was quite surprised to see your rather pessimistic view that sea dumping of sewage sludge will continue for 10 years. If you look at perhaps the reality of the situation it is going to end a lot sooner than that. Is that not going to affect your calculations a lot more and would not it be better to look at a more realistic time frame on sewage sludge dumping?

ANSWER: C. B. POWLESLAND.
In terms of costs, if you move the stopping of sea disposal forward to, say, a five year period, then certainly your costs would increase quite significantly. Obviously there is quite a lead time to put in plant to take over from sea disposal and certainly my feeling is that in a lot of cases if you were forced to go for incineration then probably six or seven years lead time would not be unrealistic, particularly bearing in mind current planning processes.

COMMENT: R. C. RAMSAY, DEPT. OF THE ENVIRONMENT.
At the North Sea Conference in 1987, the declaration which the Ministers put their name to said, yes, there would be drive to reduce the amount of sewage sludge going to sea but that reduction would not be made at the expense of environmental disaster to other parts of the environment. It is one thing to make an environmental impact study which includes all environmental inputs to the sea, to the land, to the air, for each particular case, it is another thing to make an arbitrary decision that one option is not available. We have to remember that we are looking after the future for our children and great grandchildren and that includes looking after the land as well as looking after the sea.

LIST OF PARTICIPANTS

Adams, Mr M
Member Solid Waste Section,
Ministerie Van Vrom, PO Box 450,
Leidschendam, Holland

Addison, Mr R
Principal Engineer Processes,
Anglian Water, Chivers Way,
Histon, Cambs, CB4 4ZY

Apps, Mr P G
Business Development Manager,
Foster Wheeler Energy Limited,
Station Road, Reading,
Berkshire, RG1 1LY

Arnot, Mr J M
Area Sewage Treatment Controller,
Strathclyde Regional Council,
Department of Sewerage,
Regional Offices, Hamilton,
ML3 0AJ

Austin, Mr E P
Consultant, Simon-Hartley Ltd,
Etruria, Stoke-on-Trent, ST4 7BH

Barber, Mr K G
Business Development Manager, OSC
Process Engineering Ltd. PO Box 57,
Stockport, Cheshire, SK3 0TJ

Barnert-Wiemer, Dr H
Kernforschungsanlage Julich GmbH
(KFA), Postfach 1913,
D-5170 Jülich, West Germany

Bayes, Mr C D
WRc Scotland, Unit 16, Beta Centre,
Stirling Univ. Innovation Park, Stirling

Beker, Ir. D
RIVM, PO Box 1, 3720 BA
Bilthoven, The Netherlands

Bendall, Dr R A
Int. Marketing Director, Ag Chem
Equipment Company Inc., 5720
Smetana Drive Metonka, Minnesota,
USA

Berg, van den, Dr J J
Grontmij, De Holle Bilt 22,
De Bilt, Postbus 203, 3730 AE
de Bilt, Netherlands

Best, Mr A A
Senior Operations Officer,
North West Water, Dawson House,
Liverpool Road, Gt. Sankey,
Warrington, WA5 3LW

Bevan, Mr D H
Quality Control Officer,
Welsh Water, Northern Division,
Penrhosgarnedd, Bangor,
Gwynedd, LL57 2DJ

Black, Mr B
Scientific Officer,
Grampian Regional Council,
Department of Water Services,
Woodhill House, Westburn Road,
Aberdeen, AB9 2LU

Blakey, Mr N C
Water Research Centre, PO Box 16
Marlow, Bucks, SL7 2HD

Border, Mr D J
Technical Director,
Hensby Biotech Ltd, Woodhurst,
Huntingdon, Cambs. PE17 3BS

Brown, Mr J M
Yorkshire Water, Spenfield,
182 Otley Road, West Park,
Leeds, LS16 5PR

Bruce, Mr A M
Water Research Centre, Elder Way,
Stevenage, Herts, SG1 1TH

Bull, Mr F
Supervisor, Southern Water,
58 St Johns Road, Newport,
Isle of Wight, PO30 7LT

Burrell, Mr A J
Assistant Regional Land Manager,
British Coal Corporation,
Opencast Exec, Farm Road,
Aberaman, Aberdare,
Mid Glamorgan

Bush, Mr J
Group Scientist, Anglian Water,
Northern Group, Waterside House,
Waterside North, Lincoln,
LN2 5HA

Byrom, Dr K
West Yorkshire Waste Management
55 Kirkstall Road, Leeds, LS4 2AG

Caffoor, Dr M I
R & D Co-ordinator,
Yorkshire Water, Broadacre House,
Vicar Lane, Bradford, BD1 LP2

Clark, Mr W
Senior Technician,
Fife Regional Council, Eng Dept
Water Drainage Div, Fife House,
North Street, Glenrothes, Fife, KY7 5LT

Collinson, Mr D I
Business Development Officer,
Ontario Hydro,
700 University Avenue, Toronto,
Ontario, Canada, M5G 1X6

Connor, Mr P
Planning Support Engineer,
Wessex Water South Division,
2 Nuffield Road, Poole, Dorset,
BH17 7RL

Cossu, Prof R
Instituto di Idraulica,
Universita di Cagliari, Piazzi
d'Armi, 09100 Cagliari, Sardinia,
Italy

Crossland, Mr I
Sludge Disposal Technologist,
Thames Water, New River Head,
173 Rosebery Avenue, London,
EC1R 4TP

Cutts, Mr D
Drainage Manager,
Southern Water, Kent Div,
Luton House, Capstone Road,
Chatham, Kent, ME5 7QA

Davies, Mr I W
Environmental Scientist,
Wardell-Armstrong,
Lancaster Building, High St,
Newcastle-under-Lyme, Staffs,
ST5 1PQ

Davies, Mr P W
Process Development Manager,
Biwater Treatment Ltd. Gregg St,
Heywood, Lancs. OL10 2DX

Davis, Dr R D
Water Research Centre, PO Box 16,
Marlow, Bucks. SL7 2HD

Dawson, Mr B
Yorkshire Water, Spenfield,
182 Otley Road, West Park,
Leeds, LS16 5PR

Dixon, Mr P J
Engineer,
Brian Colquhoun & Partners,
Elizabeth House, St Peter's Square,
Manchester, M2 3DF

Dover, Mr P
Northumbrian Water,
Northumbria House,
Manor Walks Shopping Centre,
Cramlington, Northumberland,
NE23 6UP

Earthy, Mr R P
Sludge Disposal Manager,
Wessex Water,
Avonmouth Treatment Works,
Eton Lane, Avonmouth, Avon

Edwards, Prof R
Deputy Chairman, Welsh Water,
Plas-y-ffynnon, Cambrian Way,
Brecon, Powys. LD3 7HP

Engers, Duvoort van, Mrs L E
RIVM, Postbus 1, 3720 BA
Bilthoven, Netherlands

Entwistle, Mr R
Project Manager,
Energy and Waste Systems Ltd.,
EWS House, Station Road,
Westbury, Wilts. BA13 4HT

Fairnell, Mr D F
Operations Development Engineer,
Northumbrian Water, N & T Div.,
Northumbrian House, Town Centre,
Cramlington, Northumberland,
NE23 6UP

Ferry, Mr J M
Senior Scientist - Public Health,
Anglian Water, Ambury Road,
Huntingdon, Cambs. PE18 6NZ

Field, Mr D
Director, WRc Swindon, P O Box 85,
Blagrove, Swindon, Wilts. SN5 8YR

Flint, Mr J G
Principal Inspector,
Her Majesty's Inspectorate of Pollution,
Room AG21B, Romney House,
43 Marsham Street, Westminster
SW1P 3PY

Fleming, Dr G A
The Agricultural Institute,
Johnstown Castle Research Centre,
Wexford, Republic of Ireland

Foster, Mr D M
Senior Research Engineer,
Thames Water, Nugent House,
Vastern Road, Reading, Berkshire

Frost, Dr R C
Water Research Centre,
P O Box 85, Blagrove, Swindon,
Wilts. SN5 8YR

Fry, Mr D
Area Manager, Anglian Water,
Southern Group, 33 Sheepen Road,
Colchester, Essex, CO3 3LB

Fullam, Mr P
Senior Executive Engineer,
Dublin Corporation, Civic Offices,
Fishamble St, Dublin 8,
Republic of Ireland

Garnett, Mr P
Southern Group, Anglian Water,
33 Sheepen Road, Colchester,
Essex, CO3 3LB

Gayler, Mr N D
Principal Environmental Health
Officer, Essex County Council,
Beehive Lane, Chelmsford, Essex,
CM2 9SY

Glegg, Dr G A
Environmental Scientist,
Greenpeace,
30-31 Islington Green, London,
N1 8XE

Grant, Mr J
Senior Project Engineer,
M C O'Sullivan Consulting Engs,
Clara House, Glenageary,
Co Dublin, Republic of Ireland

Gross, Mr T S C
Principal Engineer,
Sir William Halcrow & Ptnrs,
Burderop Park, Swindon,
Wilts. SN4 0QD

Gunn, Mr A S
Consultant,
Edwards and Jones Ltd.
Whittle Road, Meir,
Stoke-on-Trent, ST3 7QD

Gupta, Mr S K
Head of Soil,
Swiss Fed. Res. Station for
Environment,
Schwarzenburgstr. 155, CH-3097
Leibefeld, Switzerland

Hall, Mr J E
Water Research Centre,
PO Box 16, Marlow, Bucks.
SL7 2HD

Hawkins, Dr J E
Quality Control - Monitoring
Manager,Thames Water,
New River Head,
173 Rosebery Avenue,
London, EC1R 4TP

Haynes, Mr N T
Assistant Reclamation Controller,
Severn Trent Water, Northern Div,
PO Box 51, Raynesway, Derby

Haywood, Mr S P
Marketing & Sales Manager,
OSC Process Engineering Ltd.
PO Box 57, Stockport, Cheshire,
SK3 0TJ

l'Hermite, Mr P
Commission of the European
Communities,
rue de la Loi 200,
B-1049, Brussels, Belgium

Hiley, Dr P
SSO Biological Services,
Yorkshire Water,
Broad Acre House, Bradford,
BD1 5PZ

Hill, Mr C P
West Yorkshire Waste Management,
Chantry House, 123 Kirkgate,
Wakefield, WF1 1YG

Hill, Mr D M
Corporate Data Scientist,
South West Water Services Ltd.
Penninsula House, Rydon Lane,
Exeter

Holdhus, Mr O
Agricultural University,
Dept of Soil Sciences, PO Box 28,
N-1432 Aas-NLH Norway

Holton, Mr D J
Development Co-ordinator,
Welsh Water, Cambrian Way,
Brecon, Powys, LD3 7HP

Horsman, Mr P V
Director-Toxics Division,
Greenpeace,
30-31 Islington Green,
London, N1 8XE

Hydes, Mr P
Assistant Waste Disposal Officer,
Southern Water, Sparrowgrove,
Otterbourne, Winchester,
SO21 2SW

Jackson, Mr P C
Water Quality Officer,
Welsh Water, Pentwyn Road,
Nelson, Treharris,
Mid Glamorgan, CF46 6LY

James, Mr M
Business Development Manager,
Crown House Energy,
Crown House,
550 Mauldreth Road West,
Chorlton-cum-Hardy, Manchester,
M21 2RX

Jarrin, Mrs F J
Engineer, A.N.R.E.D., 2 Square
Lafayette, BP 406, 49004 Angers
Cedex, France

Keil, Dr L
Assistant Scientist, Welsh Water,
86 The Kingsway, Swansea,
SA1 5JL

Keith, Mr L B
Senior Agricultural Officer,
DAFS, Room 340, Pentland House,
47 Robbs's Loan, Edinburgh,
EH14 1TW

Keyes, Mr J
Senior Executive Engineer,
Dublin Corporation, Civic Offices,
Fishamble St, Dublin 8,
Republic of Ireland

King, Mr E C
Director, Universal Compost
Corp. Ltd.
Low Farm, Runhall, Norwich.

Klessa, Dr D A
Research Scientist,
West of Scotland College,
John F Niven Building,
Auchincruive, Ayr,
KA6 5HW

Kuhlman, Dr L R
President,
Resource Recovery Systems
of Nebraska Inc, Route 4,
Sterling, Colorado, USA - 80751

Larre-Larrouy, Dr M C
Agronomist,
Chambre D'Agriculture Du Vaucluse,
Cantarel - BP734,
84034 - Avignon, Cedex-France

Lavergne, Mr G
ociete du Metro de Marseilles,
44 Ave Alexandra Dumas, 13272
Marseilles, France

Lavin, Mr J C
Principal Ecological Advisory Officer,
West Yorkshire Ecological
Advisory Service,
Cliffe Castle, Keighley,
West Yorkshire, BD20 6LH

Leach, Dr M J
Process Engineering Manager,
Birwelco Ltd. Mucklow House,
Mucklow Hill, Halesowen,
West Midlands, B62 8DG

Lee-Harwood, Mr B
Friends of the Earth Ltd.
Water Pollution & Toxics
Campaigner,
26-28 Underwood Street,
London, N1 7JQ

Lenton, Mr K W
Senior Area Sales Manager,
E & A West, Parcel Terrace,
Derby, DE1 1NA

Livesey, Mr B
ICI Fertilisers,
PO Box 1, Billingham,
Cleveland, TS23 1LB

Loine, Mr S
Tayside Regional Council,
Water Services Dept.
Clatto Laboratories,
Dalmahoy Drive, Dundee DD3 9PR

Love, Mr S C P
HPTO Sewerage, D.O.E (NI)
Water Service, PO Box 8,
Altnagelvin, Londonderry, NI

Lowe, Mr P
Regional Support Manager,
Yorkshire Water, 38 Southgate,
Wakefield, WF1 1TR

Lubben, Dr S
Institute for Plant & Soil Research,
Bundesallee 50, D-3300
Braunschweig, FRG

Lune, van, Mr P
Agriculture Research Worker,
Inst.for Soil Fertility, (IB),
PO Box 30003, 9750 RA Haren,
Netherlands,

MacKenzie, Mr J M
Forest District Manager,
North York Moors,
Forestry Commission, 42 Eastgate,
Pickering, North Yorkshire,
YO10 7DU

Mann, Mr J
Senior Engineering Planner,
Thames Water, Maple Lodge
Sewage Works, Denham Way,
Rickmansworth

Martel, Mr J L
Technical Manager,
Agro-Developpement,
2 rue Stephenson, 78181 St Quentin
Yuelines Cedex, France

Matthews, Dr P J
Director of Quality,
Anglian Water, Ambury Road,
Huntingdon, Cambs. PE18 6NZ

McCrum, Mr W A
Senior Scientist (Quality Assess.)
North West Water,
Newtown House, PO Box 30,
Buttermarket Street, Warrington,
WA1 2QG

McNeill, Mr J D
Research Forest Officer,
Forestry Commission,
Northern Research Station,
Roslin, Midlothian, Scotland, EH25 9SY

McTavish, Mr G
Process Control & Treatment
Officer,
Welsh Water, Northern Division,
Penrhosgarnedd, Bangor,
Gwynedd, LL57 2DJ

Metcalfe, Miss B
Biologist, Mitchel Laithes STW,
Yorkshire Water, Dewsbury, York

Metcalfe, Mr B
Treatment Co-ordinator,
Yorkshire Water, Central Division,
"Spenfield", 182 Otley Road,
West Park, Leeds, LS16 5PR

Mitchell, Mr D J
Scientific Officer,
Warren Spring Laboratory,
Gunnels Wood Road, Stevenage,
Herts. SG1 2BX

Moore, Mr B J
Mott MacDonald, Demeter House,
Station Road, Cambridge, CB1 2RS

Moore, Mr J
Scottish Development Dept.
Room 252, 27 Perth Street,
Edinburgh, EH3 5DW

Mullett, Dr J
Science Director,
Secondary Resources,
Midland Bank Chambers,
The Square, Chipping Norton,
Oxon, OX7 5NE

Nichols, Mr C G
Reclamation Reuse Conservation,
Administrator, County Sanitation
Districts of Orange County,
PO Box 8127, Fountain Valley,
California, 92728-8127, USA

O'Sullivan, Mr J J
Senior Engineer,
Cork Corporation, City Hall,
Cork, Republic of Ireland

O'Sullivan, Mr M J
Senior Project Engineer,
M C O'Sullivan Consulting Engs.
Innishmore, Ballincollig,
Co Cork, Republic of Ireland

Oake, Mr R J
Environmental Policy Manager,
Thames Water, Nugent House,
Vastern Road, Reading, Berkshire,
RG1 8DB

Olesen, Dr S E
Head of Research,
Danish Land Development
Service, The Research Dept,
PO Box 110, 8800 Viborg,
Denmark

Oster, Ms B M
Engineer, Stockholm Water and
Wastewater Works,
Box 6407, 113 82, Stockholm,
Sweden

Palm, Mr O
Swedish Environment Protection
Board,
Box 1302, 17 125 Solna,
Sweden

Palmer, Mr R J
Proprietor, Flotag, PO Box 62,
Rugby, Warks. CV22 7EF

Panter, Mr K M
General Manager,
Thamesgro Land Management Ltd.
Slough WPC Works, Wood Lane,
Slough, SL1 9EB

Papadopoulos, Dr P
Ministry of Agriculture,
28 Blessa Street, 15669 Papago,
Athens, Greece

Parsons, Mr N
Contracts & Marketing Officer,
Thames Water, New River Head,
173 Rosebery Avenue,
London, EC1R 4TP

Paulsrud, Mr B
Director, Aquateam-Norwegian Water,
Tech Centre, PO Box 6326,
Etterstad, O604 Oslo 6, Norway

Pearce, Mr M J
Business Development Manager,
Biwater Supply Ltd. 46 New Road,
Bromsgrove, Worcestershire,
B60 2JT

Pearce, Mr P A
Research Scientist,
Thames Water, Nugent House,
Vastern Road, Reading, Berkshire

Persson, Mr J
Swedish University of Soil Sciences,
Box 7014, 75007
Uppsala, Sweden

Pini, Dr R
Soil Scientist,
C.N.R Inst for Soil Chemistry,
Via Corridoni 78, 56100 Pisa,
Italy

Pomares, Dr F
Inst Valenciano de Investigaciones
Agrarias, Apartado Oficial,
E-Moncada, Valencia, Spain

Powlesland, Dr C B
Scientist,
Water Research Centre, PO Box 16,
Marlow, Bucks. SL7 2HD

Prisum, Mr J
Department Manager, Kruger,
363 Gladsaxevej, 2860 Soborg,
Denmark

Pulford, Dr I D
Lecturer, University of Glasgow,
Chemistry Department, Glasgow,
G12 8QQ

Puolanne, Dr J
National Board of Waters and
Environment,
PB 250, 00101 Helsinki, Finland

Purdy, Mr J
Manager, Resource Development Unit,
Dept of Environment (NI),
Water Service, 3a Frederick St,
Belfast, BT1 2NS

Rae, Dr P A S
Ecologist, British Gas plc,
London Research Station,
Michael Road, London, SW6 2AD

Ramsay, Mr R C
Water Technical Division,
Dept of Environment, Room B4.52,
Romney House, 43 Marsham Street,
London, SW1P 3PY

Ramsay, Dr R J
Principal Scientific Officer,
Environmental Protection Div,
DoE, Calvent House,
23 Castle Place, Belfast, DT1 1FY

Reaston, Dr P
Consultant,
Laurence Gould Consultants Ltd.
Birmingham Road, Saltisford,
Warwick, CV34 4TT

Riddell-Black, Miss D
Water Research Centre,
PO Box 16, Marlow, Bucks.
SL7 2HD

Riley, Mr D
Process Engineer,
Yorkshire Water (North & East
Division), 32/34 Monkgate, York,
YO3 7RH

Roblin, Mr M G
Treatment Officer,
South Western Division,
Welsh Water, DWR
Cymrll, Hawthorn Rise,
Haverfordwest, Dyfed

Rowlands, Dr C
STW - Reclamation Officer,
Severn Trent Water,
Abelson House, 2297 Coventry Road,
Sheldon, Birmingham, B26 3PU

Royle, Miss S M
Advisory Soil Scientist, ADAS,
MAFF, Woodthorne, Wolverhampton,
WV6 8TQ

Scherer, Dr H W
Institute of Agricultural Chemistry,
Meckenheimer Allee 176, D-5300
Bonn 1, West Germany

Sercombe, Mr D C W
Principal Process Scientist,
Anglian Water, Chivers Way,
Histon, Cambridge, CB4 4ZY

Sharp, Mr C H
Project Engineer,
West Yorkshire Waste Management,
Chantry House, 123 Kirkgate,
Wakefield

Sidwick, Mr J M
Consultant,
Sir Alexander Gibb & Ptnrs,
Earley House, London Road,
Reading, Berks. RG6 1BL

Sopper, Dr W E
Professor of Forest Hydrology,
Pennsylvania State Univ,
University Park, PA 16802, USA

Stokes, Mr R
Thames Water, New River Head,
173 Rosebery Avenue,
London, EC1R 4TP

Suss, Dr A
Bayerische Landesanstalt f.
Bodenkultur u. Pflanzenbau
Vottinger Strasse 38, 8050
Freising, West Germany

Taylor, Mr C M A
Project Leader,
Forestry Commission, Northern
Research Station, Roslin,
Midlothian, Scotland, EH25 9SY

Taylor, Mr J
Director of Operations,
Yorkshire Water, 32/34 Monkgate,
York, YO3 7RA

Turner, Mr C J
Restoration Engineer,
Wimpey Waste Management,
Matthews St, Ardwick, Manchester,
M12 5BB

Unwin, Mr R J
Pollution Specialist,
Agricultural Development
Advisory Service, MAFF,
Burghill Road, Westbury-on-Tyme,
Bristol, BS10 6LT

Vigerust, Mr E
Agricultural University, Inst for
jordfag, Dept of Soil Sciences,
Box 28, 1432 As-NLH, Norway

Vigna-Guidi, Dr G
Director of Institute,
C.N.R Inst for Soil Chemistry,
Via Corridoni 78, 56100 Pisa,
Italy

Walker, Mr J
Sludge Disposal Manager,
Yorkshire Water, Central Division
"Spenfield", 182 Otley Road,
West Park, Leeds, LS16 5PR

Wallis, Mr B F J
Sewage Treatment Manager,
Borders Regional Council,
Water & Drainage Services,
West Grove, Melrose,
Roxburghshire, TD6 9SJ

Walsh, Mr J
Projects Manager,
E G Pettit & Co
Springville House, Blackrock Road
Cork, Republic of Ireland

Ward, Mr M D
Divisional Manager - Process
Systems Division, Birwelco Ltd.
Mucklow House, Mucklow Hill,
Halesowen, West Midlands,
B62 8DG

Webber, Dr M D
Research Scientist,
Environment Canada,
Wastewater Technology Centre,
PO Box 5050, Burlington,
Ontario, L7R4A6, Canada

Wheale, Mr G
Process Development Manager,
Yorkshire Water, PO Box 500,
Western House, Western Way,
Halifax Road, Bradford, BD6 2LZ

Whipps, Mr A P
Senior Engineer,
South West Water,
Peninsula House, Rydon Lane,
Exeter, Devon, EX2 7HR

Wigfull, Miss S D
PhD student,
Polytechnic of East London,
Romford Road, London, E15 4LZ

Wilcock, Mr K G
Watson Hawksley, Terriers House,
Amersham Road, High Wycombe,
Bucks. HP13 5AJ

Wild, Mr S R
PhD student,
Inst Envir & Bio Sci,
Environmental Science Div,
Univ of Lancaster, Bailrigg,
Lancaster, Lancs, LA1 4YQ

Williamson, Mrs S M
Projects Manager (Sewage Treatment),
Yorkshire Water,
PO Box 500 Western House,
Western Way, Halifax Road,
Bradford, BD6 2LZ

Woolhouse, Mr A J
Forestry Officer, ARC Eastern,
Regional Head Office, Ashby Road,
East Shepshed, Loughborough,
Leics. LE12 9BU

Woosey, Mr L A
Engineering Training Consultant,
Water Training, Burn Hall,
Sutton-on-Forest, York, YO6 1JB

Wray, Mr E C
Regional Manager-Waste,
Alfa-Laval Sharples Ltd,
Dorman Road, Camberley, Surrey,
GU15 3DN